Feature Papers of Ruminants 2024–2025

Feature Papers of Ruminants 2024–2025

Guest Editors

Leilson Rocha Bezerra
Phillip Lancaster

MDPI

Basel • Beijing • Wuhan • Barcelona • Belgrade • Novi Sad • Cluj • Manchester

Guest Editors
Leilson Rocha Bezerra
Department of Animal Science
Federal University of
Campina Grande
Patos
Brazil

Phillip Lancaster
Beef Cattle Institute
Kansas State University
Manhattan
USA

Editorial Office
MDPI AG
Grosspeteranlage 5
4052 Basel, Switzerland

This is a reprint of the Special Issue, published open access by the journal *Ruminants* (ISSN 2673-933X), freely accessible at: https://www.mdpi.com/journal/ruminants/special_issues/67093585KH.

For citation purposes, cite each article independently as indicated on the article page online and as indicated below:

Lastname, A.A.; Lastname, B.B. Article Title. *Journal Name* **Year**, *Volume Number*, Page Range.

ISBN 978-3-7258-6990-9 (Hbk)
ISBN 978-3-7258-6991-6 (PDF)
https://doi.org/10.3390/books978-3-7258-6991-6

Contents

About the Editors

Leilson Rocha Bezerra

Leilson Rocha Bezerra is a Titular Professor at the Federal University of Campina Grande (UFCG), Brazil, affiliated with the Center for Health and Rural Technology (CSTR). He has a PhD in Animal Science and has been teaching and researching since 2006. His scientific interests focus on ruminant nutrition, biomaterials and bioactive compounds for animal feeding, feed efficiency, meat quality, and the development of sustainable livestock production systems adapted to semi-arid environments. He has coordinated and participated in national and international research projects addressing nutritional strategies, slow-release technologies, and innovation in animal production, including collaborative networks in Latin America, Europe, and the United States. His achievements include more than 280 peer-reviewed publications in high-impact journals, authorship of books and book chapters, deposits of patents related to animal nutrition technologies, supervision of numerous MSc and PhD students, and recognized contributions as a reviewer, editor, and scientific consultant for national and international journals and research funding agencies.

Phillip Lancaster

Phillip Lancaster is an Assistant Professor at the Beef Cattle Institute at Kansas State University in Manhattan, Kansas, USA. He has been teaching and researching since 2008. His scientific interests focus on ruminant nutrition, energy metabolism, and gastrointestinal physiology. He has led and collaborated on research projects to improve the feed efficiency of grazing cows and growing cattle, identifying risk factors for gastrointestinal-related diseases such as liver abscess, and used model simulations to enhance the sustainability of beef cattle production systems. He has led or collaborated on over 30 funded grant proposals and 70 peer-reviewed articles and has mentored multiple MSc and PhD students.

Preface

This Reprint was developed to provide a comprehensive and updated overview of the role of ruminant animals within modern food production systems, with an emphasis on efficiency, sustainability, and responsible resource use. The motivation for assembling this collection arises from ongoing scientific and societal discussions regarding the use of feed resources, particularly the balance between human-edible and non-human-edible inputs, and the environmental footprint of ruminant production.

The scope of this Reprint encompasses experimental and applied research addressing ruminant nutrition, physiology, production systems, and environmental interactions, highlighting how dietary strategies and management practices influence productivity, nutrient utilization, and system resilience. By bringing together contributions from different regions and production contexts, this Reprint aims to foster a deeper understanding of the biological mechanisms and system-level outcomes that define ruminant efficiency.

This Reprint is primarily addressed to researchers, graduate students, educators, and professionals in animal science, agriculture, and sustainability, offering evidence-based insights to support innovation, informed decision-making, and the development of resilient and sustainable ruminant production systems.

Leilson Rocha Bezerra and Phillip Lancaster
Guest Editors

ruminants

MDPI

Editorial

Feature Papers of Ruminants 2024–2025

Leilson R. Bezerra [1,*] and Phillip Lancaster [2,*]

[1] Department of Animal Science, Federal University of Campina Grande, Patos 58708110, Paraíba, Brazil
[2] Beef Cattle Institute, Kansas State University, Manhattan, KS 66506, USA
* Correspondence: leilson@ufpi.edu.br (L.R.B.); palancaster@vet.k-state.edu (P.L.)

1. Introduction

Ruminants continue to play a pivotal role in sustainable food production by converting abundant resources, such as rangelands, pastures, crop residues, and agro-industrial byproducts, into human-edible products of high nutritional value (meat, milk, and other dairy products). Although the contribution of ruminants to food security is historically well established and remains substantial, debate persists regarding the efficiency of converting feed resources into animal products and the environmental externalities associated with livestock production.

Contemporary discussions increasingly focus on the use of potentially human-edible feeds in ruminant diets, the need to optimize feeding systems, and the challenge of mitigating environmental impacts without compromising productivity. This Special Issue was designed to assemble cutting-edge research on these topics, offering an integrated perspective on recent scientific advances and the knowledge gaps that remain to be resolved in ruminant science.

2. Materials and Methods

The *Special Issue Feature Papers of Ruminants 2024–2025* brings together a diverse collection of original research articles, reviews, and communications that reflect scientific progress and emerging directions in ruminant research. The contributions span themes ranging from physiology and nutrition to management technologies, animal welfare, genetics, and sustainability.

Collectively, the papers provide a broad and cohesive view of the multiple dimensions that currently shape ruminant production, underscoring that sustainability, productive efficiency, animal welfare, technological innovation, and socioeconomic context are inseparable components of modern livestock systems. From a socioeconomic standpoint, the study conducted in Türkiye [Varalan et al., Contribution 1] indicates that sheep farming remains largely grounded in extensive systems, primarily managed by middle-aged or older producers with limited formal education and a strong reliance on direct marketing. This profile reveals structural constraints but also delineates clear pathways for sectoral sustainability, including generational renewal, expanded technical training, and the adoption of more modern production practices. Complementing this structural perspective, the morphological characterization of the Sanmartinero Creole bovine breed using geometric morphometrics [Salamanca-Carreño et al., Contribution 2] highlights the value of advanced quantitative tools for genetic conservation by demonstrating consistent sexual dimorphism in head size and shape, information that is strategically relevant for preservation programs, population standardization, and the valorization of locally adapted breeds.

In the domain of growth and development, the assessment of the relationship between stature and live weight in dairy cows from birth to maturity advances understanding of body growth trajectories and the development of productive capacity, showing that linear measures exhibit strong associations between early life and adulthood. In contrast, traits related to thoracic capacity and live weight remain environmentally sensitive for more extended periods [Gibson et al., Contribution 3]. These findings reinforce the importance of adequate nutritional management up to later ages to maximize future productive potential, with direct implications for reproductive efficiency and lactational performance. This emphasis on nutrition is also mirrored in experimental work with growing cattle, in which replacing urea-treated rice straw with silages derived from maize–soybean or grass–soybean intercrops produced consistent improvements in performance and feed efficiency without altering dry matter intake or body conformation, demonstrating the potential of integrated forage strategies to improve the energy–protein balance in tropical production systems [Nguyen et al., Contribution 4].

The interface among nutrition, behavior, and animal welfare is further strengthened by studies that address both pathological conditions and innovative feeding strategies. Research on the behavior of cows affected by Johne's disease [Charlton et al., Contribution 5] identifies subtle yet meaningful changes in lying and rumination patterns around peak lactation, suggesting that behavioral metrics can serve as sensitive tools for monitoring health status and welfare. However, the underlying mechanisms require further clarification. In a complementary approach, the study evaluating the inclusion of technical cashew nutshell liquid combined with vegetable oils differing in fatty acid profiles in sheep diets shows that dietary lipid composition influences feed selectivity and rumination patterns without adverse physiological or thermoregulatory effects, highlighting the role of monounsaturated fatty acids in optimizing energy intake and feeding efficiency [Andrade et al., Contribution 6].

More integrative and technology-oriented contributions extend this discussion to large-scale production systems. The invited review on total mixed ration (TMR) feeding [Dey et al., Contribution 7] consolidates evidence that TMR enhances rumen fermentation stability, improves microbial activity, and increases productive performance across ruminant species, while also emphasizing challenges related to formulation errors, mycotoxin risks, and environmental outcomes, reinforcing the need for nutritional precision and sound management practices. Similarly, the review of virtual fencing technology underscores its promise as a core tool within precision livestock farming, capable of increasing grazing-management flexibility and reducing labor demands, while also recognizing technical, economic, and welfare-related constraints that must be addressed to enable broader adoption across diverse production contexts [Gadzama et al., Contribution 8].

The ethical and societal dimensions of ruminant production are robustly addressed through the analysis of enhanced animal welfare labeling schemes for cattle, sheep, and goats in Europe [Papageorgiou et al., Contribution 9]. This contribution demonstrates that growing demands for transparency and sustainability have accelerated the development of more stringent welfare standards grounded in animal-based measures, with a strong emphasis on pasture access, comfort, environmental enrichment, and, in some cases, mother–young bonding. Such labeling systems not only reflect societal expectations but also function as strategic instruments for market differentiation and for rewarding more responsible production systems.

Finally, the review on slow-release urea technologies [Bezerra et al., Contribution 10] synthesizes essential advances in the use of non-protein nitrogen sources in small ruminants, showing that microencapsulation and other controlled-release systems can improve nitrogen-use efficiency, enhance microbial protein synthesis, reduce environmental nitrogen losses, and, under specific dietary and management conditions, improve animal performance and product quality. At the same time, the review highlights the need for

 https://doi.org/10.3390/ruminants6010013

greater methodological standardization, long-term evaluations, and comparative economic assessments to support large-scale adoption of these technologies.

Taken together, these studies demonstrate that the future of ruminant production depends on integrated, interdisciplinary approaches that simultaneously address biological efficiency, economic viability, animal welfare, conservation of genetic resources, and societal acceptance. The contributions in this Special Issue reinforce the central role of science in building more resilient, efficient, and sustainable production systems capable of responding to global challenges related to food security, climate change, and evolving societal demands.

3. Conclusions

The Special Issue Feature Papers of Ruminants 2024–2025 consolidate studies that advance understanding of the mechanisms influencing production efficiency, animal welfare, and the sustainability of ruminant production systems. Collectively, the contributions emphasize the value of integrated approaches that consider not only nutritional and productive variables but also technological, behavioral, and socioeconomic factors. Despite these advances, significant knowledge gaps remain, including (i) the integration of nutritional efficiency with reduced environmental impacts, for which robust predictive models are still needed to optimize diets that maximize output while minimizing emissions; (ii) improved connectivity between emerging technologies (e.g., sensors and precision livestock farming) and on-farm practices to translate data streams into operational decisions that create measurable value; and (iii) a deeper understanding of the interactions among genetics, management, and nutrition across contrasting climates and production systems.

To accelerate progress, future research should prioritize the development of feeding systems that maximize the use of resources that do not compete with the human food supply while reducing environmental losses; the application of big data analytics and artificial intelligence to improve predictions of performance and animal health; and interdisciplinary studies integrating genetics, the rumen microbiome, precision nutrition, and public policies to advance sustainable production practices at a global scale. This body of work not only brings a relevant thematic issue to a close but also opens promising avenues for investigation that may further strengthen the role of ruminants in sustainable agriculture in the twenty-first century.

Acknowledgments: We gratefully acknowledge the authors who contributed their work to this Special Issue. We also express our sincere appreciation to the peer reviewers and academic editors for their valuable expertise and constructive evaluations, which played a crucial role in enhancing the quality and rigor of the submitted manuscripts.

Conflicts of Interest: The authors declare no conflicts of interest.

List of Contributions:

1. Varalan, A.; Barit, B.; Mat, B.; Çevrimli, M.B.; Günlü, A. Socio-Economic Structure of Sheep Enterprises in Türkiye: A Micro-Level Analysis. *Ruminants* **2025**, 5, 39. https://doi.org/10.3390/ruminants5030039.
2. Salamanca-Carreño, A.; Parés-Casanova, P.M.; Vélez-Terranova, M.; Rangel-Pachón, D.E.; Martínez-Correal, G.; Rosero-Alpala, J. Head Sexual Characterization of Sanmartinero Creole Bovine Breed Assessed by Geometric Morphometric Methods. *Ruminants* **2025**, 5, 33. https://doi.org/10.3390/ruminants5030033.
3. Gibson, M.J.; Sneddon, N.W.; Rogers, C.W.; Back, P.J.; Dittmer, K.E.; Martín, N.P. The Relationship Between Stature and Live Weight of Dairy Cows Between Birth and Maturity. *Ruminants* **2025**, 5, 7. https://doi.org/10.3390/ruminants5010007.
4. Nguyen, D.V.; Penrose, B.; Tran, N.B.T.; Le, H.T.T.; Trinh, H.T.; Ives, S. Feed Intake and Growth Performance of Vietnamese Yellow Calves Fed Silages from Intercropped Maize–Soybean and Guinea Grass. *Ruminants* **2024**, 4, 602–612. https://doi.org/10.3390/ruminants4040041.

5. Charlton, G.L.; Churches, J.; Bleach, E.C.L.; Thorup, V.M. Behaviour of Cows with Johne's Disease (Paratuberculosis). *Ruminants* **2024**, *4*, 533–542. https://doi.org/10.3390/ruminants4040037.
6. Andrade, É.L.G.; Pereira Filho, J.M.; Lucena, K.H.D.O.S.D.; Barreto, Y.C.S.; Oliveira, R.L.; Sousa, B.B.D.; Vaz, A.F.D.M.; de Oliveira, J.P.F.; Fonseca, M.A.; Bezerra, L.R. Changes in the Fatty Acid Composition of Vegetable Oils Affect the Feeding Behavior, Feed Preference, and Thermoregulatory Responses of Sheep. *Ruminants* **2024**, *4*, 433–447. https://doi.org/10.3390/ruminants4030031.
7. Dey, M.C.; Jairath, G.; Gadzama, I.U.; Alves, S.P.; Ponnampalam, E.N. Impact of Mixed Rations on Rumen Fermentation, Microbial Activity and Animal Performance: Enhancing Livestock Health and Productivity—Invited Review. *Ruminants* **2025**, *5*, 42. https://doi.org/10.3390/ruminants5030042.
8. Gadzama, I.U.; Asadi, H.; Hina, Q.; Ray, S. Influence of Virtual Fencing Technology in Cattle Management and Animal Welfare. *Ruminants* **2025**, *5*, 21. https://doi.org/10.3390/ruminants5020021.
9. Papageorgiou, M.; Karageorgou, A.; Tzamaloukas, O.; Simitzis, P. Enhanced Animal Welfare and Labeling in Cattle, Sheep, and Goats. *Ruminants* **2025**, *5*, 19. https://doi.org/10.3390/ruminants5020019.
10. Bezerra, L.R.; Mazza, P.H.S.; Lucena, K.H.D.O.S.D.; Amorim, M.A.F.; Moura, J.F.P.D.; Oliveira, J.P.F.D.; Oliveira, R.L.; Barbosa, A.M.; Fonseca, M.A.; Pereira Filho, J.M. Slow-Release Urea Technologies: Effects on Nitrogen Utilization and Small Ruminant Performance. *Ruminants* **2025**, *5*, 50. https://doi.org/10.3390/ruminants5040050.

ruminants

MDPI

Article

Socio-Economic Structure of Sheep Enterprises in Türkiye: A Micro-Level Analysis

Alperen Varalan [1,*], Burak Barit [2], Burak Mat [3], Mustafa Bahadır Çevrimli [3] and Aytekin Günlü [3]

1 Department of Animal Health Economics and Management, Faculty of Veterinary Medicine, Ataturk University, TR-25240 Erzurum, Türkiye
2 Department of Animal Health Economics and Management, Faculty of Veterinary Medicine, Hatay Mustafa Kemal University, TR-31060 Hatay, Türkiye; burak.barit@mku.edu.tr
3 Department of Animal Health Economics and Management, Faculty of Veterinary Medicine, Selçuk University, TR-42250 Konya, Türkiye; burakmat@selcuk.edu.tr (B.M.); bahadir.cevrimli@gmail.com (M.B.Ç.); agunlu@selcuk.edu.tr (A.G.)
* Correspondence: alperenvaralan@atauni.edu.tr

Simple Summary: This study explores the structure of sheep farming enterprises in Türkiye. Surveys were conducted with 201 farmers during the 2023 production period to collect data on farm size, production goals, marketing methods, and economic performance. The results show that most sheep farmers are middle-aged or older and have low education levels. They usually use traditional (extensive) production systems and sell their products directly to consumers. To make sheep farming more sustainable, it is important to attract younger farmers, offer better training, and support the use of modern farming methods.

Abstract: This study aims to analyze the technical and economic infrastructure of sheep farming enterprises operating in Türkiye. It assesses the demographic characteristics of enterprise owners, enterprise scales, production objectives, marketing strategies, and economic performance. Primary data were collected through face-to-face surveys conducted with 201 sheep farming enterprises during the 2023 production period. The sample was selected based on information provided by the Provincial Directorates of Agriculture and Forestry and the Breeding Sheep and Goat Breeders' Associations. Data analysis was performed using SPSS 27. Categorical data related to enterprise characteristics and the demographic profiles of enterprise owners were examined. The findings indicate that the majority of enterprise owners are middle-aged or older individuals, have a low level of education, and operate predominantly within an extensive production system. The producers' marketing methods rely mainly on direct sales. In conclusion, ensuring the sustainability of the sheep farming sector requires encouraging young producers to enter the industry, expanding educational programs, and adopting modern production techniques.

Keywords: enterprise structure; marketing; production systems; sheep farming; socio-economic structure

1. Introduction

The sheep husbandry sector represents a significant sub-branch of the livestock industry in Türkiye. This economic activity enables the production of various products, including meat, milk, wool, and leather [1,2]. Owing to their high adaptability, sheep can be raised across a wide range of regions by adapting to diverse geographical conditions.

The fact that sheep farming requires lower maintenance/feeding conditions and operating costs than cattle farming makes sheep farming an attractive economic activity, especially for small-scale enterprises [3]. In addition, sheep husbandry contributes to economic development by providing employment in rural areas [4]. In addition, the sheep husbandry sector is strategically important in meeting the demand for red meat because of the very limited production and consumption of pork in Türkiye. Sheep husbandry contributes to an increase in red meat consumption per capita [5,6].

Sheep husbandry is a prominent livestock activity in Türkiye, and is particularly prevalent in inland regions and areas with low precipitation. Despite its economic contributions, the sector's development is constrained by the low yield capacity of domestic sheep breeds and suboptimal management practices [3]. Following the establishment of the Republic in 1923, scientific research and policy support emphasized the development of the sector, initially prioritizing wool and meat production. However, with the global proliferation of synthetic fibers—particularly from the mid-20th century onward—competition in wool production diminished. Consequently, efforts have increasingly focused on enhancing meat and milk productivity [5].

At present, the state of sheep and goat husbandry in Türkiye remains below targeted levels, as indicated by expert reports on animal husbandry [7]. As of 2024, per capita red meat consumption (including beef, sheep, and goat meat) in Türkiye is about 16.6 kg, whereas the average for the European Union (EU-28), including pork, is 54 kg—39 kg of which is accounted for by pork alone [8]. In this context, strengthening the ovine livestock sector is essential for raising red meat consumption levels. Furthermore, the continued reliance on red meat imports underscores the sector's underutilized potential and its ongoing relevance to national food security and economic sustainability [5].

When the last 10-year period, independent of the past, is examined, a general increase in the number of sheep can be seen. The farming data for Ankara, Balıkesir, Erzurum, Eskişehir, and Konya Provinces, representing Türkiye and Türkiye in general for the years 2014–2023, are presented in Table 1 [9].

Table 1 reveals that there is a general upward trend in the number of sheep throughout Türkiye and in the provinces where the study was conducted between 2014 and 2023. The number of sheep in Türkiye has increased by about 35%, and differences in these rates are noteworthy on the basis of individual provinces. Konya stands out as the province with the highest number of sheep, while Ankara has recorded the largest increase in its sheep population. This growth is primarily attributed to imports rather than structural improvements or transformation, with most imports driven by investments from capitalists outside the sector [5]. Furthermore, there is evidence of a transition from extensive to intensive production methods in ovine farming, spurred by capital inflows [10]. In order to ensure the sustainability of this transformation, it is essential to conduct accurate and comprehensive analyses of sheep husbandry activities.

In recent years, various structural problems have been addressed [11–15], and socio-economic conditions have been examined [16–19] in studies on sheep husbandry in different districts and provinces of Türkiye. However, these studies are largely limited to the provincial or district level.

The main purpose of this study is to reveal the socio-economic, technical, and economic infrastructure of sheep husbandry enterprises operating in Ankara, Balıkesir, Erzurum, Eskişehir, and Konya Provinces in different geographical regions of Türkiye and the sectoral profile of the enterprises. The analysis is conducted at the micro level, based on enterprise-level characteristics such as demographics, production scale, and income structure.

Table 1. Total sheep (2014–2023) in provinces and in Türkiye and changes within the scope of the study, x1000 (2014 = 100).

Years	Ankara	Index	Balıkesir	Index	Erzurum	Index	Eskişehir	Index	Konya	Index	Türkiye	Index
2014	817	100.00	816	100.00	633	100.00	602	100.00	1896	100.00	31,140	100.00
2015	879	107.59	793	97.18	698	110.27	693	115.12	1862	98.21	31,508	101.18
2016	984	120.44	797	97.67	617	97.47	651	108.14	1827	96.36	30,984	99.50
2017	1233	150.92	984	120.59	615	97.16	644	106.98	1895	99.95	33,678	108.15
2018	1319	161.44	1006	123.28	650	102.69	706	117.28	2001	105.54	35,195	113.02
2019	1455	178.09	1112	136.27	689	108.85	927	153.99	2191	115.56	37,276	119.70
2020	1718	210.28	1300	159.31	799	126.22	1071	177.91	2557	134.86	42,127	135.28
2021	1940	237.45	1495	183.21	810	127.96	1153	191.53	2769	146.04	45,178	145.08
2022	1680	205.63	1272	155.88	912	144.08	1154	191.69	2771	146.15	44,688	143.51
2023	1620	198.29	1183	144.98	800	126.38	995	165.28	2793	147.31	42,060	135.07

2. Materials and Methods

The primary data of the study were obtained from field visits, and the survey method was applied to the enterprises carrying out sheep husbandry activities in Ankara, Balıkesir, Erzurum, Eskişehir, and Konya Provinces during the production cycle of 2023, which represented Türkiye. These provinces were selected because they are among the most active sheep production regions in the country, covering diverse geographical and climatic conditions and accounting for a significant share of the national sheep population. The questionnaire development process was based on the form designed by Gambelli et al. (2021) [20], with adaptations made to align with local conditions and the specific objectives of this study. In this context, the validity of the survey questions was first evaluated through a review of the relevant literature. Subsequently, the scope of the content was expanded based on feedback from industry experts and academics. In the final stage, a draft version of the survey was pilot-tested on a small sample group, and necessary revisions were made in light of the feedback received. The content of the questionnaires prepared for sheep husbandry enterprises included questions tailored to the statistical analysis methods employed. These questions aim to determine the current status and structural characteristics of sheep husbandry enterprises in the selected provinces. Face-to-face interviews were conducted with enterprise owners engaged in sheep husbandry using the questionnaires developed during the research process. Additionally, data were drawn from the five-year development plans of the Turkish Statistical Institute (TSI), the United Nations Food and Agriculture Organization (FAO), and the Ministry of Development, alongside secondary data compiled from relevant national and international literature.

In five provinces where the primary data of the study were obtained, a sample with high representation power was selected via a stratified random sampling method among the enterprises that are members of the Brood Sheep and Goat Breeders Association. The sample size was determined as a 95% confidence interval and 5% margin of error in line with the method reported by Güneş & Arıkan (1985) [21]. The calculation for this is presented below:

$$n_0 = \frac{Nt^2PQ}{d^2(N-1)+t^2PQ} = \frac{65265(1.96)^2 0.9 \times 0.1}{(0.05)^2(65265-1)+(1.96)^2 0.9 \times 0.1} \cong 195$$

[Table value t corresponding to confidence level (95%) = 1.96; probability of selecting an enterprise: P = 0.9; Q = 0.1].

N: Number of units in the population

P: Frequency of occurrence of the event to be examined

Q: Frequency of non-occurrence of the event to be examined

t: Theoretical value found in the t-table

d: Frequency of occurrence of the event

$Strata\ weight = \frac{195}{65265} = 0.003\ (whole\ sample)$ [22,23].

It is important to clarify that the value "195/65,265 = 0.003" refers to the overall sampling fraction, which indicates the proportion of the selected sample to the total population. This value is distinct from the t-value used in the sample size determination formula. In this context, the t-value (1.96) represents the critical value from the t-distribution corresponding to a 95% confidence level, whereas the sampling fraction is used solely for proportional allocation in the stratified sampling design.

Table 2 shows the distribution of enterprises by province and stratum.

In Table 3, information on the number of enterprises determined according to the scale after each province is presented.

Of the 195 enterprises included in the study, 72 were categorized as small, 79 as medium, and 45 as large based on their flock size, as shown in Table 3.

Table 2. The number of samples determined according to stratified provinces.

Provinces	Number of Enterprises	W_h *	n_h **
Ankara	10,121	0.16	30
Balıkesir	20,821	0.32	62
Eskişehir	8670	0.13	26
Erzurum	7045	0.11	21
Konya	18,608	0.29	56
Total	65,265	1	195

* Wh: strata weight; ** nh: number of individuals per strata.

Within the scope of the study, about 10% of the total sample size, that is, 20 enterprises, were determined as reserves. In this context, 215 enterprises were visited. After removing those enterprises that refused to participate in the questionnaires or whose data were not reliable, the number of valid enterprises surveyed was determined to be 201. This process was carried out carefully to ensure the accuracy of the data. Consequently, the actual number of enterprises visited in each province (as presented in Figure 1) may slightly differ from the planned sample sizes ('n_h') shown in Table 2.

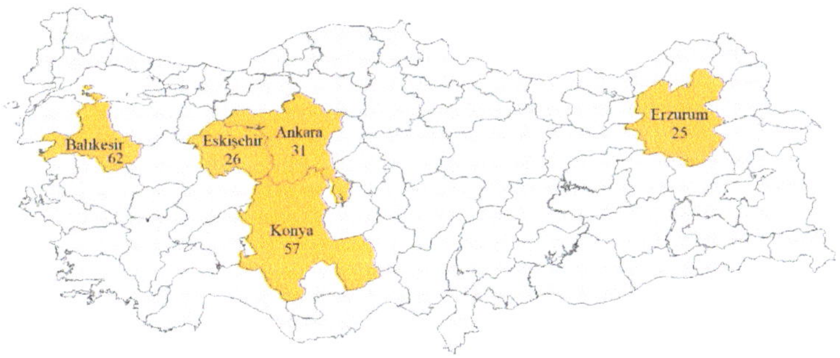

Figure 1. Provinces where sheep husbandry enterprise owners carry out their activities.

Sheep husbandry enterprises were identified through consultations with the Brood Sheep and Goat Breeders' Associations, as well as the District Directorates of Agriculture and Forestry in the relevant provinces and districts. Time and budget constraints were taken into consideration to ensure the collection of accurate and reliable data.

The survey data collected from sheep husbandry enterprises were entered into the computer using Microsoft Office 365 programs and subsequently processed using the Statistical Package for the Social Sciences (SPSS) version 27 (IBM Corp., Armonk, NY, USA), making them ready for analysis.

Within the scope of the study, enterprise characteristics such as enterprise type, enterprise scale, and questions such as the age, education level, and experience period of the enterprise owners were designed as categorical data. The number (n) and percentage (%) values were determined for the categorical data obtained through the questions. The arithmetic mean, standard deviation, median, minimum, and maximum values were used to evaluate the income of enterprise owners from activities such as milk sales, livestock sales, brood sales, and fleece sales. Findings on the market type of sheep husbandry enterprises were evaluated via "Multiple Response" analysis since enterprise owners can mark more than one option.

Table 3. Distribution of enterprises by scale in selected provinces.

Provinces	Enterprises			Number of Enterprises								Example Total
	Small *	Medium *	Large *		W_h ** (Small)	W_h ** (Medium)	W_h ** (Large)		n_h *** (Small)	n_h *** (Medium)	n_h *** (Large)	
Ankara	2600	3703	3818	10,121	0.26	0.37	0.38	1	8	11	11	30
Balıkesir	10,121	9189	1511	20,821	0.49	0.44	0.07	1	30	27	5	62
Eskişehir	3257	3370	2043	8670	0.38	0.39	0.24	1	10	10	6	26
Erzurum	2722	2615	1708	7045	0.39	0.37	0.24	1	8	8	5	21
Konya	5238	7531	5839	18,608	0.28	40	0.31	1	16	23	17	56
Total	23,938	26,408	14,919	65,265					72	79	45	195

* Enterprises are classified based on flock size: Small (1–50 animals), Medium (51–200 animals), and Large (201 and more animals). ** Wh: strata weight; *** nh: number of individuals per strata.

3. Results

The number (n) values of the provinces where sheep husbandry enterprise owners, who contributed to the survey questions in the study with their answers, carried out sheep husbandry activities are presented in Figure 1.

Figure 1 shows that the majority of the participants carried out sheep husbandry activities in Balıkesir and Konya Provinces, with rates of 30.8% and 28.4%, respectively.

In the present study, 200 (99.5%) of the 201 participants were men, and only 1 (0.5%) was a woman.

The number (n) and percentage (%) value findings regarding the age and educational status of sheep husbandry enterprise owners are given in Table 4.

Table 4. Demographic characteristics of the study participants.

	Statistics	
Characteristics	**Number (n)**	**Percentage (%)**
Breeder's Age		
18–25	2	1.0
26–40	37	18.4
41–50	57	28.4
51 and above	105	52.2
Total	201	100.0
Education level		
Literate	7	3.5
Primary Education	123	61.2
High School	50	24.9
University	19	9.4
Postgraduate	2	1.0
Total	201	100.0
Course/training status on sheep husbandry		
Those who have not taken a course/training on sheep husbandry	114	56.7
Those who took courses/trainings on sheep husbandry	87	43.3
Total	201	100.0

An examination of Table 4 reveals that the majority of sheep husbandry enterprise owners are aged 51 and above, accounting for 52.2% of the sample. In terms of educational attainment, primary school graduates constitute the largest group at 61.2%, followed by high school graduates at 24.9% and university graduates at 9.4%. The proportion of enterprise owners with postgraduate education is notably low, at just 1%. Additionally, 43.3% of the participants reported having received courses or training related to sheep husbandry, while 56.7% indicated that they had not received any such training.

The number (n) and percentage (%) values concerning the agricultural and livestock sector experience of sheep owners, as well as their specific experience in ovine livestock, are presented in Table 5.

According to Table 5, when the duration of experience in the agriculture and livestock sector is examined, the majority of the participants (66.7%) have 21 years or more of experience. This rate is 49.8% in ovine livestock. Those with 11–20 years of experience are 22.9% and 21.9%, those with 6–10 years of experience are 7.5% and 17.4%, and those with 1–5 years of experience are 3.0% and 10.9%, respectively.

Table 5. Experience-related characteristics of the study participants.

	Statistics	
	Number (n)	**Percentage (%)**
Years of experience in agriculture and livestock sector		
1 to 5	6	3.0
6 to 10	15	7.5
11 to 20	46	22.9
21 and above	134	66.7
Total	201	100.0
Duration of experience in the ovine livestock sector (Years)		
1 to 5	22	10.9
6 to 10	35	17.4
11 to 20	44	21.9
21 and above	100	49.8
Total	201	100.0

The number (n) and percentage (%) findings regarding the type of enterprise, scale of the enterprise, production purpose of the enterprise, and lambing per litter status of sheep husbandry enterprises, in which the characteristics of the enterprise are examined, are given in Table 6.

Table 6. Enterprise characteristics of the study participants.

	Statistics	
	Number (n)	**Percentage (%)**
Enterprise Scale		
Small (1–50)	72	35.8
Medium (51–200)	79	39.3
Large (201 and above)	50	24.9
Total	201	100.0
Enterprise type		
Extensive	125	62.2
Semi-intensive	70	34.8
Intensive	6	3.0
Total	201	100.0
Purpose of enterprise owners for raising animals		
For fatlings	134	66.7
For both fatling and milk production (combined)	51	25.3
For breeding purposes	10	5.0
For milk production	6	3.0
Total	201	100.0
Lambing/litter status		
Once a year	116	57.7
Three times in two years	65	32.3
Twice a year	20	10.0
Total	201	100.0

When the findings related to Table 6 are evaluated, the scale distributions of enterprises are 35.8%, 39.3%, and 24.9%, respectively, from small to large. When the type of enterprise is examined, the percentage of enterprises that produce extensively is 62.2%, and the percentage of semi-intensive enterprises is 34.8%. The rate of intensive enterprises is quite low, at only 3%. For the purpose of raising livestock, 66.7% of the enterprise owners prefer to breed fatlings, while the percentage of enterprises aiming at both fatling and milk production is 25.3%, and the percentage of enterprises operating for breeding purposes constitute 5% of the sample, while those focused on milk production account for just 3%. Regarding lambing frequency, 57.7% of enterprises reported lambing once per year. Additionally, 32.3% of enterprises lamb three times every two years, and 10% report lambing twice a year.

Findings related to the dairy and livestock marketing practices of sheep farm owners, as well as their marketing methods, are presented in Table 7.

Table 7. Milk and livestock marketing status and marketing methods used in sheep husbandry enterprises *.

	Statistics	
	Frequency (n)	**Percentage (%)**
Milk marketing status (processed or unprocessed)		
Marketed	96	47.8
Not marketed	105	52.2
Total	201	100.0
Milk marketing methods * (processed or unprocessed)		
Direct sale by hand	89	70.1
Small shops	27	21.2
Contract with a retail company	8	6.3
Online sales	3	2.4
Total	127	100.0
Live animal marketing status		
Marketed	198	98.5
Not marketed	3	1.5
Total	201	100.0
Livestock marketing methods *		
Direct sale by hand	188	70.7
Animal Market	57	21.4
Contract with a retail company	8	3.0
Online sales	13	4.9
Total	266	100.0

* Participants selected more than one option.

As shown in Table 7, various aspects of the technical infrastructure of sheep husbandry enterprises are presented. It is observed that the majority of enterprises (52.2%) do not engage in milk marketing, while a very high proportion (98.5%) are involved in the sale of live animals. In both product categories, direct hand-to-hand sales emerge as the most commonly used marketing method.

The findings regarding the arithmetic mean ($\overline{X}$), standard deviation (SD), median (M), minimum (min), and maximum (max) values of the enterprise owners to evaluate their income from activities such as annual milk sales, live animal sales (butchery/sacrifice), brood sales, and fleece sales are shown in Table 8. In Figure 2, the average income levels of enterprise owners according to sales income elements are presented as percentages.

Table 8. Distribution of sales revenues of the participants (total Turkish Lira (TRY) per year).

	Statistics			
Sales Income Elements	$\overline{X}$	*SD*	*Median*	*Min–Max*
Milk sales (TRY Thousand)	27.3	106.9	0	0–1000
Livestock sales—sacrifice (TRY Thousand)	155.8	241.3	75	0–2000
Livestock sales—butchery (TRY Thousand)	327.3	499.3	125	0–3000
Brood sales (TRY Thousand)	94.8	269.5	0	0–1700
Fleece sales (TRY Thousand)	0.4	1.5	0	0–10
Total sales (TRY Thousand)	605.6	700.8	362	30–4309

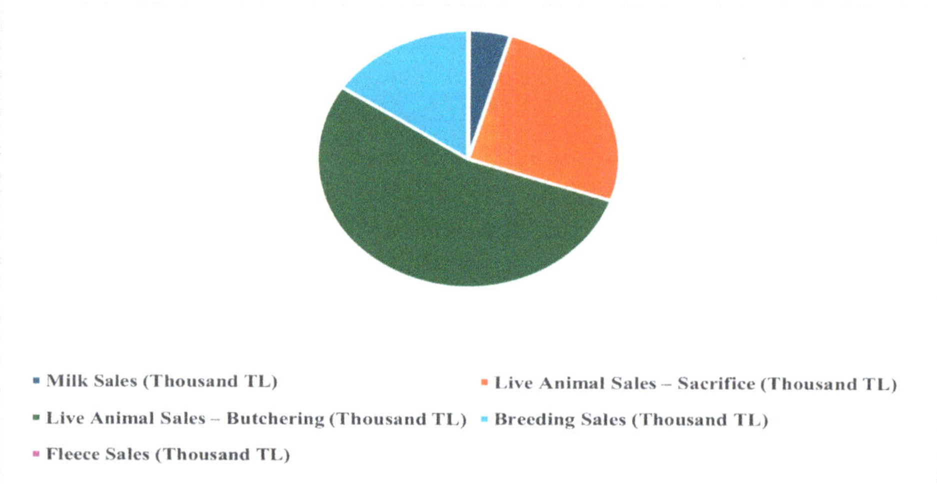

Figure 2. Percentage income averages of the participants according to sales income elements.

An examination of the data in Table 8 reveals that different sales income items in the livestock sector vary widely. Fleece sales are the lowest source of income, with an annual average of TRY 0.41 thousand (0.07%), whereas butchery animal sales have the highest average, with an annual average of TRY 327.32 thousand (54.05%). Sacrificial animal sales represent a significant income component, with an average annual revenue of TRY 155.77 thousand (25.77%), followed by brood animal sales at TRY 94.79 thousand (15.65%) (Figure 2). Total sales revenues exhibit considerable variation, with an annual average of TRY 605.61 thousand (100.00%) (SD: 700.81; Min–Max: 30–4309), indicating substantial income disparities among livestock enterprises.

4. Discussion

In this study, it was determined that the majority of the producers in the sheep husbandry sector were 41 years old or older (Table 4). Similarly, studies conducted in Konya, Siirt, and Çanakkale have shown that enterprise owners are generally concentrated in middle-aged and older groups [14,24,25]. Other studies across Türkiye have revealed similar results that are consistent with these findings. This shows that the sector is based on more traditional production methods and that the demand for ovine breeding by young populations is insufficient. In studies conducted in Greece, South Wales, and South Australia, the sheep breeding sector has been carried out mainly by middle-aged and older groups [26–28]. However, there are supportive policies in developed countries that encourage the participation of young people in the sector. The implementation of similar practices in Türkiye may be critical for the modernization and sustainability of the sector.

The study found that the majority of sheep husbandry enterprise owners (61.2%) were primary school graduates (Table 4). Similar findings have been reported in studies conducted in Konya, Selçuklu, and Aksaray [14,19,29]. In contrast, the majority of enterprise owners in countries such as Greece, Norway, and Chile possess at least a secondary level of education [26,30,31]. The relatively lower education levels observed among enterprise owners in Türkiye may pose limitations in areas such as financial planning, adoption of modern technologies, and effective enterprise management.

In this study, 43.3% of the enterprise owners received courses or training on sheep husbandry (Table 4). This rate is higher in Muş, Konya, and Ardahan Provinces than in previous studies [16,19,32,33]. This finding, which is higher than the rates reported in Greece (15.1%) and Brazil (30.6%), shows that participation in vocational training in Türkiye is similar to that reported in European countries in some regions [26,34]. However, considering the regional differences throughout the country, it is important to disseminate educational activities at the national level in the sheep husbandry sector.

The study determined that most sheep husbandry enterprise owners had 21 years or more of experience in the sector (Table 5). Similar findings have been reported in studies conducted in the provinces of Konya, Siirt, and Çanakkale [14,24,25]. In Greece and Chile, the average experience ranges around 24 years, while in Norway it falls between 21 and 25 years [25,30,31], aligning closely with the findings from Türkiye. These results highlight that sheep husbandry is a sector that is globally grounded in long-term knowledge and experience. However, a high level of experience does not always correspond with parallel development in modernization. In many countries, including Türkiye, it is observed that business owners who have been engaged in the profession for many years tend to adhere to traditional methods. This tendency may hinder the adoption of modern technologies and the integration of younger generations into the sector. Therefore, alongside experience, it is essential to provide continuous professional development opportunities and to expand training programs focused on digital technologies and innovative production methods. Such efforts would not only enhance the capacities of experienced producers but also support the overall modernization of the sector.

In this study, it was determined that enterprise owners conducted substantially extensive production (62.2%), followed by semi-intensive production (34.8%) (Table 6). In a study conducted in Aksaray, 94.2% of enterprises preferred semi-intensive production, and 96% preferred semi-intensive production in Mersin [29,35]. This difference may be due to differences in breeding habits, enterprise structure, pasture facilities, and regional conditions. In this study, which was conducted in five provinces, the wide range of pasture opportunities in some regions may have caused enterprises to turn to extensive production. It has been reported that 78% of enterprises in northeastern Brazil operate under semi-intensive systems, while 14% utilize extensive systems [34]. A study conducted in Europe found that 37.8% of enterprises employed extensive systems, and 34.4% adopted semi-intensive methods [36]. In Greece, the rate of semi-intensive production is 38.06% [26]. Findings from Türkiye indicate that production systems vary across regions due to differing local conditions; however, semi-intensive production appears to be a widely adopted approach on a global scale. Structural factors such as the regional distribution of agricultural support policies, the level of infrastructure investment, and the effectiveness of producer organizations play a decisive role in shaping producers' preferences for production systems in Türkiye. In particular, the inadequacy of support mechanisms and the limited marketing channels in the Eastern and Southeastern Anatolia regions may lead to a preference for extensive production systems. Conversely, in more developed regions such as Central Anatolia and the Aegean, the tendency toward semi-intensive systems can provide producers with higher added value. This situation highlights the need for a comprehensive approach to address the structural and policy factors underlying regional differences in production systems.

This study found that the vast majority of sheep husbandry enterprises raise animals primarily for fatling purposes (Table 6). In a study conducted in Aksaray, 47.8% of enterprises engaged in the production of meat, milk, and wool, while in Mersin, 72% of enterprises preferred the combined production of meat and milk [29,35]. The dominance of fatling cultivation is thought to be due to factors such as regional differences, producer preferences, pasture use, and market conditions. In a study conducted in Muğla, 46% of the enterprises were engaged in meat production only, while 34% were involved in combined production of both meat and milk [37]. In Brazil, 49% of enterprises produce milk, 37% produce meat, and 14% have combined production; in Europe, 67.8% produce milk and 32.2% produce meat [34,36]. The observed variations in results may be attributed to factors such as climatic conditions, pasture facilities, consumer demands, support policies, and regional production traditions.

In this study, it was seen that lambing occurs once a year in 67.7% of the enterprises and twice in 10% (Table 6). In a study conducted in Konya, it was determined that 83.2% of the enterprises had a single lambing per year, while 8.4% had lambing twice a year, and in Bursa, these rates were 68.1% and 6.4%, respectively [14,38]. A single lambing per year is common throughout Türkiye, and farmers are hesitant due to maintenance and feeding difficulties associated with more than one lamb. In contrast, the frequency of lambing is higher in Spain and Egypt, with an average of 1.5 in Spain and 8–10 months in Egypt [39,40]. This shows that the range of lambing in Türkiye is longer than that in international samples.

In this study, 47.8% of the enterprises marketed milk and dairy products, and 70.1% made these sales directly to consumers (Table 7). In a similar study conducted in Aksaray, located in the Central Anatolia Region of Türkiye, the milk marketing rate was reported as 71.6% [29], while in Ardahan, from the Eastern Anatolia Region, it was 45.5% [33]. Direct sales are common throughout Türkiye, although regional and sectoral differences significantly influence marketing methods. To provide an international perspective and contextualize the findings, comparable trends can be observed in other countries as well. For example, in Greece, 33.1% of milk producers sell to local milk processing plants, and 24.8% to national milk companies [41]. Similarly, in Italy, 73.68% of traditional cheese producers prefer direct sales to companies [42]. In Sicily, 70% of dairy products are sold in wholesale and 30% are sold in retail [43]. These international examples help to highlight that while direct sales are prevalent across different contexts, the structure of marketing channels may vary depending on regional production dynamics, policy environments, and market organization—parallels that can also be observed in Türkiye.

It was observed that 98.5% of the enterprise owners marketed live animals, and 70.7% of them used the direct sale method (Table 7). Similarly, in studies conducted in Aksaray and Niğde, which are located in the Central Anatolia Region of Türkiye, direct sales were also commonly practiced. These provinces are frequently used for comparison in the literature due to their similar regional, structural, and socio-economic characteristics to the study area [13,29]. While marketing methods among enterprises in Türkiye exhibit similar trends, data from Brazil show that 74.4% of producers market their animals for economic purposes, with 38.5% selling directly by hand and 33.3% marketing through unions [44]. In Greece, 56.7% of enterprises sell directly to wholesalers, while 21.7% sell to retailers [41]. Although the prevalence of direct sales in Türkiye is comparable to that in Brazil and Greece, notable differences exist in the use of sales channels. This suggests that the livestock sector in Türkiye remains heavily reliant on traditional marketing practices and that modern cooperative structures have yet to be fully developed.

Most annual sales revenues in sheep husbandry enterprises in Türkiye are derived from the sale of butchery and sacrificial animals, while the share of revenue from brood animal and milk sales remains relatively low (Table 8). Studies based on provincial-level data from Konya Province (Selçuklu district) in Türkiye's Central Anatolia Region and Muş Province (Korkut district) in the Eastern Anatolia Region reported that the share of lamb and red meat sales in total household income were 72.2% and 63.5%, respectively. These findings are consistent with the present study [16,19]. In contrast, the high productive asset increase (PDKA) rate of 56.4% observed in Yozgat indicates that enterprises there tend to adopt a herd growth strategy [17]. In general, butchery and sacrificial animal sales are the main sources of income in Türkiye. In studies conducted in other countries, milk sales in Spain constitute 78.6% of revenues, whereas the income distribution of sheep husbandry enterprises in France, Spain, and the United Kingdom varies according to the production system [45,46]. The dominance of meat sales in Türkiye can largely be attributed to the economic impact of Eid al-Adha, whereas the prominence of milk revenues in some other countries reflects variations in enterprise structures and consumer preferences. This indicates that sheep husbandry in Türkiye is predominantly shaped by periodic and traditional demand patterns. To ensure long-term sustainability, it is essential to develop a more balanced income model within the sector.

In addition to the academic and sectoral implications of the findings, it is crucial to consider how these results can be effectively communicated to the wider public. Given the growing prevalence of misinformation, particularly in animal science, social media platforms such as Twitter (X), Facebook, and Instagram offer valuable opportunities to share accurate, research-based knowledge with a broad audience. Transparent and accessible dissemination of information can enhance public understanding of the socio-economic dynamics of sheep husbandry and combat misconceptions. In a parallel vein, a study focused on utilizing Instagram illustrates how social media can serve as an effective tool for agricultural outreach [47]. This study underscores the power of digital platforms in conveying complex topics, such as the socio-economic structure of sheep enterprises, to a diverse audience. Promoting engagement through modern communication channels not only increases the visibility of research but also fosters trust and awareness among consumers, producers, and policymakers alike.

5. Conclusions

As a result, this research makes an important contribution to the literature by addressing the sheep husbandry sector in Türkiye not only with economic indicators but also with a multidimensional perspective in the context of producer profiles and structural practices. This study presents the structural profile of the sheep farming sector in Türkiye based on a sample drawn from five key provinces that are among the most active sheep production regions across different geographical areas of the country. These provinces not only represent diverse regional conditions but also account for a significant share of national sheep production, thereby enhancing the relevance of the findings for understanding sectoral dynamics at the national level. The selected regions provide a robust basis for gaining a broader perspective on the structural characteristics of sheep farming in Türkiye. In this respect, it not only captures the on-the-ground realities of the field but also reinforces the link between scientific knowledge and sectoral policymaking.

The dominance of traditional production and marketing methods in the sheep husbandry sector reveals that the development potential of the sector cannot be adequately evaluated. This clearly underscores the need for structural transformations aimed at enhancing efficiency, sustainability, and profitability in the sector. The inclusion of the young population in the sector, strengthening producers' access to educational opportunities, and comprehensive policies that encourage the adoption of modern production techniques are expected to form the basis of this transformation.

Limitations and Future Research

This study focuses on five key provinces where sheep husbandry is prevalent in Turkey, which, while not entirely limiting, may somewhat constrain the generalizability of the findings to the national level. Moreover, due to time and budgetary constraints, the analysis primarily relied on descriptive statistics. Consequently, the study's ability to uncover causal relationships and deeply explore socio-economic interactions remains limited. Another limitation is the lack of an integrated theoretical framework, which future studies should address to strengthen academic contributions.

Future research is recommended to expand the geographical scope to include additional regions and to employ more advanced quantitative methods, such as regression and econometric modeling, alongside qualitative approaches. Such comprehensive analyses would facilitate a more nuanced understanding of the interrelations between economic performance and social factors—such as gender roles and knowledge transfer—within the sector. This, in turn, would contribute to better-informed policy-making aimed at supporting the sustainable development of sheep husbandry.

Author Contributions: Conception—B.M. and M.B.Ç.; Design—B.M. and M.B.Ç.; Supervision—A.G., B.M. and M.B.Ç.; Data Collection and/or Processing—A.V. and B.B.; Analysis and/or Interpretation—B.B.; Literature Search—A.V. and B.B.; Writing Manuscript—A.V.; Critical Review—A.G. All authors have read and agreed to the published version of the manuscript.

Funding: This research was supported by the Selçuk University Scientific Research Projects Coordination Unit, Project No. 23401052.

Institutional Review Board Statement: Ethics committee approval was received for this study from the ethics committee of Selçuk University (Approval number: 2023/28).

Data Availability Statement: The data that support the findings of this study are available from the corresponding author (A. Varalan) on reasonable request.

Conflicts of Interest: The authors declare no conflicts of interest.

References

1. İşler, H.; Ünlü Ören, H.G. Examination and SWOT analysis of the livestock sector in Burdur province. *J. Appl. Soc. Sci. Fine Arts* **2021**, *3*, 166–190.
2. TİGEM. 2023 yılı Hayvancılık Sektör Raporu. 2023. Available online: https://www.tigem.gov.tr/Folder/CarouselDosyasi/d722366d-7a4d-4929-ab06-10bc2614778e.pdf (accessed on 4 August 2024).
3. Gökmener, H.; Öztürk, A. Ovine livestock activities and general problems in Uzundere district of Erzurum province. *J. Bahri Dagdas Anim. Res.* **2022**, *11*, 21–29.
4. Skapetas, B.; Kalaitzidou, M. Current status and perspectives of sheep sector in the world. *Livest. Res. Rural. Dev.* **2017**, *29*, e2000.

5. Günlü, A.; Mat, B. Türkiye ekonomisinde koyun keçi yetiştiriciliğinin yeri ve önemi. In *Kuzu ve oğlak Kayıplarının Önlenmesinde Koyun Keçi Sağlığı ve Yetiştiriciliği*; Erdem, H., Çiftci, E., Işık, M.K., Yorgancılar, M.Ü., Eds.; Akademisyen Yayınevi: Ankara, Turkey, 2021; pp. 3–14.
6. Ertaş, N. Türkiye's place in the world red meat market from past to present in terms of production, consumption and marketing. *J. Acad. Soc. Resour.* **2023**, *8*, 2191–2214. [CrossRef]
7. Başkanlığı, V.B.B. Specialized Commission Reports. 2025. Available online: https://www.sbb.gov.tr/ozel-ihtisas-komisyonu-raporlari/ (accessed on 22 January 2025).
8. FAOSTAT. Food and Agriculture Data. 2024. Available online: https://www.fao.org/faostat/en/#home (accessed on 21 January 2025).
9. Turkish Statistical Institute (TURKSTAT). Turkish Statistical Institute. 2024. Available online: https://biruni.tuik.gov.tr/medas/?kn=101&locale=tr (accessed on 21 January 2025).
10. Cedden, F.; Cemal, İ.; Daşkıran, İ.; Esenbuğa, N.; Gül, S.; Kandemir, Ç.; Karaca, O.; Kaymakçı, M.; Keskin, M.; Koluman, N.; et al. Türkiye küçükbaş hayvancılığında mevcut durum ve gelecek. In Proceedings of the Türkiye Ziraat Mühendisliği IX. Teknik Kongresi, Ankara, Türkiye, 13–17 January 2020.
11. Kılıç, İ.; Bozkurt, Z.; Tekerli, M.; Koçak, S.; Çelikeloğlu, K. The perceptions of employees in Afyonkarahisar sheep enterprises on factors affecting animal 0welfare. *Lalahan Hayvancılık Araştırma Enstitüsü Derg.* **2013**, *53*, 29–38.
12. Şahinli, M.A. Sheep flock management: Case of Karaman Province. *Anadolu J. Agric. Sci.* **2014**, *29*, 113–120. [CrossRef]
13. Ceyhan, A.; Şekeroğlu, A.; Ünalan, A.; Cingi, C.C.; Erdoğan, S. A research on structural characteristics and problems of sheep breeding in Nigde. *KSÜ J. Nat. Sci.* **2015**, *18*, 60–68. [CrossRef]
14. Arıtunca, D.; Karabacak, A. The status of sheep farm in central district of Konya. *J. Bahri Dagdas Anim. Res.* **2020**, *9*, 13–24.
15. Gürer, B.; Ulutaş, Z. Analysis of the factors affecting the farms' probability continuing sheep farming in the TR71 Region. *Anadolu J. Agric. Sci.* **2021**, *36*, 63–72. [CrossRef]
16. Kaymak, K.; Sarıözkan, S. Socio-economic structure and production costs of sheep breeding enterprises in Korkut district of Muş province. *Van Vet. J.* **2016**, *27*, 141–146.
17. Tamer, B.; Sarıözkan, S. Socio-economic structure and production costs of sheep farms in central district of Yozgat province. *Erciyes Univ. J. Vet. Fac.* **2017**, *14*, 39–47.
18. Bakırtaş, Ş.; Günlü, A. Technical and socio-economics analysis of goat breeding enterprises in region of Aksaray province. *Eurasian J. Veter- Sci.* **2018**, *34*, 134–141. [CrossRef]
19. Mohamud, A.H.; Çevrimli, M.B. Technical and economic analysis of sheep breeding enterprises in Selcuklu district of Konya province. *Kocatepe Vet. J.* **2023**, *16*, 342–356. [CrossRef]
20. Gambelli, D.; Solfanelli, F.; Orsini, S.; Zanoli, R. Measuring the economic performance of small ruminant farms using balanced scorecard and importance-performance analysis: A European case study. *Sustainability* **2021**, *13*, 3321. [CrossRef]
21. Güneş, T.; Arıkan, R. *Tarım Ekonomisi Istatistiği*, 1st ed.; Ankara Üniversitesi Ziraat Fakültesi Yayınları: Ankara, Turkey, 1988.
22. Botev, Z.; Ridder, A. Variance reduction. In *Wiley StatsRef: Statistics Reference Online*; Balakrishnan, N., Colton, T., Everitt, B., Piegorsch, W., Ruggeri, F., Teugels, J.L., Eds.; John Wiley & Sons: Hoboken, NJ, USA, 2017; pp. 1–6.
23. Esfahani, S.M.; Dougherty, E.R. Effect of separate sampling on classification accuracy. *Bioinformatics* **2013**, *30*, 242–250. [CrossRef]
24. Bakır, G.; Mikail, N. Structural characteristics of sheep and goat farms in Siirt province. *Atatürk Univ. J. Agric. Fac.* **2019**, *50*, 66–74. [CrossRef]
25. Özsayin, D.; Everest, B. Socio-economic structure of farmers that make sheep breeding and practices related to their sheep breeding activities. *Ksu Tarim Ve Doga Dergisi-Ksu J. Agric. Nat.* **2019**, *22*, 440–448. [CrossRef]
26. Lianou, D.T.; Fthenakis, G.C. Dairy sheep and goat farmers: Socio-demographic characteristics and their associations with health management and performance on farms. *Land* **2021**, *10*, 1358. [CrossRef]
27. Kopp, K.; Hernandez-Jover, M.; Robertson, S.; Abuelo, A.; Friend, M. A survey of New South Wales sheep producer practices and perceptions on lamb mortality and ewe supplementation. *Animals* **2020**, *10*, 1586. [CrossRef]
28. Bates, A.L.; McGrath, S.R.; Allworth, M.B.; Robertson, S.M.; Refshauge, G. A cross-sectional study of commercial ewe management practices for different sheep breeds across Southern Australia. *Animals* **2023**, *13*, 388. [CrossRef]
29. Gül, S.; Oflaz, N.; Keskin, M.; Behrem, S. The structural characteristics of sheep farming in Aksaray province. *ETHABD* **2022**, *5*, 1–11. [CrossRef]
30. Muri, K.; Tufte, P.A.; Coleman, G.; Moe, R.O. Exploring work-related characteristics as predictors of Norwegian sheep farmers' affective job satisfaction. *Sociol. Rural.* **2020**, *60*, 574–595. [CrossRef]
31. Larrondo, C.; Bustamante, H.; Gallo, C. Sheep farmers' perception of welfare and pain associated with routine husbandry practices in Chile. *Animals* **2018**, *8*, 225. [CrossRef]
32. Şahinli, M.A. Konya Ilinde Koyunculuk Faaliyetine yer Veren Tarım Işletmelerinin Ekonomik Analizi ve Koyunculuk Faaliyetinde Etkili Olan Unsurların Saptanması. Ph.D. Thesis, Ankara Üniversitesi Fen Bilimleri Enstitüsü, Ankara, Turkey, 2011.

33. Ayvazoğlu Demir, P.; Işık, S.A.; Aydın, E.; Yazıcı, K.; Ayvazoğlu, C. Socio-economic importance of sheep breeding farms in Ardahan province. *Van Vet. J.* **2015**, *26*, 141–146.
34. Guilherme, R.D.F.; Lima, A.M.C.; Alves, J.R.A.; da Costa, D.F.; Pinheiro, R.R.; Alves, F.S.F.; de Azevedo, S.S.; Alves, C.J. Characterization and typology of sheep and goat production systems in the State of Paraíba, a semi-arid region of northeastern Brazil. *Semin. Agrar.* **2017**, *38*, 2163–2178. [CrossRef]
35. Tüney Bebek, D.; Keskin, M. Current status, some yield and structural properties of sheep breeding in Mersin. *Mustafa Kemal Univ. J. Agric. Fac.* **2018**, *23*, 315–323.
36. Belanche, A.; Martín-Collado, D.; Rose, G.; Yáñez-Ruiz, D. A multi-stakeholder participatory study identifies the priorities for the sustainability of the small ruminants farming sector in Europe. *Animal* **2021**, *15*, 100131. [CrossRef]
37. Aydin, M.K.; Keskin, M. Muğla ilinde küçükbaş hayvan yetiştiriciliğinin yapısal özellikleri. *Mediterr. Agric. Sci.* **2018**, *31*, 317–323. [CrossRef]
38. Dönmez, O. Bursa Ili Koyunculuk Işletmelerinin Yetiştiricilik Açısından Yapısı. Master's Thesis, Namık Kemal Üniversitesi Fen Bilimleri Enstitüsü, Tekirdağ, Turkey, 2008.
39. Toro-Mujica, P.; García, A.; Gómez-Castro, A.G.; Acero, R.; Perea, J.; Rodríguez-Estévez, V.; Aguilar, C.; Vera, R. Technical efficiency and viability of organic dairy sheep farming systems in a traditional area for sheep production in Spain. *Small Rumin. Res.* **2011**, *100*, 89–95. [CrossRef]
40. Abdallah, S. Assessment of the village sheep production systems under the prevalent management practices in the Nile Delta of Egypt. *Egypt. J. Veter Sci.* **2024**, *55*, 517–530. [CrossRef]
41. Tsourgiannis, L.; Errington, A.; Eddison, J. *Marketing Strategies of Agricultural Producers in Objective One Greek Regions: The Factors Affecting the Selection of Marketing Channels of Sheep and Goat Producers*; University of Plymouth, School of Geography: Plymouth, UK, 2002.
42. Bonadonna, A.; Peira, G.; Giachino, C.; Molinaro, L. Traditional cheese production and an EU labeling scheme: The Alpine cheese producers' opinion. *Agriculture* **2017**, *7*, 65. [CrossRef]
43. Schimmenti, E.; Viola, E.; Funsten, C.; Borsellino, V. The contribution of geographical certification programs to farm income and rural economies: The case of pecorino siciliano PDO. *Sustainability* **2021**, *13*, 1977. [CrossRef]
44. Matte, A.; Waquil, P.D. Changes in markets for lamb in livestock family farming in Brazil. *Small Rumin. Res.* **2021**, *205*, 106535. [CrossRef]
45. Milán, M.; Frendi, F.; González-González, R.; Caja, G. Cost structure and profitability of Assaf dairy sheep farms in Spain. *J. Dairy Sci.* **2014**, *97*, 5239–5249. [CrossRef]
46. Theodoridis, A.; Vouraki, S.; Morin, E.; Rupérez, L.R.; Davis, C.; Arsenos, G. Efficiency analysis as a tool for revealing best practices and innovations: The case of the sheep meat sector in Europe. *Animals* **2021**, *11*, 3242. [CrossRef]
47. Lamanna, M.; Muca, E.; Buonaiuto, G.; Formigoni, A.; Cavallini, D. From posts to practice: Instagram's role in veterinary dairy cow nutrition education—How does the audience interact and apply knowledge? A survey study. *J. Dairy Sci.* **2025**, *108*, 1659–1671. [CrossRef]

ruminants

MDPI

Article

Head Sexual Characterization of Sanmartinero Creole Bovine Breed Assessed by Geometric Morphometric Methods

Arcesio Salamanca-Carreño [1,*], Pere M. Parés-Casanova [2], Mauricio Vélez-Terranova [3], David E. Rangel-Pachón [1], Germán Martínez-Correal [4] and Jaime Rosero-Alpala [5]

1 Facultad de Medicina Veterinaria y Zootecnia, Universidad Cooperativa de Colombia, Villavicencio 50001, Colombia; david.rangelp@campusucc.edu.co
2 Departamento de Bromatologia, Universitat Oberta de Catalunya, 08035 Catalonia, Spain; ppares@uoc.edu
3 Facultad de Ciencias Agropecuarias, Universidad Nacional de Colombia Sede Palmira, Palmira 763531, Colombia
4 Asociación de Criadores de Bovinos de Razas Criollas y Colombianas de los Llanos Orientales, Villavicencio 50001, Colombia
5 Corporación Colombiana de Investigación Agropecuaria-Agrosavia, Villavicencio 50001, Colombia
* Correspondence: asaca_65@yahoo.es

Simple Summary: The study of sexual dimorphism of the head can provide information on natural selection and adaptation to different environments. The head sexual dimorphism of the Sanmartinero Creole bovine through geometric morphometric was studied. Research on sexual dimorphism of the head in Creole cattle is limited. In Colombia, the Sanmartinero Creole cattle are known, an animal adapted to the foothills and high plains of the Meta department (Colombia). The aim was to compare, using geometric morphometrics, the head shape (size + shape) of the Sanmartinero Creole breed bovine according to the animals' sex. The heads of males and females differed in size and shape. Females had smaller heads. This study provides valuable insights into the morphological characteristics of a Creole bovine breed, contributing to veterinary anatomy, biology, and domestic breed phylogeny.

Abstract: Geometric morphometrics is performed on different species in different contexts. Here, the aim was to investigate morphological differences in the head of the Sanmartinero Creole bovine to examine head shape variations between sexes using geometric morphometric methods. A sample of cranial pictures of 43 animals (13 males and 30 females) was obtained, and form (size + shape) was studied by means of geometric morphometric techniques using a set of 14 landmarks. This approach eliminated potential dietary effects, ensuring that the observed shape variations were primarily due to intrinsic morphological differences. Sexual dimorphism was found in form (for both size and shape) of the head of the Sanmartinero Creole bovine breed. Males had significantly larger heads based on centroid size ($U = 714$, $p = 0.0004$), confirming true sexual size differences, and Principal Component Analysis revealed overlapping head shapes with sexual dimorphism concentrated at midline sagittal landmarks (between the most rostral and caudal orbit points) and paired lateral points, indicating that males have broader and longer heads. The two evaluated characters (head size and shape) are of special interest for the conservation of the breed, especially in those cases whose objectives are to maintain the uniqueness, distinctiveness, and uniformity of the populations. This study analyzed animals subjected to the same feeding program, ensuring the elimination of additional variables.

Keywords: *Bos taurus*; cephalometry; natural selection; sexual dimorphism; skull

1. Introduction

Colombia is home to several distinctive cattle breeds, adapted to specific regional conditions and valued for their unique characteristics. A notable Creole breed is the Sanmartinero, which was adapted on the foothills of the Colombian Orinoquia plains, through natural selection, with the bovine introduced by the Spaniards in the 15th century. This Creole bovine has developed economically important traits under adverse environmental conditions and extensive management, including rusticity, resistance, and tolerance to ectoparasites and extreme conditions of temperature and humidity, as well as the ability to use fibrous forage. Thus, Sanmartinero constitutes a biological and economic heritage for the efficient provision of food (meat, milk), skins, and work [1].

Sanmartinero cattle are classified as mesolinean (medium size) and eumetric (medium weight) due to their conformation and size, respectively. The height (measured in the vertical of the withers) is 130 and 135 cm for females and males, respectively. Large and strong head with a straight profile; small, rounded ears with an abundant amount of hair. Strong and well-implanted horns with light colors at the root and darker tones at the tips. In males, horns are thick and directed forward in a crown shape, while in females the horns are lyre-shaped (twisted forward, backward, and upward) [1].

Morphometry can be defined as the study of shape variation and its covariation with other variables [2]. Its objective is to define the conformation of an individual and obtain a stable representation of physiological variations in size, and seeks to eliminate the variation in size that results from individual growth, which frequently has environmental causes, and differentiate this change in size from the change that results from evolutionary genetic causes [3].

Two types of morphometric techniques are recognized: traditional morphometry, based on distances between anatomical points, and geometric morphometrics, which uses the coordinates of these points [2,4]. Geometric morphometrics is more efficient than traditional morphometry because landmarks are digitized from scans or images and the spatial coordinates are automatically saved in special programs [4,5]. The advantage of the latter is that coordinates not only contain size information as distances, but also configuration and relative position. In addition, they allow the visualization of the conformation and its variation between individuals or groups. Unlike the point-to-point approach of traditional morphometry, geometric morphometrics allows the holistic characterization of the biological specimens under study and a graphical output of shape variation [6,7].

Different studies have observed variations in head morphology within the same species [8]. Shape analysis is an approach to explain the morphological diversity of biological structures [7,9]. Geometric morphometrics is a method used to quantitatively evaluate shape variation with biological foundations and sexual dimorphism, especially in scientific areas such as taxonomy and veterinary anatomy [10–12].

Various biological processes, mutations, and genetic developments cause differences in shape between individuals [7]. The evolutionary changes in body size that accompany the domestication process, and the morphological radiation of domestic breeds, imply a differentiation between sexes [13]. These phenotypic differences between females and males of the same species are observed throughout the animal kingdom and are manifested in morphological, behavioral, and life cycle traits [14].

Sexual dimorphism is the existence of phenotypic differences between individuals of the same species but of different sex [15]. Significant differences are manifested in body size, color, and morphology [16,17]. What is known about the patterns and causes of dimorphism comes from studies focusing on sexual dimorphism in body size [18]. It is not appropriate to study only sexual dimorphism in size since it is not clear whether selection (sexual or natural) operates on body size or on a specific part such as head size [18–20].

The degree of sexual dimorphism usually varies greatly from one species to another. In the case of domestic cattle, although not very pronounced [21], these differences may include the following [13,22]: size and weight (males are usually larger and heavier than females), and horns (in some breeds, males have larger, more prominent horns than females). The head is crucial in reflecting morphological features specific to species and breeds. In cattle, the skull is broad and short, with a square shape, and these characteristics may vary among different breeds. Another feature in some breeds is the presence of cornual processes, which form as epidermal elevations from the frontal bone. Identifying these differences provides more specific data for breed description and contributes to the body of knowledge in the zoo ethnological literature [23,24].

Most studies have been conducted for qualitative (size form of horn, prominence of hump) or quantitative traits assessed by traditional methods (e.g., lineal morphometry), but few studies have been conducted (in the Sanmartinero creole bovine) applying geometric morphometric methods, which, as said before, provide information on size and also on shape [6,25]. The study of sexual dimorphism of the head can be an indicator of adaptation to the environment, adaptation to types of forage, reproductive function, and evolutionary processes. The aim of this research was to use geometric morphometrics to compare the head form (size + shape) of the Sanmartinero Creole bovine breed, according to the animal's sex. This study is based on the hypothesis that cattle possess distinctive morphological features in their skulls (besides the horns) according to sex.

2. Materials and Methods

2.1. Study Site

The research was carried out at the Libertad Research Centre of the Colombian Agricultural Research Corporation (with AGROSAVIA being its acronym in Spanish), Meta department, Colombia. It is in the municipality of Villavicencio, kilometer 17 of the road that leads to Puerto López, Meta (4.0574386, −73.4676386), with an average annual rainfall of 2.933 mm, average temperature of 26 °C, height above sea level of 336 m, and acidic and low-fertility soils. Its landscape consists of high, medium, and low terraces, typical of the "Piedemonte Llanero" sub region.

2.2. Sampling

A sample of 43 Sanmartinero Creole bovines (13 males and 30 females), aged 2 years for bulls, and from 2 to 4 years for cows, was studied. Animals were raised at the AGROSAVIA Research Centre. The sample was randomly taken from the 10 families that make up the Sanmartinero Creole bovine breed's in vivo conservation animal germplasm bank, during 2024. A cranial digital photograph was taken of each animal using a Nikon P530 42X camera, Tokyo, Japan. The camera was positioned at a sufficient distance to ensure that the head occupied only a portion of the visual field, free from distortion. An inch ruler was included in the photograph. The images were stored in JPEG extension and then transferred to the computer. To facilitate the photograph, the animals were immobilized in livestock traps.

However, the ideal sample size can vary depending on the specific research question, the complexity of the shape variation and, effectively, the number of landmarks used. Our sample was of 42 animals. We know that the complexity of shape variation being studied was considered low. However, the results are relevant to this study in the Sanmartinero Creole bovine.

This study analyzed animals subjected to the same feeding program, to ensure a homogeneous sample, and avoiding potential dietary effects. This ensures that the observed shape variations were primarily due to intrinsic morphological differences.

2.3. Pictures

Once the photographs were collected, they were imported into the TPSdig program ("Thin-Plate Spline digitizer") v. 1.40 [5], which made it possible to obtain and assemble a matrix of the coordinates of the landmarks or homologous points (Figure 1). Landmarks were located on classical corporal points used in ethnological research [26]. To ensure the accuracy of landmark placements, landmarking was performed by the same researcher (PMPC) with two independent replicas. The description of the reference points is shown in Table 1.

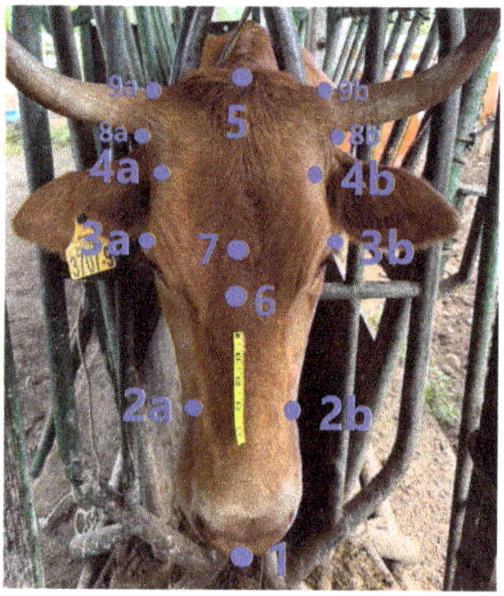

Figure 1. Set of 1 landmarks used in geometric morphometric analysis on the head (cranial plane) for the Sanmartinero bovine breed (Photography CI AGROSAVIA).

Table 1. Characterization of reference points in the Sanmartinero Creole bovine.

1	Most Rostral Point of the Head ("Point of the Muzzle")
2a and 2b	Narrowest points of the muzzle
3a and 3b	Most lateral margins of the orbits
4a and 4b	Narrowest points of the front
5	Most caudal point of the head (*protuberantia intercornualis*)
6	Midline between most rostral points of the orbitas
7	Midline between most caudal points of the orbitas
8a and 8b	Ventral point of horn base
9a and 9b	Dorsal point of horn base

2.4. Geometric Morphometric Analysis

A Procrustes ANOVA was used to determine the measurement error (whether the value of individuals mean squares was greater than the value of the error) to reduce the handling error of the researcher and statistically verify whether the digitization of the landmarks was suitable for further statistical analysis.

In geometric morphometrics, the isometric component of shape (i.e., the shape that changes in a 1:1 proportion with size) is removed from the dataset through the scaling procedure of the Procrustes superimposition. This step brings all specimens to the same size, producing "isometry-free" shape coordinates and a centroid size. Centroid size (CS) can be used subsequently as a proxy for specimen size. Allometric effect refers to how the size of the structure influences the percentage or rate of change in the shape. The effect of allometry was verified using the multivariate regression of shape (Procrustes coordinates) on size ($\log_{10}$-transformed CS) which was treated here as a proxy for head size, while the information about the shape variation was extracted from the Procrustes superimposition. Comparison of allometries was performed with an ANCOVA.

Once these procedures were performed, a *U* Mann–Whitney test was applied to check the differences between sizes (expressed as centroid size). Principal Component Analysis (PCA) was then performed, to detect which changes are presented, related to the two first components. Following this, a discriminant function (DF) with 1000 runs was carried out in order to test for statistically significant differences in shape of the individuals.

The analysis was carried out with the statistical programs MorphoJ v. 1.07a [27] and PAST v. 4.03c [28], establishing the confidence level at 95%. All these computer programs are free to use.

3. Results

When carrying out image processing for geometric morphometric project, the Procrustes ANOVA established that the value of the mean square of individuals was greater than the error (MS Individual 0.0005194767 >> MS Error 0.0000097978) (Table 2), meaning that landmark digitization was performed correctly.

Table 2. Procrustes ANOVA to verify the intra-observer error and calculate the measurement error for the head dataset through the head shape of Sanmartinero Creole animals (n = 43).

Effect	Sum of Squares	Mean Square	Degrees of Freedom	F	*p*
Individual	0.305452	0.000519	588	4.84	<0.0001
Side	0.004745	0.000395	12	3.69	<0.0001
Individual * Side	0.063045	0.000107	588	10.94	<0.0001
Error	0.011757	9.8×10^{-6}	1200		

* Individual by size.

3.1. Sexual Dimorphism in Size

The centroid size of the head significantly differed between the sexes, with males showing the higher value ($U = 714$; $p = 0.0004$).

3.2. Allometry

The allometric effect was 14.74% of shape variation explained by size. This represents the proportion of variation for which the regression accounts as a percentage of the total variation.

We tested for differences in allometric slopes between sexes and found no differences ($F = 0.484$; $p = 0.488$). This means that our tests on allometry in this study are not missing informative differences among the sexes related to differing allometric regression slopes. As allometry appeared to be significative, ulterior analyses were performed with residual regression.

3.3. Sexual Dimorphism in Shape

The PCA showed that the clusters formed by both sexes superimposed themselves on each other without a marked separation between the groups (Figure 2). The first two components explained a 58.89% of the total variation, each of them contributing 36.39% and 22.49%, respectively. This lack of differentiation between groups in PC1/PC2 space does not mean that the groups are undifferentiated. In fact, PCA does not assume *a priori* division and concentrates on the relation between variables and individuals of the same sample, reducing the dimensions of the original data to a new group of variables with axis or orthogonal components [21].

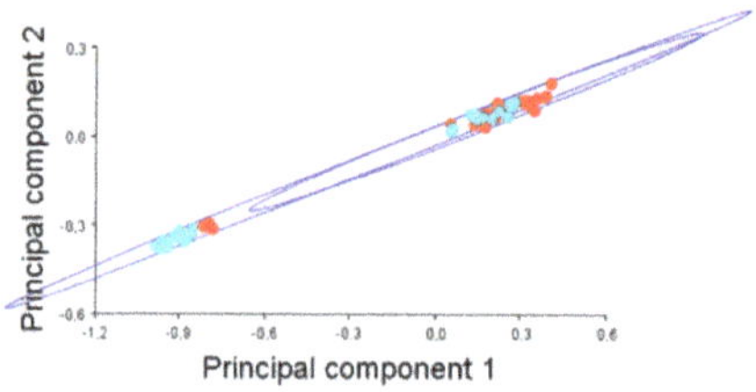

Figure 2. PCA for the first two principal components explained a 58.89% of the total variation, each of them contributing 36.39% and 22.49%, respectively. Clusters formed by both sexes superimposed themselves on each other without a marked separation between the groups (Figure 2). Males in blue and females in red. Presented are 95% confidence ellipses.

Sexual dimorphism is primarily related to variation at anatomical landmarks 6 and 7 (sagittal), and 2, 3, and 4 (with counterparts), so males tended to present more massive heads (wider head and longer cranium than females) (Figure 3).

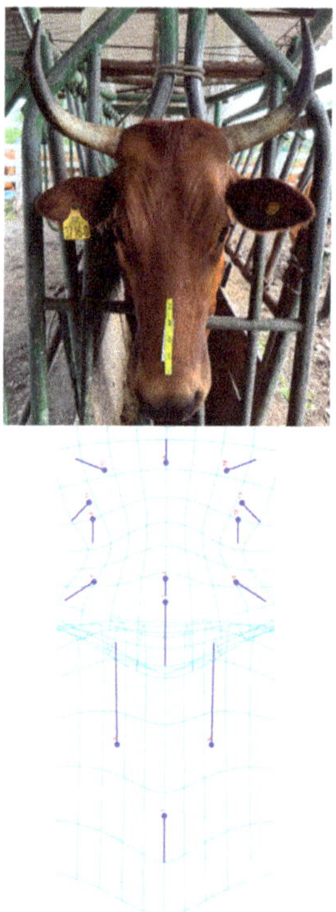

Figure 3. Deformation grid indicating the shape variation of the head according to sex in the Sanmartinero breed. Variation at anatomical landmarks. Main changes were centered on sagittal landmarks and widths.

In the discriminant function, the head shape differed significantly between males and females (Procrustes distance = 0.812; $p < 0.01$; Mahalanobis distance = 5.644; $p < 0.001$), with a correct classification percentage of 100% for both sexes (Figure 4).

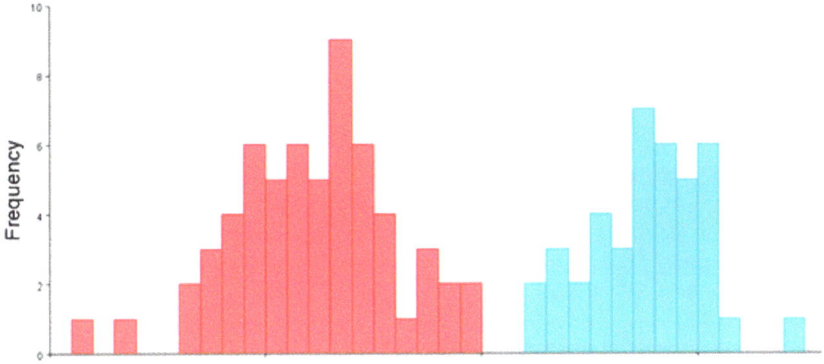

Figure 4. Discriminant function. The shape of the skulls differed significantly between males (blue) and females (red) (Procrustes distance = 0.812; $p < 0.01$; Mahalanobis distance = 5.644; $p < 0.001$). The correct classification percentage was 100% for both sexes.

4. Discussion

Geometric morphometrics has emerged as a powerful tool in recent research for elucidating the relationship between shape and size within samples and explaining variations between groups [6,10,29]. Unlike linear morphometry, which relies on length, depth, and width measurements, shape analysis focuses on the spatial relationships between landmarks, and while linear morphometry provides only limited information about the overall shape of a structure, often with overlapping measurements, geometric morphometry techniques offer a more comprehensive approach as they contain information related to form [11,25,29].

Sexual dimorphism is an important physiological characteristic for understanding animal adaptation to the natural and social environment [30]. Numerous studies have focused on it since Darwin's theory of evolution [15,22,31–33], and it is considered to be a consequence of natural adaptation and sexual selection [31]. The head is the fundamental biological and skeletal structure for distinguishing morphological differences between species, breeds, and sexes, offering valuable information about their functional and environmental adaptations [24,34].

Sexual dimorphism in head size and shape was observed in the Sanmartinero Creole bovine, with a greater manifestation in size. Males have wider heads and longer skulls. Although sexual dimorphism in head size has been little studied by geometric morphometrics in domestic Creole bovines, a study conducted on the Casanare Creole bovine found that head length and width are greater in males [35], while in the Limonero Creole bovine, sexual dimorphism is evident due to greater head width in males [21]. In other animal species such as the Araucan Creole pig, males presented slightly greater head length and width, without any statistical differences [36].

The main mechanisms responsible for the evolution of sexual dimorphism in size and shape are sexual selection and natural selection [37,38], fertility, and intrasexual competition [39]. In several mammals, circulating testosterone concentrations are associated with cranial bone growth [40]. Studies in two species of lizards (*Podarcis muralis* and *P. siculus*) have shown that sexual dimorphism in head shape can arise from head segregation or competitive displacement in response to competition between the sexes [41]. In *Canis lupus familiaris*, males had a wider and longer head than females [42]; in *Hynobius maoershanensis*, sexual dimorphism is expressed in terms of head width with a bias towards males [17]. Other research carried out on several species of lizards (Reptilia: Squamata: Sauria) showed that males had longer heads [20]. In a population of domestic cats (*Felis silvestris catus*), males had a larger head circumference [43].

Studies report that sexual dimorphism in head width is explained by sexual selection, ecological selection, and fertility [17]. Sexual selection favors large-headed males, winning male competition and obtaining more mating opportunities [20,44]. A wide head facilitates the intake of greater amounts of food, providing more energy to carry out the corresponding activities [17]. In males, a large head may be a consequence of intrasexual selection, translating into the ability to carry out confrontations (fights) and aggressive encounters to maintain a territory [45,46]. In the case of the Sanmartinero, we have observed that Creole bovines are animals which become aggressive when their territory or herd is invaded.

The mechanisms of sexual selection modify the morphological properties of each sex; therefore, sexual dimorphism can be amplified by natural selection resulting in a natural segregation of both sexes [19,47]. Environmental adaptation, a complex multi-level structure, and a distinctive feeding system allow the evolution of sexual dimorphism [48]. These natural processes may be the cause of the sexual dimorphism expressed in the Sanmartinero Creole bovine, but the causes still need to be explained more deeply.

We found sexual dimorphism in form (for both size and shape) of the head of the Sanmartinero cattle breed, so not only was size bigger in males, as would be expected for the species, but they are also bigger in shape: bulls have bigger but shorter heads, with a wider forefront and a major base of the horns, compared to cows.

This variation is related to different reproductive functions and adaptations of the sexes and can be a consequence of different sexual strategies of the group, once the energy for growth is used from distinct shape in males and females [49]. Sexual dimorphism can be a direct consequence of sex-specific hormones that provide a distinct morphology related to the reproductive success of the species [50]. The variation in the shape of the head between the sexes occurs because the differentiated growth relative to head dimensions during ontogenesis, and this can be regulated by the presence or absence of androgen hormones [51]. In most mammals, anatomical differences between sexes are due to a trait associated with the action of sex hormones and, therefore, gonad maturation, which marks the beginning of sexual differentiation and the end of the period of sexual or pregonadal undifferentiation that precedes it. This differentiation would seem to follow a different "speed" in each sex [50,51].

Information on head dimorphism in bovines is limited. Potential studies are still lacking on the effect of environmental conditions, genetic factors, dietary habits, and housing type on the presence of sexual dimorphism in head size in the Sanmartinero Creole bovine. The breed's uniqueness stems from various factors, including its origin, physical characteristics, and its temperament. Moreover, it is a rare breed due to limited populations and distribution, as well as its distinctive appearance. The consideration of all these evaluated characters is of special interest for the conservation of the breed, especially in those cases whose aims are to maintain the uniqueness, distinctiveness, and uniformity of the populations.

5. Conclusions

Sexual dimorphism in head size and shape was observed in the Sanmartinero Creole bovine. Males have wider heads and longer skulls. These results bring new character information on the form of the head of the Sanmartinero Creole bovine. In addition, the technique of geometric morphometry has proven to be efficient in recognizing cattle cephalic sexual dimorphism with refined detailing.

Author Contributions: Writing—original draft preparation, methodology, and formal analysis, P.M.P.-C. and A.S.-C.; writing—review and editing: P.M.P.-C., M.V.-T., G.M.-C., D.E.R.-P. and J.R.-A. All authors have read and agreed to the published version of the manuscript.

Funding: This research was funded by the Research Committee—CONADI (ID3640)—of the Universidad Cooperativa de Colombia.

Institutional Review Board Statement: Not applicable. No request for ethical approval was considered since in no case were animals sacrificed, nor were traumatic manipulations that caused pain performed.

Informed Consent Statement: Not applicable. Only photographs were taken without causing trauma or pain to the animals. The owners knew the photographs that were used for this study.

Data Availability Statement: Data are available upon reasonable request to the second author. The data have not been published because we are going to carry out further analysis.

Acknowledgments: We would like to thank the students who helped with data collection and the anonymous referees for their valuable comments on the manuscript. All persons included in this section have consented to their acknowledgment.

Conflicts of Interest: Author Jaime Rosero-Alpala was employed by the company Corporación Colombiana de Investigación Agropecuaria. The remaining authors declare that the research was conducted in the absence of any commercial or financial relationships that could be construed as a potential conflict of interest.

References

1. Martínez, C.G.; Flórez, D.H.; Martínez-Villate, G.C. El ganado criollo colombiano Sanmartinero, su conservación y aporte sostenible a la producción bovina en la Orinoquia colombiana. *Arch. Latinoam. Prod. Anim.* **2020**, *28*, 69–86.
2. Bookstein, F.L. Reworking geometric morphometrics into a methodology of transformation grids. *Evol. Biol.* **2023**, *50*, 275–299. [CrossRef]
3. Parés-Casanova, P.M. (Ed.) Introductory Chapter—Morphometric Studies: Beyond Pure Anatomical Form Analysis. In *New Insights into Morphometry Studies*; IntechOpen: Tokyo, Japan, 2017; Chapter 1, p. 96.
4. Mitteroecker, P.; Gunz, P. Advances in Geometric morphometrics. *Evol. Biol.* **2009**, *36*, 235–247. [CrossRef]
5. Rohlf, F.J. The tps series of software. *Hystrix It. J. Mamm.* **2015**, *26*, 9–12.
6. Klingenberg, C.P. Size, shape, and form: Concepts of allometry in geometric morphometrics. *Dev. Genes Evol.* **2016**, *226*, 113–137. [CrossRef]
7. Cooke, S.B.; Terhune, C. Form, function, and geometric morphometrics. *Anat. Rec.* **2015**, *298*, 5–28. [CrossRef]
8. Lawing, A.M.; Polly, P.D. Geometric morphometrics: Recent applications to the study of evolution and development. *J. Zool.* **2010**, *280*, 1–7. [CrossRef]
9. Gündemir, O.; Koungoulos, L.; Szara, T.; Duro, S.; Spataru, M.C.; Michaud, M.; Onar, V. Cranial morphology of Balkan and West Asian livestock guardian dogs. *J. Anat.* **2023**, *243*, 951–959. [CrossRef]
10. Bookstein, F.L. Combining the tools of geometric morphometrics. In *Advances in Morphometrics*; Springer: Boston, MA USA, 1996; pp. 131–151.
11. Günay, E.; Szara, T.; Çakar, B.; Deveci, E.I.; Coşkun, A.S.; Gün, G.; Yiğit, F.; Gündemir, O.; Duro, S.; Spataru, M.C. Geometric Morphometric Analysis of Sexual Dimorphism in the Bill of the White Stork (*Ciconia ciconia*). *Animals* **2025**, *15*, 1312. [CrossRef]
12. Ağaç, K.D.; Onuk, B.; Gündemir, O.; Kabak, M.; Manuta, N.; Çakar, B.; Janeczek, M.; Crampton, D.A.; Szara, T. Comparative Cranial Geometric Morphometrics among Wistar Albino, Sprague Dawley, and WAG/Rij Rat Strains. *Animals* **2024**, *14*, 1274. [CrossRef]
13. Polák, J.; Frynta, D. Patterns of sexual size dimorphism in cattle breeds support Rensch's rule. *Evol. Ecol.* **2010**, *24*, 1255–1266. [CrossRef]

14. Cordeschi, G.; Canestrelli, D.; Porretta, D. Sex-biased phenotypic plasticity affects sexual dimorphism patterns under changing environmental conditions. *Sci. Rep.* **2024**, *14*, 892. [CrossRef] [PubMed]
15. Berns, C.M. The evolution of sexual dimorphism: Understanding mechanisms of sexual shape differences. In *Sexual Dimorphism;* Moriyama, H., Ed.; IntechOpen: Tokyo, Japan, 2013; pp. 1–15.
16. Turcotte, C.M.; Mann, E.H.J.; Stock, M.K.; Villamil, C.I.; Montague, M.J.; Dickinson, E.; Surratt, S.B.; Martinez, M.; Williams, S.A.; Anton, S.C. The ontogeny of sexual dimorphism in free-ranging rhesus macaques. *Am. J. Biol. Anthr.* **2022**, *177*, 314–327. [CrossRef] [PubMed]
17. Chen, H.; Bu, R.; Ning, M.; Yang, B.; Wu, Z.; Huang, H. Sexual Dimorphism in the Chinese Endemic Species *Hynobius maoershanensis* (Urodela: Hynobiidae). *Animals* **2022**, *12*, 1712. [CrossRef]
18. Cox, R.M.; Kahrl, A.F. Sexual Selection and Sexual Dimorphism. In *Reproductive Biology and Phylogeny of Lizards and Tuatara*, 1st ed.; Rheubert, J.L., Siegel, D.S., Trauth, S.E., Eds.; CRC Press: Boca Raton, FL, USA, 2015; pp. 78–108.
19. Kaliontzopoulou, A.; Carretero, M.A.; Llorente, G.A. Sexual dimorphism in traits related to locomotion: Ontogenetic patterns of variation in Podarcis wall lizards. *Biol. J. Linn. Soc.* **2010**, *99*, 530–543. [CrossRef]
20. Scharf, I.; Meiri, S. Sexual dimorphism of heads and abdomens: Different approaches to 'being large' in female and male lizards. *Biol. J. Linn. Soc.* **2013**, *110*, 665–673. [CrossRef]
21. Chirinos, Z.; Contreras, G.; Zambrano, S.; Molero, E.; Paéz, A. Caracterizacion del dimorfismo sexual en ganado criollo limonero mediante medidas corporales. *Rev. Fac. Agron. (LUZ)* **2011**, *28* (Suppl. S1), 554–564.
22. Lang, D.R.; Hansel, W. A Sexual Dimorphism in Three Somatic Tissues of Cattle. *J. Dairy Sc.* **1959**, *42*, 1330–1337. [CrossRef]
23. Çakar, B.; Tandir, F.; Can Güzel, B.; Bakıcı, C.; Ünal, B.; Duro, S.; Szara, T.; Spataru, C.; Spataru, M.-C.; Gündemir, O. Comparison of Skull Morphometric Characteristics of Simmental and Holstein Cattle Breeds. *Animals* **2024**, *14*, 2085. [CrossRef]
24. Gündemir, O.; Manuta, N.; Can Güzel, B.; Bakıcı, C.; Duro, S.; Ünal, B.; de Cakar, C.; Szara, T. Skull morphology in native and non-native cattle breeds in Türkiye. *J. Anat.* **2025**, *247*, 179–188. [CrossRef]
25. Adams, D.C.; Rohlf, F.J.; Slice, D.E. A field comes of age: Geometric morphometrics in the 21st century. *Hystrix Ital. J. Mammal.* **2013**, *24*, 7–14.
26. Sañudo Astiz, C. *Valoración Morfologica de los Animales Domésticos. Ministerio de Medio Ambiente y Medio Rural y Marino;* Artegraf, Industrias Graficas S.A: Madrid, Spain, 2009; p. 865.
27. Klingenberg, C.P. MorphoJ: An integrated software package for geometric morphometrics. *Mol. Ecol. Res.* **2011**, *11*, 353–357. [CrossRef]
28. Hammer, Ø.; Harper, D.A.T.; Ryan, P.D. PAST v. 4.17c. Paleontological Statisticts Software Package for education and data analysis. *Palaeontol. Electron.* **2001**, *4*, 1–229.
29. Zelditch, M.L.; Swiderski, D.L.; Sheets, H.D. *Geometric Morphometrics for Biologists: A Primer;* Elsevier Academic Press: Amsterdam, The Netherlands, 2012.
30. Clutton-Brock, T. Sexual selection in males and females. *Science* **2007**, *318*, 1882–1885. [CrossRef]
31. Clutton-Brock, T.H.; Harvey, P.H.; Rudder, B. Sexual Dimorphism, Socionomic Sex Ratio and Body Weight in Primates. *Nature* **1977**, *269*, 797–800. [CrossRef] [PubMed]
32. Kamalakannan, M. Characterisations of Hair of Hoolock Gibbon Hoolock Hoolock (Harlan, 1834) (Hylobatidae: Primates: Mammalia). *J. Entomol. Zool. Stud.* **2017**, *5*, 986–988.
33. Cassini, M.H. Role of Fecundity Selection on the Evolution of Sexual Size Dimorphism in Mammals. *Anim. Behav.* **2017**, *128*, 1–4. [CrossRef]
34. Gündemir, O.; Szara, T.; Yalin, E.E.; Karabagli, M.; Mutlu, Z.; Yilmaz, O.; Büyükünal, S.K.; Blagojevic, M.; Parés-Casanova, P.M. Examination of Shape Variation of the Skull in British Shorthair, Scottish Fold, and Van Cats. *Animals* **2023**, *13*, 614. [CrossRef]
35. Salamanca Carreño, A.; Crosby Granados, R.A. Estudio fenotípico del bovino criollo Casanare biotipo Araucano. Análsisi zoométrico. *Zootec. Trop.* **2013**, *31*, 201–208.
36. Salamanca-Carreño, A.; Parés-Casanova, P.M.; Vélez Terranova, O.M.; Castro Rosa, G.; Jáuregui, R. Valoración morfométrica de una población de cerdos criollos araucanos (Colombia). *Rev. Investig. Vet. Perú* **2022**, *33*, 1–12. [CrossRef]
37. Liao, W.B.; Liu, W.C.; Merilä, J. Andrew meets Rensch: Sexual size dimorphism and the inverse of Rensch's rule in Andrew's toad (*Bufo andrewsi*). *Oecologia* **2015**, *177*, 389–399. [CrossRef]
38. Yan, C.; Ma, H.; Yang, Y.; Mi, Z. Sexual Dimorphism in the Limb Bones of Asiatic Toad (*Bufo gargarizans*) in Relation to Sexual Selection. *Animals* **2023**, *13*, 2638. [CrossRef]
39. Dursun, C.; Gül, S.; Özdemir, N. Sexual size and shape dimorphism in Turkish common toads (*Bufo bufo* Linnaeus 1758). *Anat. Record* **2022**, *305*, 1548–1558. [CrossRef]
40. Stirrat, M.; Perrett, D.I. Valid Facial Cues to Cooperation and Trust: Male Facial Width and Trustworthiness. *Psychol. Sci.* **2010**, *21*, 349–354. [CrossRef] [PubMed]

41. Sacchi, R.; Mangiacotti, R.; Scali, S.; Storniolo, F.; Zuffi, M.A.L. Species-Specific Spatial Patterns of Variation in Sexual Dimorphism by Two Lizards Settled in the Same Geographic Context. *Animals* **2023**, *13*, 736. [CrossRef] [PubMed]
42. Carrasco, J.J.; Georgevsky, D.; Valenzuela, M.; McGreevy, P.D. A pilot study of sexual dimorphism in the head morphology of domestic dogs. *J. Veter. Behav.* **2014**, *9*, 43–46. [CrossRef]
43. Sagrero-Del Moral, M.; Castañeda-Ortega, J.C. Dimorfismo sexual mediante el uso de la morfometría en *Felis silvestris catus* (Schreber, 1775). *Rev. Dig. Act. Med. Vet. Zoot. México* **2020**, *en línea*. Available online: https://acmevez.mx/dimorfismo-sexual-mediante-el-uso-de-la-morfometria-en-felis-silvestris-catus-schreber-1775/ (accessed on 29 May 2025).
44. Xiong, J.; Zhang, B.; Liu, Q.; Pan, T.; Gou, J. Sexual dimorphism in the Chinese endemic species *Pachyhynobius shangchengensis* Fei, Qu and Wu, 1983 (Urodela: Hynobiidae). *PeerJ* **2019**, *7*, e6408. [CrossRef]
45. Aguilar-Moreno, M.; Rodríguez-Romero, F.J.; Aragón-Martínez, A.; Muñoz-Manzano, J.A.; Granados-González, G.; Hernández-Gallegos, O. Dimorfismo sexual de *Aspidoscelis costata costata* (Squamata: Teiidae) en el sur del Estado de México, México. *Rev. Chil. Hist. Nat.* **2010**, *83*, 585–592. [CrossRef]
46. Borczyk, B.; Kusznierz, J.; Paśko, L.; Turniak, E. Scaling of the sexual size and shape skull dimorphism in the sand lizard (*Lacerta agilis* L.). *Vertebr. Zool.* **2014**, *64*, 221–227. [CrossRef]
47. Punzalan, D.; Hosken, D.J. Sexual dimorphism: Why the sexes are (and are not) different. *Curr. Biol.* **2010**, *20*, 972–973. [CrossRef]
48. Li, Y.-P.; Huang, Z.-P.; Yang, Y.; He, X.-B.; Pan, R.-L.; He, X.-M.; Yang, G.-W.; Wu, H.; Cui, L.-W.; Xiao, W. Ontogenetic Development of Sexual Dimorphism in Body Mass of Wild Black-and-White Snub-Nosed Monkey (*Rhinopithecus bieti*). *Animals* **2023**, *13*, 1576. [CrossRef]
49. Székely, T. Evolution of reproductive strategies: Sex roles, sex ratios and phylogenies. *Biol. Futur.* **2023**, *74*, 351–357. [CrossRef]
50. Shingleton, A.W.; Vea, I.M. Sex-specific regulation of development, growth and metabolism. *Semin. Cell Dev. Biol.* **2023**, *138*, 117–127. [CrossRef]
51. Milella, M.; Franklin, D.; Belcastro, M.G.; Cardini, A. Sexual differences in human cranial morphology: Is one sex more variable or one region more dimorphic? *Anat. Rec.* **2021**, *304*, 2789–2810. [CrossRef]

ruminants

MDPI

Article

The Relationship Between Stature and Live Weight of Dairy Cows Between Birth and Maturity

Michaela J. Gibson [1,*], Nick W. Sneddon [2], Chris W. Rogers [1], Penny J. Back [1,2], Keren E. Dittmer [1] and Natalia P. Martín [2]

1 School of Veterinary Sciences, Massey University, Private Bag 11-222, Palmerston North 4442, New Zealand; c.w.rogers@massey.ac.nz (C.W.R.); p.j.back@massey.ac.nz (P.J.B.); k.e.dittmer@massey.ac.nz (K.E.D.)
2 School of Agriculture Environmental Sciences, Massey University, Private Bag 11-222, Palmerston North 4442, New Zealand; n.w.sneddon@massey.ac.nz (N.W.S.); n.p.martin@massey.ac.nz (N.P.M.)
* Correspondence: m.gibson@massey.ac.nz

Simple Summary: Milk yield and the ability to get pregnant are strongly associated with cow live weight and size. However, there are limited data published describing the prediction of cow size at birth. Therefore, the aim of this study was to examine the relationship between cow shape and live weight of female dairy cattle between birth and maturity. Forty dairy cows with records of cow shape and live weight from birth up until two years of age, underwent follow-up measurements for live weight, height, length, girth circumference and leg length on four occasions between 42 and 52 months of age. Measures of height, leg and body length at fourth mating had a greater correlation with measures at birth compared to girth and live weight. The lower correlation between birth and maturity measures for girth and live weight indicates growth of these measures is affected by environmental factors and emphasises the importance of adequate nutrition to maximise cow size.

Abstract: Lactational and reproductive performance are strongly associated with cow live weight and capacity. However, there are limited data published describing capacity (thoracic) growth and the prediction of final stature and capacity from measurements at birth. Therefore, the aim of this study was to examine the relationship between stature and live weight of female dairy cattle between birth and maturity. Forty dairy cows, with records of stature and capacity from birth up until two years of age, underwent follow-up measurements for live weight, height at withers, wither-rump length, girth circumference and leg length on four occasions between 42 and 52 months of age. Measures of wither height, leg length and wither rump length at fourth mating had the strongest association with measures at birth ($R^2 > 0.90$) compared to girth and live weight ($R^2 = 0.88$ and 0.82, respectively). The weaker association between birth and maturity measures for girth is likely a reflection of the stronger relationship with live weight resulting in a later maturity (approximately 810 days) compared to linear measures such as height (approximately 730 days). Therefore, to maximise capacity, adequate nutrition is required until approximately 810 days of age when capacity growth is most sensitive to environmental input.

Keywords: bone; girth; height; nutrition; growth; mature size

1. Introduction

Both reproductive and lactational performance are strongly associated with cow live weight. Cows that are heavier at puberty and first calving are reported to subsequently have greater pregnancy rates and reproductive performance compared to lighter cows in the herd [1,2]. Most studies that use cow live weight as a predictor of milk production do not take into account the stature or shape of the cow [1,3]. Therefore, a cow may be "tall and skinny" but have the same live weight as a "short fat cow". One approach to account for this is the use of body condition score (BCS) to estimate fat deposition. Within the literature, BCS is also positively correlated with lactation and reproductive performance [4,5]. However, BCS is a measure at one time point and changes throughout the season based on stage of lactation and seasonal pasture growth. Stature measures of height and girth (capacity) are not affected by seasonal changes and reflect the frame size of the cow, thus removing the effect body condition can have on live weight measures [6].

It has been hypothesised that the height of the calf is associated with subsequent lactational performance, with taller calves producing more milk across multiple lactations than shorter calves when adjusted for live weight [7,8]. Stature growth in cattle up until 23 months follows the typical pattern of growth in mammals [9]. Similar to other herbivorous mammals [10], the bones in the distal limb of cattle, such as the metacarpus, have limited capacity for longitudinal growth and cease longitudinal growth at one year of age [11]. Whereas bones in the proximal limb and thorax related to capacity, such as the humerus, are capable of longitudinal and appositional growth after two years of age [11]. It has been estimated that the humerus ceases longitudinal growth (estimated to maturity by physeal closure) at approximately 3.5–4 years of age [12]. Thus, the growth and development of the humerus could still be affected by changes in nutrient supply during the cow's second or even third year of life [9,11,12].

Potential mature size can also be affected by nutritional supply over different growth phases, whereby times of nutritional deficit affect mature size through growth pauses prior to physeal closure. The dairy systems of New Zealand have a heavy reliance on grazed pasture which can result in heifers (cows less than 2 years old) undergoing a growth check during winter due to poor pasture quality and availability [13]. This seasonal growth check was described by Handcock, et al. [14], whereby heifers that had a seasonal growth check in winter had a subsequent period of compensatory growth that resulted in a taller animal than those that grew in a linear growth pattern. This growth check has been identified as a risk factor for spontaneous humeral fractures in first lactation dairy heifers in New Zealand. Studies examining humeri from heifers affected by these fractures have reported the presence of growth arrest lines and reduced bone density suggesting a recent period of restricted bone growth [15,16]. Therefore, the winter prior to first calving has been identified as a key period for humeral growth [15,17].

The effect of nutrition has also been examined in calves whereby calves exposed to a period of undernutrition in utero had a smaller stature size and lower bone density at birth, compared with calves that were not exposed to a period of undernutrition [18]. The relative magnitude and timing of the nutritional deficit could independently and synergistically alter the extent of the growth check and how stature growth is impacted.

Stature growth after first calving of dairy cows has not been described in detail and is required to estimate the timing of bone growth in relation to nutrition due to changes in on-farm seasons. In addition, measures are required after first calving to determine if the shape and stature of a cow in early life can predict the stature and shape of the cow at maturity and indicate when a mature (cessation of growth) state has been achieved.

Therefore, the aim of this study was to examine the relationship between stature and live weight of female dairy cattle between birth and maturity among different breeds.

2. Materials and Methods

This experiment was conducted with the approval of Massey University Animal Ethics Committee (MUAEC 19/81- birth to 23 months and MUAEC 23/01 from three to four years of age).

The cohort of animals consisted of 40 Holstein–Friesian (F), Jersey (J) and crossbred (HxJ) heifers that remained in the herd from the cohort of 57 heifers measured in Gibson, et al. [9]. Heifers were born during the 2019 spring calving season at Massey University's Dairy One farm and 40 were still present on the same farm during the 2023 milking season. There was attrition due to failure to conceive and low production (n = 17).

Calf measure data from birth to 23 months of age were retrieved from the dataset used by Gibson et al. [9]. For analysis, heifers were classified into one of three breed groups based on pedigree records for each calf. The breeds were defined as follows: Heifers that were at least 15/16 (93.8%) F were classed as F (n = 8); heifers that were at least 15/16 (93.8%) J were classified as J (n = 9); heifers that did not meet the requirements of either F or J were classified as HxJ (n = 23).

2.1. Sampling and Measurements

Cows were measured on 4 occasions during 2023: at pregnancy diagnosis of their 3rd pregnancy (1280 days/42 months), dry-off of their 2nd lactation (1370 days/45 months of age), calving of their 3rd calf and start of 3rd third lactation (1460 days/48 months of age) and 4th mating (1540 days/52 months of age). Their 4th mating was considered maturity for all measures. On each measure day, cows were weighed using a Tru-test weigh bridge (Tru Test, Auckland, New Zealand) to the nearest kilogram. When contained in the weigh crate, single linear measures to the nearest one centimetre were obtained using a flexible tape measure (Korband, Lincolnshire, UK) as outlined in Gibson, et al. [9]. Linear measurements were obtained for wither height (from ground to point of wither), heart girth, wither-rump length (wither to tuber ischii) and leg length (left front leg, from ground to point of olecranon). Thoracic height was calculated by subtracting leg length from wither height.

2.2. Statistical Analysis

Statistical analysis was conducted using SAS version 9.4 (SAS Institute Inc., Cary, NC, USA). Using the NLMixed procedure, a von Bertalanffy growth function was applied to the data to produce growth curves for weight and stature measurements (data not presented).

The general equation was:

$$Y = L\infty(1 - b \times \exp(-k \times \text{age}))^3$$

where Y was the trait measured at a certain age (months), exp is the base of the natural logarithms, and L, b and k are parameters to be estimated. Starting parameters were b = 0.61, k = 0.0041 and L∞ was set to 450, 140, 75, 120, 180, for live weight, wither height, leg length, crown-rump length, heart girth, respectively.

Fourth-order Legendre polynomials were constructed to predict weight and stature measures at a given age excluding wither-rump length which were modelled using a fifth-order Legendre polynomial. Individual polynomials were constructed using the solutions for each individual animal fitted as a random function. Predictions were made for given ages were based on key milestones in cow production which were; birth (0 days), weaning (100 days), 1 year of age (365 days), 1st mating (450 days), 1st calving (730 days), 2nd mating (810 days), 1st dry off (1000 days), 2nd calving (1100 days), 3rd mating (1180 days), 2nd dry off (1370 days), 3rd calving (1460 days) and 4th mating (1540 days). Data were also analysed using the Proc GLM procedure for each milestone with the fixed effect of breed group. Correlations between stature measurements were evaluated using the Corr procedure.

3. Results

Live weight and all stature measures at birth were highly correlated with measures at 3rd calving (Table 1). Height, wither rump length and leg length at birth were strongly correlated with their respective measures at maturity ($R^2 \geq 0.90$).

Table 1. Regression values for stature measures for Holstein–Friesian (F) heifers, Holstein–Friesian-cross–Jersey (FxJ) and Jersey (J) cows from between birth and maturity.

Maturity / Birth	Live Weight	Heart Girth	Wither Height	Wither-Rump Length	Leg Length
Live weight	**0.82**	0.77	0.79	0.79	0.84
Heart girth	0.85	**0.88**	0.80	0.72	0.83
Wither height	0.80	0.75	**0.90**	0.73	0.95
Wither-rump length	0.86	0.74	0.77	**0.94**	0.84
Leg length	0.77	0.70	0.85	0.75	**0.95**

At one year of age, cows were only 60–62% of their mature live weight and approximately 82–84% of their mature girth circumference (Table 2). Measures of wither height and leg length were almost at mature size by one year of age with measures at or above 89% of maximal measurement.

Table 2. Proportion of estimated mature size at key production stages by breed for leg length, height, wither rump length, girth and liveweight for Holstein–Friesian (F), Holstein–Friesian-cross–Jersey (FxJ) and Jersey (J) dairy cows from birth to 4th mating. A value of 1 indicates mature size was achieved.

		Leg Length			Wither Height			Wither-Rump Length			Heart Girth			Live Weight		
Key Milestone	**Age (Days)**	**J**	**FxJ**	**F**	**J**	**FxJ**	**F**	**J**	**FxJ**	**F**	**J**	**FxJ**	**F**	**J**	**FxJ**	**F**
Birth	1	0.67	0.67	0.66	0.55	0.55	0.56	0.44	0.44	0.45	0.40	0.40	0.42	0.06	0.06	0.07
Weaning	100	0.81	0.80	0.79	0.69	0.68	0.70	0.61	0.60	0.61	0.56	0.56	0.56	0.20	0.20	0.20
1 year old	365	0.97	0.95	0.94	0.90	0.89	0.90	0.88	0.86	0.85	0.84	0.83	0.82	0.62	0.62	0.60
1st mating	450	0.98	0.97	0.96	0.93	0.92	0.93	0.92	0.90	0.90	0.89	0.88	0.87	0.72	0.71	0.69
1st calving	730	1	0.99	0.99	0.98	0.98	0.98	0.98	0.97	0.97	0.96	0.96	0.95	0.90	0.90	0.88
2nd mating	810		1	0.99	0.99	0.98	0.99	0.98	0.98	0.98	0.97	0.97	0.97	0.93	0.92	0.91
1st dry off	1000			1	0.99	0.99	0.99	0.99	0.99	0.99	0.99	0.99	0.98	0.96	0.96	0.95
2nd calving	1100				1	1	1	1	0.99	0.99	0.99	0.99	0.99	0.98	0.97	0.97
3rd mating	1180								1	0.99	0.99	0.99	0.99	0.98	0.98	0.98
2nd dry off	1370									1	1	1	1	0.99	0.99	0.99
3rd calving	1460													0.99	0.99	0.99
4th mating	1540													1	1	1

Leg length growth slowed at the youngest age at just over a year (377–431 days, Table 3), and reached maximal measurement first; in contrast, live weight continued to increase across the measurement period.

Table 3. Inflection points (in days of age) for liveweight, girth, height, wither rump length and leg length growth curves for Holstein–Friesian (F), Holstein–Friesian-cross–Jersey (FxJ) and Jersey (J) cows.

Knee of Curve	J	FxJ	F
Live weight	609	612	631
Heart girth	503	516	527
Wither-rump length	469	493	500
Wither height	473	484	476
Leg length	377	409	431

4. Discussion

All measures at the 4th mating (maturity) were strongly correlated with birth values which reflects earlier observations for these cows when examined until first calving (23 months) [9]. Measures that effectively relate to height (wither height, leg length and wither to rump length) had the strongest associations between birth and 4th mating values. These strong associations reflect the high heritability estimates published for these values within the literature [8,19] and the limited impact environmental variables such as nutrition, either early in life or later post-puberty, have on these values. Cows in the current study grew at a greater rate than industry standards [9], hence nutrition would not have limited the attainment of mature height.

Across many mammalian species, where even in situations of severe and sustained under nutrition, individuals will always achieve predicted height. Predicted height or final height occurs earlier than final capacity and in this population of pasture-managed dairy cows, was achieved around the time of first calving at 730 days old (24 months). The majority of contributions to wither height is driven by the longitudinal growth in the limbs and subsequently in the proximal limb and thorax. The attainment of the threshold of 96–98% of final length in the limb at 450 days (15 months), earlier than any of the other measurements, reflected the earlier closure of the growth plates in the long bones within the limb and that wither height early in life is predominantly due to leg length and then increases in thoracic capacity [9]. The strong association of final height with birth values, and the reported high heritability for wither height [8], indicates that there is limited opportunity to manipulate this aspect of cow size via nutrition. Thus, if milk production is associated with body size and live weight, the potential for increased capacity (body size) relies on the promotion of growth associated with the thoracic limb and thorax.

Final wither-rump length, which equates to the human equivalent of sitting height, was achieved at an older age than wither height and reflects that the growth plates in the vertebrae are some of the last bones to cease growth [10]. The relative increase in wither-rump length was small and thus, cow length is unlikely to be a variable with great plasticity or capability to contribute to greater cow capacity and relative efficiency of milk production.

Final heart girth and live weight measures occurred later than stature measures and had the lowest association with measures at birth. These data demonstrate that heart girth and live weight are not only plastic and sensitive to environmental input until approximately 810 days, but the magnitude of the response may also be temporally sensitive at different periods of growth. In dairy cattle, girth has a greater association with live weight than wither height [20], as girth describes both the width and thoracic height of the cow. The close relationship between girth and live weight is also reflected in the later timing of bone maturity in the proximal limb and thoracic cavity. For example the humerus is a late maturing bone located in the thoracic cavity which is highly correlated girth size [21], versus bones in the distal limb, such as the metacarpus, which are associated with increases in height.

Growth in the metacarpus (distal limb) in cows as measured by changes in length and bone morphology has been estimated to cease at approximately one year of age [11]. In the current study, the slowing of leg length and wither height at one year of age is in agreement with this estimation. The humerus (proximal limb), in comparison, continues to grow until at least two years of age, as evidenced by increases in girth size. The difference in maturity of these bones reflects their location and function within the limb. The humerus is surrounded by muscle and is located within the proximal limb/thoracic cavity. Therefore, increases in live weight and capacity would provoke a greater relative bone response in the humerus than in the metacarpus [11]. The current study estimated that the heifers were approximately 50% of their mature height at their birth but less than 10% of their mature live weight. In early life, increases in live weight are driven by increases in long bone growth resulting in increases in height [9,22]. At puberty, longitudinal bone growth is slowed and increases in live weight is driven by increases in capacity (axial skeleton) [23]. This growth model is reflective of the current results where increases in height after birth were rapid, resulting in 89–90% of mature height being attained but only 60–62% of mature weight by one year of age. This is similar to other mammalian species such as horses where they are born at approximately 60% of their mature height and 10% of their mature live weight and are 92% of their mature height but only 61% of their mature live weight at one year of age [24].

Measures of leg length, wither height and wither-rump length at birth were highly correlated with the respective measures at maturity, indicating a relatively conserved growth pattern [25]. With appendicular bones being a greater proportion of their mature size at birth compared to axial bones, the growth trajectory is well conserved and less affected by perturbations in live weight gain from environmental factors such as nutrition and disease. This trend has also been observed in human studies where birth length and mature height had a greater association than birth weight and mature weight [26].

It is well established that nutrition plays a crucial role in bone growth and development [27,28]. However, the magnitude of any nutritional deficit is dependent on the stage of growth, severity of the restriction and its duration. In the current study, there were no obvious signs of a growth check in the first two years of life when the majority of growth occurs. However, if the heifers had undergone a significant growth check, the mature live weight and stature shape may have differed. In New Zealand's pasture-based systems, winter is often associated with a period of reduced growth due to decreased pasture growth, quality and supply. It is common for heifers to have a period of decreased growth during winter which tends to be more severe in the second winter than the first [29]. In the first winter (10–12 months of age), long bones, such as the metacarpus, reach maturity, so deficits in nutrition will have negligible effects on their growth trajectory. However, increases in live weight after one year of age continues to drive increases in capacity (e.g., humerus), so perturbations in live weight gain are likely to impact longitudinal growth of proximal bones. In a study by Handcock, et al. [14], heifers that were grown in a seasonal manner at 6–15 months of age, were lighter and attained puberty later than heifers that grew in a linear manner. At 21 months of age, heifers grown in a seasonal manner were lighter than heifers grown in a linear manner. However, when adjusted for live weight, heifers that grew in a seasonal manner were taller at 21 months of age. This result suggests a compensatory growth mechanism by heifers in the seasonal group whereby skeletal growth was achieved, but the ability to achieve the genetic potential for live weight was constrained. Conversely, in a study by Barash, et al. [30] where heifers were fed a "stair-step" regime consisting of a restricted diet followed by a compensatory diet, heifers were a similar live weight compared to heifers fed a constant diet. However, heifers fed the "stair-step" diet had a shorter hip height, suggesting a restriction in longitudinal bone growth and the promotion of fat deposition over muscle gain. Although the results of these studies differ, the overall trend suggests that animals that experience a check in growth are likely to have an alteration in the development of bone, muscle and fat composition. The effect of nutrition is likely to be dependent on the timing of the restriction within the developmental window, as well as the severity and length of the restriction.

A condition in New Zealand that has been linked to the seasonal nature of pasture supply is spontaneous humeral fractures in first lactation dairy heifers. The second winter has been identified as a risk factor for this type of fracture [31]. When examining bone morphology from affected heifers, the humerus shows signs of osteoporosis and growth arrest lines [15,17], whereas the metacarpus does not show the same signs. Observations of the early timing of bone maturity and lack of environmental impacts on growth in the metacarpus reinforces the lack of pathology in metacarpals from affected heifers. Conversely, the presence of the growth arrest lines and osteoporosis in the humerus highlights the importance of the second winter nutrition to achieve peak bone mass in the humerus. Given the importance of the second winter for heifer growth, future research should aim to examine how the growth check in the second winter affects stature growth.

5. Conclusions

The results of this study indicate that measuring calves at birth is a good predictor of these values at maturity (52 months old) in dairy cattle. Measures of capacity, such as heart girth and live weight, while still associated with measures at birth, had lower associations with measures at maturity. The lower associations of live weight and heart girth with values at birth reflect the greater opportunity for improvement in these capacity values with positive management strategies. Optimising nutrition, particularly prior to maturity, could increase capacity and therefore increase potential milk yield given the observation that milk yield appears to be influenced by measures of capacity. However, additional research is required to quantify the magnitude of the strength of the association of stature measures at birth and maturity with future lactational and reproductive performance.

Author Contributions: Conceptualization, M.J.G. and C.W.R.; methodology, M.J.G.; formal analysis, M.J.G. and N.W.S.; investigation, M.J.G., N.W.S. and N.P.M.; data curation, M.J.G.; writing—original draft preparation, M.J.G., N.W.S. and C.W.R.; writing—review and editing, M.J.G., N.W.S., N.P.M., C.W.R., K.E.D. and P.J.B.; project administration, M.J.G. and N.P.M.; funding acquisition, N.P.M. All authors have read and agreed to the published version of the manuscript.

Funding: This research was funded by College of Sciences Massey University Research Fund.

Institutional Review Board Statement: This experiment was conducted with the approval of Massey University Animal Ethics Committee (MUAEC 19/81- Birth to 23 months and MUAEC 23/23 from three to four years of age).

Informed Consent Statement: Not applicable.

Data Availability Statement: Data available upon request from corresponding author.

Acknowledgments: Authors acknowledge Massey University's Dairy One Farm and Keeble farm and technical staff for managing the animals. Many thanks to Trevor and Stephanie Rudman for their managing of animals and assistance in data collection.

Conflicts of Interest: No authors declare no real or perceived conflicts of interest.

References

1. Handcock, R.; Lopez-Villalobos, N.; McNaughton, L.; Back, P.; Edwards, G.; Hickson, R. Positive relationships between body weight of dairy heifers and their first-lactation and accumulated three-parity lactation production. *J. Dairy Sci.* **2019**, *102*, 4577–4589. [CrossRef] [PubMed]
2. Archbold, H.; Shalloo, L.; Kennedy, E.; Pierce, K.; Buckley, F. Influence of age, body weight and body condition score before mating start date on the pubertal rate of maiden Holstein–Friesian heifers and implications for subsequent cow performance and profitability. *Animal* **2012**, *6*, 1143–1151. [CrossRef] [PubMed]
3. Martín, N.P.; Hickson, R.E.; de Clifford, R.P.; Tulley, W.; Lopez-Villalobos, N.; Back, P.J. Production benefits from meeting liveweight targets in dairy heifers. *N. Z. J. Agric. Res.* **2020**, *63*, 220–232. [CrossRef]
4. Berry, D.; Buckley, F.; Dillon, P. Body condition score and live-weight effects on milk production in Irish Holstein-Friesian dairy cows. *Animal* **2007**, *1*, 1351–1359. [CrossRef]
5. Roche, J.; Macdonald, K.; Burke, C.; Lee, J.; Berry, D. Associations among body condition score, body weight, and reproductive performance in seasonal-calving dairy cattle. *J. Dairy Sci.* **2007**, *90*, 376–391. [CrossRef] [PubMed]
6. Sloniewski, K.; Mao, I.; Jensen, J.; Madsen, P. Changes in body weight and frame and its genetic variation during the productive life of dairy cows. *Acta Agric. Scand Sect. A* **2005**, *55*, 128–136. [CrossRef]
7. Van De Stroet, D.; Díaz, J.C.; Stalder, K.; Heinrichs, A.J.; Dechow, C.D. Association of calf growth traits with production characteristics in dairy cattle. *J. Dairy Sci.* **2016**, *99*, 8347–8355. [CrossRef] [PubMed]
8. Weik, F.; Hickson, R.E.; Morris, S.T.; Garrick, D.J.; Archer, J.A. Genetic parameters for maternal performance traits in commercially farmed New Zealand beef cattle. *Animals* **2021**, *11*, 2509. [CrossRef]
9. Gibson, M.J.; Adams, B.R.; Back, P.J.; Hickson, R.E.; Dittmer, K.E.; Rogers, C.W. Live Weight and Bone Growth from Birth to 23 Months of Age in Holstein-Friesian, Jersey and Crossbred Heifers. *Dairy* **2022**, *3*, 333–344. [CrossRef]
10. Rogers, C.W.; Gee, E.K.; Dittmer, K.E. Growth and bone development in the horse: When is a horse skeletally mature? *Animals* **2021**, *11*, 3402. [CrossRef]
11. Gibson, M.; Hickson, R.; Back, P.; Dittmer, K.; Schreurs, N.; Rogers, C. The Effect of Sex and Age on Bone Morphology and Strength in the Metacarpus and Humerus in Beef-Cross-Dairy Cattle. *Animals* **2021**, *11*, 694. [CrossRef] [PubMed]
12. Greenough, P.R.; Weaver, A.D.; MacCallum, F.J. *Lameness in Cattle*; Oliver & Boyd: Edinburgh, UK, 1972.
13. Handcock, R.; Lopdell, T.; McNaughton, L. *More Dairy Heifers Are Achieving Liveweight Targets*; New Zealand Society of Animal Production: Adelaide, Australia, 2016; pp. 3–7.
14. Handcock, R.C.; Jenkinson, C.M.; Laven, R.; McNaughton, L.R.; Lopez-Villalobos, N.; Back, P.J.; Hickson, R.E. Linear versus seasonal growth of dairy heifers decreased age at puberty but did not affect first lactation milk production. *N. Z. J. Agric. Res.* **2021**, *64*, 83–100. [CrossRef]
15. Wehrle-Martinez, A.; Lawrence, K.; Back, P.J.; Rogers, C.W.; Gibson, M.; Dittmer, K.E. Osteoporosis is the cause of spontaneous humeral fracture in dairy cows from New Zealand. *Vet. Pathol.* **2023**, *60*, 88–100. [CrossRef] [PubMed]
16. Dittmer, K.; Hitchcock, B.; McDougall, S.; Hunnam, J. Pathophysiology of humeral fractures in a sample of dairy heifers. *N. Z. Vet. J.* **2016**, *64*, 230–237. [CrossRef]
17. Gibson, M.; Dittmer, K.; Hickson, R.; Back, P.; Wehrle-Martinez, A.; Rogers, C. The mid-diaphysis is a poor predictor of humeral fracture risk indicating that predisposing factors are recent. *Ruminants* **2021**, *1*, 23–30. [CrossRef]
18. Gibson, M.J.; Rogers, C.W.; Back, P.J.; Dittmer, K.E.; Wehrle-Martinez, A.; Dalley, D.E.; Woods, R.R. The intrauterine effects of a maternal winter diet of either kale or fodder beet on measures of calf stature and bone morphology at birth. *N. Z. J. Agric. Res.* **2024**, *67*, 1–16. [CrossRef]
19. Dib, M.; Van Vleck, L.; Spangler, M.L. Genetic parameters for cow weight and height using a repeatability model in American Angus cattle. In Proceedings of the Western Section, American Society of Animal Science, Fort Collins, CO, USA, 16–18 June 2009; pp. 40–41.
20. Nelsen, T.; Short, R.; Reynolds, W.; Urick, J. Palpated and visually assigned condition scores compared with weight, height and heart girth in Hereford and crossbred cows. *J. Anim. Sci.* **1985**, *60*, 363–368. [CrossRef]
21. Gibson, M.; Dittmer, K.; Hickson, R.; Back, P.; Rogers, C. Bone Morphology and Strength in the Mid-Diaphysis of the Humerus and Metacarpus in Dairy Calves Prior to Weaning. *Animals* **2020**, *10*, 1422. [CrossRef] [PubMed]
22. Guilbert, H.; Gregory, P. Some features of growth and development of Hereford cattle. *J. Anim. Sci.* **1952**, *11*, 3–16. [CrossRef]
23. Hammack, S.P.; Gill, R.J. Texas Adapted Genetic Strategies for Beef Cattle X: Frame Score, Frame Size, and Weight. Texas FARMER Collection. 2009. Available online: https://oaktrust.library.tamu.edu/items/5ce5ff1d-3950-46d1-a19c-58201842731b (accessed on 20 January 2025).
24. Huntington, P.; Brown-Douglas, C.; Pagan, J. Growth and development of Thoroughbred horses. *Anim. Prod. Sci.* **2020**, *60*, 2093–2102. [CrossRef]

25. Crichton, J.; Aitken, J.; Boyne, A. The effect of plane of nutrition during rearing on growth, production, reproduction and health of dairy cattle.* I. Growth to 24 months. *Anim. Prod.* **1959**, *1*, 145–162. [CrossRef]
26. Eide, M.G.; Øyen, N.; Skjærven, R.; Nilsen, S.T.; Bjerkedal, T.; Tell, G.S. Size at birth and gestational age as predictors of adult height and weight. *Epidemiology* **2005**, *16*, 175–181. [CrossRef]
27. Gibson, M.J.; Adams, B.R.; Back, P.J.; Dittmer, K.E.; Herath, H.M.G.P.; Pain, S.J.; Kenyon, P.R.; Morel, P.C.H.; Blair, H.T.; Rogers, C.W. Increased Dietary Protein to Energy Ratio in Pre-Weaning Lambs Increases Average Daily Gain and Cortical Bone Thickness in the Tibia. *Ruminants* **2022**, *2*, 478–486. [CrossRef]
28. Turner, J.; Bulsara, M.; McDermott, B.; Byrne, G.; Prince, R.; Forbes, D. Predictors of low bone density in young adolescent females with anorexia nervosa and other dieting disorders. *Int. J. Eat. Disord.* **2001**, *30*, 245–251. [CrossRef] [PubMed]
29. Handcock, R.C.; Lopez-Villalobos, N.; McNaughton, L.R.; Back, P.J.; Edwards, G.R.; Hickson, R.E. Live weight and growth of Holstein-Friesian, Jersey and crossbred dairy heifers in New Zealand. *N. Z. J. Agric. Res.* **2019**, *62*, 173–183. [CrossRef]
30. Barash, H.; Bar-Meir, Y.; Bruckental, I. Effects of a low-energy diet followed by a compensatory diet on growth, puberty and milk production in dairy heifers. *Livest. Prod. Sci.* **1994**, *39*, 263–268. [CrossRef]
31. Gibson, M.J.; Dittmer, K.E.; Back, P.J.; Wehrle-Martinez, A.S.; Lawrence, K.E.; Rogers, C.W. Humeral fracture in first lactation dairy heifers: Temporal effects—Why first lactation? In Proceedings of the Australasian Dairy Science Symposium, Christchurch, New Zealand, 25–28 November 2024; pp. 87–91.

ruminants

MDPI

Article

Feed Intake and Growth Performance of Vietnamese Yellow Calves Fed Silages from Intercropped Maize–Soybean and Guinea Grass

Don V. Nguyen [1,2,*], Beth Penrose [3,4], Ngoc B. T. Tran [2,5], Huyen T. T. Le [5], Hong T. Trinh [6] and Stephen Ives [7]

1 School of Environmental and Rural Science, University of New England, Armidale, NSW 2351, Australia
2 Faculty of Animal Science, Vietnam National University of Agriculture, Hanoi 12406, Vietnam; bichngocnias@gmail.com
3 Tasmanian Institute of Agriculture, University of Tasmania, Locked Bag 98, Hobart, TAS 7000, Australia; beth.penrose@utas.edu.au
4 Research Institute for Northern Agriculture, Charles Darwin University, Darwin, NT 0909, Australia; beth.penrose@cdu.edu.au
5 National Institute of Animal Science, Hanoi 11913, Vietnam; lehuyen1973@yahoo.com
6 Dien Bien Department of Agriculture and Rural Development, Dien Bien 32110, Vietnam; hongtt.dbi@gmail.com
7 Tasmanian Institute of Agriculture, University of Tasmania, Locked Bag 1325, Newnham, TAS 7248, Australia; stephen.ives@utas.edu.au
* Correspondence: don.nguyen@une.edu.au

Simple Summary: Combining legumes with maize biomass or cultivated grass in silage offers a balanced source of carbohydrates and proteins, and can serve as a high-quality roughage during forage shortages. This study investigated the feed intake and growth responses of indigenous Yellow calves to diets with Guinea grass, Guinea grass–soybean, and maize–soybean silages. The results showed that replacing 30% of urea-treated rice straw with these silages significantly improved feed conversion ratios, total weight gain, and average daily gain in growing Yellow calves, although there were no significant effects on body conformation. These findings indicate that incorporating Guinea grass–soybean or maize–soybean silage into urea-treated rice straw diets can enhance growth in Yellow calves, supporting more sustainable mixed crop–livestock production systems.

Abstract: Combining soybeans with grass or biomass maize in silage holds promise in addressing the nutritional limitations of individual crops, providing a roughage with a good energy–protein balance. This study evaluated the effects of replacing urea-treated rice straw (UTRS) with silages made from intercropped maize–soybean and Guinea grass (GG) in calf diets on feed intake and growth performance. Sixteen native Yellow calves (130.7 ± 16.1 kg live weight and 12.8 ± 2.6 months old) were used; the experiment had a randomised complete block design with four dietary treatments: Treatment 1 (70% UTRS + 30% GG); Treatment 2 (40% UTRS + 30% GG + 30% Guinea grass silage); Treatment 3 (40% UTRS + 30% GG + 30% Guinea grass–soybean silage); and Treatment 4 (40% UTRS + 30% GG + 30% maize–soybean silage). The animals were fed 0.5 kg concentrate per 100 kg live weight daily, with unlimited access to forage and clean water, for 12 weeks after a two-week adaptation. The results indicate that silages containing soybean increased total weight gain and average daily gain (ADG) and decreased feed conversion ratio (FCR); however, silage replacements had no impact on dry matter intake and body conformation, suggesting that Guinea grass–soybean or maize–soybean silage can effectively enhance the ADG and decrease the FCR of growing calves fed UTRS-based diets.

Keywords: average daily gain; body conformation; maize–soybean silage; native cattle; small scale; urea-treated rice straw

1. Introduction

Yellow cattle are the predominant indigenous breed in Northwest Vietnam and are commonly raised both for beef and as a source of draught power [1]. As an indigenous breed, Yellow cattle are well adapted to the poor local feeding conditions and harsh climates and have a good reproductive capacity, but they have a low meat yield, with poor growth and a small body size [2]. In recent years, several genetic and nutritional studies have been conducted to improve those characteristics [2–5]. Both Nguyen et al. [4] and Dung et al. [5] stated that feeding protein–energy-balanced diets to Yellow cattle improved their general appearance, growth rate, carcass properties, and meat yield.

Small-scale Yellow cattle production in Northwest Vietnam has mainly relied on extensive grazing in forests and native pastures. Recently, many exotic forage species have been introduced and evaluated in traditional pasture areas of Vietnam to supplement the local feed source. The improved, higher-yielding varieties used for cut-and-carry production, such as elephant grass (*Pennisetum purpureum*) and Guinea grass (*Megathyrsus maximus*), were introduced in the last 20 years. In a trial conducted by Hue [6], Guinea grass achieved a yield of 74 tonnes dry matter per ha per year. However, the amount and (or) nutritional values of the forage vary greatly depending on season, grazing area, grazing intensity, grass species, fertiliser management, and focus of the small-scale farmer. Nguyen et al. [7] reported that, in winter or the dry season, natural and cultivated grasses only supply around 35–57% of the total forage demand. Consequently, supplemental feed including agricultural by-products from rice (*Oryza sativa* L.), maize (*Zea mays* L.) and cassava (*Manihot esculenta* Crantz) crops have been utilised to maintain ruminant nutrition [8]. In cold, dry winters, many livestock households in Northern Vietnam routinely provide their cattle with dry and/or urea-treated rice straw [7]. Many studies have tried to optimise the use of urea-treated rice straw as a substitute for fresh forage in cattle diets [4,9,10].

Maize is the predominant crop in the northwest region of Vietnam; in 2021, the largest cultivating area of maize in Vietnam was the Northern Midlands and mountains, with 414.4 thousand hectares, accounting for 45.9% of the country's total maize-growing area [11]. Currently, maize is mainly grown for use as a grain in monocropping systems, and the vegetative parts of the plant are mainly left to dry and are then burnt. The monocropping maize systems deplete soil nitrogen reserves, reduce soil fertility, decrease soil pH and, when grown on steep slopes, increase soil erosion [12]. There is opportunity to intercrop the maize in these systems with a legume such as soybean (*Glycine max* L.) to replenish soil nitrogen, improve soil health, reduce soil erosion, and increase long-term productivity in intercropped systems [13,14]. Additionally, ensiling maize–soybean from the intercropped systems could provide a highly nutritious feed for cattle [15,16], which could be used in the winter/dry season feed gap.

Ensiling the vegetative soybean with maize biomass or cultivated grass could overcome the nutritional limitations of the individual crops. Soybean has a relatively small amount of fermentable sugars, and silage made of soybean only is a high-crude-protein (CP)-content feedstuff [17]. Vegetative soybean silage usually has a higher pH value compared to maize and maize–soybean mixtures [15,18]. As a result, proteins and amino acids are decomposed, leading to an unpleasant odor and the deterioration of silage [17,18]. Conversely, grass and maize silages have a low CP content [18,19]. Maize–soybean intercropping can also improve the soil's biological and chemical properties, increase total biomass yield [20,21], and produce better quality silages compared to silages made of only grass, biomass maize, or soybean [16,18,22].

Generally, most recent studies have focused on silage characteristics, such as its microbial profile, nutritive values, and fermentation parameters [17,18,21]. Limited attention has been paid to the effects of the inclusion of maize–soybean silage on the responses of live animals, including effects on feed intake, digestibility, and growth performance. When Dorper × Santa Inês crossbred lambs were fed iso-nitrogenous (14% CP) and iso-neutral detergent fibre (16.5% NDF) diets, Bolson et al. [23] concluded that the maize–soybean intercropped silage increased the feed efficiency and growth of the lambs compared to monocultured soybean silage. Kang et al. [24] also reported that maize–soybean intercropped silage increased in vitro rumen fermentation, feed degradability, and milk quality of Holstein dairy cows; however, there is insufficient information available regarding the use of maize–soybean intercropping to make silage for cattle under small-scale production in Vietnam. Therefore, this study aimed to understand the impacts of incorporation of maize–soybean silage on growth performance of Yellow calves in comparison to other silages and fresh grass.

2. Materials and Methods

2.1. Silage and Rice Straw Preparation

In August 2021, three types of silage were prepared: Guinea grass, maize–soybean and soybean. Guinea grass was cut at 50 days old after its previous harvest. Maize–soybean was sown as an intercrop, while soybean was sown as a monocrop in May 2021, and both crops were simultaneously harvested at 90 days old. For the maize–soybean intercrop, soybean was sown in four rows with two rows of maize in between. Within the intercrop, maize was sown with 40 cm between rows whilst soybean was sown with 30 cm between rows. The distances between columns were 10 cm for soybean and 20 cm for maize. For sole-cropping soybean, row spacing was 40 cm, and the column spacing was 10 cm (Figure 1).

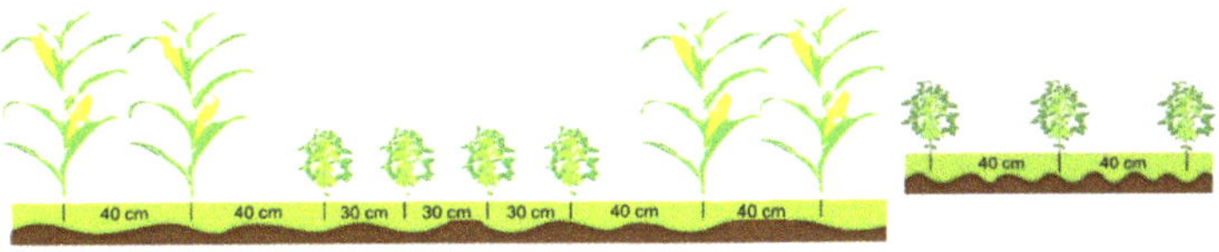

(a) (b)

Figure 1. Row and column spacings for (a) intercropped maize–soybean and (b) monocropped soybean.

At 90 days of age, maize and soybean were at the seed-filling and 2/3 milk line stages, respectively. Guinea grass, maize, and soybean were cut and wilted on the field for 4–6 h to reduce moisture content. After wilting, they were chopped into 10–15 cm pieces and mixed according to the following formulations:

1. Guinea grass silage: 97% guinea grass + 2.5% cornmeal + 0.5% salt.
2. Guinea grass–soybean silage: 80% Guinea grass + 20% soybean.
3. Maize–soybean silage: 80% maize + 20% soybean.

The three silage mixtures were sealed in individual plastic bags (approximately 50 kg per bag). In the maize–soybean silage, the 4:1 (w/w) ratio reflects the approximate proportion of biomass maize and soybean yields harvested from the intercropped crop. The same percentage of soybean (20%) was used in the Guinea grass–soybean silage to facilitate direct comparison between the two silages. Urea-treated rice straw (UTRS) was also ensiled in August 2021. Dry rice straw was treated with 4% urea (as-fed basis) with the addition of 8 L clean water for every 10 kg dry rice straw and also sealed in individual plastic bags (approximate 40 kg per bag). All ensiled forages were anaerobically preserved and fed to cattle after at least four weeks of ensilage.

2.2. Experimental Design and Feed Compositions

The feeding trial was carried out in Dien Bien district, Dien Bien province, Vietnam from September to December 2021. Sixteen indigenous Yellow calves (130.7 ± 16.1 kg of LW, 12.8 ± 2.6 months old and balanced by sex) were used in the feeding trial. They were dewormed with Ivermectin (VIAMECTIN 25, Viavet Ltd., Hanoi, Vietnam) and immunised against foot-and-mouth disease before the experiment commenced.

The experiment followed a randomised complete block design. The calves were weighed and allocated to one of four blocks based on their LW. Subsequently, the four animals in each block were randomly assigned one of the four treatments below:

Treatment 1 (Control): 70% UTRS + 30% Guinea grass.
Treatment 2: 40% UTRS + 30% Guinea grass + 30% guinea grass silage.
Treatment 3: 40% UTRS + 30% Guinea grass + 30% guinea grass–soybean silage.
Treatment 4: 40% UTRS + 30% Guinea grass + 30% maize–soybean silage.

The experimental diets were formulated using the National Research Council feed standard system [25] to achieve an expected weight gain of 400–600 g per day. The calves were housed in separate pens and fed treatment diets individually for 14 weeks, which included a two-week adaptation period and a 12-week trial period. During the experiment, the calves were given ad libitum access to their allotted roughage treatment and clean drinking water, and were also given 0.5 kg of concentrate (as-fed basis) per 100 kg of LW per day at approximately 8.00 am before feeding roughage to ensure they received all the nutrients required for maintenance and growth. The ingredients of the concentrate mixture (as-fed basis) are displayed in Table 1. The amount of concentrate offered to individual animals was updated fortnightly based the change in their live weigh. Each roughage treatment was well mixed (as-fed basis) before being fed to the animals as a mixture.

Table 1. Concentrate ingredients and feed chemical compositions.

Item	Concentrate	Urea-Treated Rice Straw	Fresh Guinea Grass	Guinea Grass Silage	Guinea Grass–Soybean Silage	Maize–Soybean Silage
Ingredient (g per kg) (as-fed basis)						
Corn flour	595					
Rice bran	260					
Soybean	140					
Premix	5					
Chemical composition						
Dry matter (DM) (%)	88.6	42.3	19.3	32.0	30.3	29.2
Organic matter (%DM)	96.0	86.6	88.5	87.4	87.2	94.4
Crude protein (%DM)	12.3	8.3	6.1	5.9	7.5	7.9
Ether extract (%DM)	6.5	2.1	2.6	2.7	3.2	3.3
Crude fibre (%DM)	10.8	29.8	30.6	38.2	35.7	26.9
Neutral detergent fibre (%DM)	17.7	68.9	49.3	48.6	49.7	53.0
Acid detergent fibre (%DM)	6.1	37.6	28.4	24.2	26.3	24.1
Total ash (%DM)	4.0	13.4	11.5	12.7	12.9	5.6
Non-fibrous carbohydrate (%DM)	59.5	7.3	30.5	30.1	25.7	28.2
Total digestible nutrient	81.0	47.4	53.6	52.0	53.9	58.2
Digestible energy ($MJ\cdot kg^{-1}$ DM)	14.9	8.7	9.9	9.4	9.9	10.7
Metabolisable energy ($MJ\cdot kg^{-1}$ DM)	12.2	7.2	8.0	7.6	8.2	8.7

Premix includes vitamin A (4100.000 $IU\cdot kg^{-1}$), vitamin D3 (350.000 $IU\cdot kg^{-1}$), vitamin E (8 $g\cdot kg^{-1}$), vitamin B1 (850 $mg\cdot kg^{-1}$), vitamin B2 (1.6 $g\cdot kg^{-1}$), vitamin B6 (1.7 $g\cdot kg^{-1}$), vitamin B12 (6 $mg\cdot kg^{-1}$), vitamin K3 (350 $mg\cdot kg^{-1}$), niacin (12 $g\cdot kg^{-1}$), folic acid (250 $mg\cdot kg^{-1}$), biotin (16 $mg\cdot kg^{-1}$), iron (30 $g\cdot kg^{-1}$), copper (30 $g\cdot kg^{-1}$), manganese (13 $g\cdot kg^{-1}$), and other minerals (Zn, Se, I, Co) (380 $mg\cdot kg^{-1}$).

2.3. Calculation of Feed Intake and Feed Analysis

Daily feed intake was calculated by weighing the offered feed and the residual feed. The concentrate mixture, ensiled mixtures, UTRS and Guinea grass were separately sampled on days 1, 42 and 84 of the 12-week trial period, and stored in a freezer at −20 °C until ready for feed analyses. At the end of the trial period, samples from each timepoint were dried at 65 °C in a fan-forced oven until their weight between two consecutive weighing times remained unchanged to ascertain the DM content. For each feed type, the three timepoint samples were pooled and then ground through a 1 mm screen using a CT 293 Cyclotec laboratory mill (Tecator, Höganäs, Sweden).

An elemental analyser (PE2400 Series II; Perkin-Elmer Corp., Waltham, MA, USA) was used to quantify total nitrogen content (N) for each feed type, which was then multiplied by 6.25 to determine CP content [26]. The samples were placed in a furnace for five hours at 550 °C to determine the total ash content according to the methods of AOAC [26]. An ANKOM XT15 fat/oil extractor (ANKOM Technology, Macedon, NY, USA) was used to quantify ether extract (EE). The neutral detergent fibre (NDF) and acid detergent fibre (ADF) content were quantified using an ANKOM 200 fibre analyser (ANKOM Technology, Macedon, NY, USA). Organic matter (OM) was computed as OM = 100 − Ash. Non-fibrous carbohydrate (NFC) was determined using equation NFC = 100 − (CP +NDF + EE + Ash) [27]. Total digestible nutrient (TDN) was computed utilising the equations established by Jayanegara et al. [28] TDN = 0.714CP + 1.594EE + 0.479NDF + 0.704NFC for forage and TDN = 0.885CP + 1.829EE + 0.323NDF + 0.883NFC for concentrate. Metabolisable energy (ME) was calculated by converting TDN todigestible energy (DE) = 0.01 × 4.4 × 4.185 × TDN), which was converted as ME = 0.82 × DE as per Nguyen et al. [4]. The concentrate ingredients and chemical compositions of feed used in the experiment are displayed in Table 1.

2.4. Measurements of Live Weight and Body Conformation

Animals were weighed fortnightly before morning feeding, by walking them over a calibrated Ruddweigh 200 electronic scale (Ruddweigh Electronic Scales, Guyra, NSW, Australia) with an increment of 100 g. Average daily gain was calculated as the total liveweight gain divided by the number of days on experimental feed. Feed conversion ratio (FCR) is an indicator of the efficiency with which an animal converts feed into body mass, and computed as the ratio of feed intake to live weight gain. The body conformation traits including chest girth, body length and wither height were also measured fortnightly using a plastic tape measure and a metal vernier equipped with two vertically sliding arms to record span. The detailed description of body measurements was provided by Nguyen et al. [4]. To maintain, accuracy, consistency and repeatability, the same researcher implemented all measurements throughout the experiment. During the measurements, the animals were restrained in a relaxed state with their four legs stabilised on flat ground and their head comfortably erect.

2.5. Statistical Analyses

The feed intake, ADG and FCR of each calf were calculated using linear regression, based on individual data collected from the start to the end of the feeding trial. Summary descriptive statistics, including means and standard errors of mean, were computed and carefully examined for any potential data entry errors. Version 16.2 of Minitab statistical software [29] was used to analyse all collected data. Analysis of variance (ANOVA) was employed to analyse the data through a general linear model, incorporating various forage mixtures as a fixed effect and blocks as a random effect, while dependent variables included growth traits, feed intake, and body measurements. Tukey's probability pairwise comparison tests were applied to identify significant differences at the $p < 0.05$ threshold.

3. Results

3.1. Feed Intake

There were no significant differences in feed intake (as-fed basis) between the four treatments ($p > 0.05$; Table 2). Similarly, DMI and DMI per 100 kg LW were not influenced by the treatments ($p > 0.05$), with values ranging from 3.4 to 3.7 kgDM per day per head and from 2.36 to 2.57 kg, respectively. Replacing 30% UTRS with different silages had no significant effects on both CP and ME intakes ($p > 0.05$). Moreover, dietary treatments did not cause any significant variation ($p > 0.05$) in concentrate to forage ratios, ranging from 0.21 to 0.28.

Table 2. Daily feed intake.

	Treatment 1 (*n* = 4)	Treatment 2 (*n* = 4)	Treatment 3 (*n* = 4)	Treatment 4 (*n* = 4)	SEM	*p*
As-fed basis						
Concentrate (kg)	0.68	0.69	0.75	0.66	0.02	0.878
Forage (kg)	9.0	9.0	9.7	10.1	0.37	0.231
Total intake (kg)	9.7	9.7	10.4	10.8	0.38	0.238
Dry matter basis						
Concentrate (kg)	0.60	0.61	0.66	0.59	0.02	0.960
Forage (kg)	3.1	2.7	2.9	3.0	0.11	0.875
DMI (kg)	3.7	3.4	3.6	3.6	0.12	0.889
DMI/100 kg LW (kg)	2.57	2.40	2.36	2.40	0.08	0.447
Daily CP intake (kg)	0.31	0.26	0.29	0.30	0.01	0.933
Daily ME intake (MJ)	30.2	28.4	30.3	30.6	1.00	0.705
Concentrate to forage ratio	0.21	0.28	0.27	0.21	0.01	0.866

Treatment 1 = 70% urea-treated rice straw (UTRS) + 30% Guinea grass; Treatment 2 = 40% UTRS + 30% Guinea grass + 30% Guinea grass silage; Treatment 3 = 40% UTRS + 30% Guinea grass + 30% Guinea grass–soybean silage; Treatment 4 = 40% UTRS + 30% Guinea grass + 30% maize–soybean silage; ADG = average daily gain; CP = crude protein; DMI = dry matter intake; LW = live weight; ME = metabolisable energy; SEM = standard error of the mean.

3.2. Live Weight Gain and Conversion Ratio

Calves fed diets containing Guinea grass–soybean silage and maize–soybean silage (treatments 3 and 4, respectively) had considerably higher total weight gain ($p = 0.035$) than those fed the control and Guinea grass silage diets (treatments 1 and 2, respectively) (Table 3). No significant differences in total weight gain were observed between the two groups fed diets without soybean (treatments 1 and 2), or between the two groups fed soybean-containing diets (treatments 3 and 4) ($p > 0.05$). Similarly, significant differences ($p = 0.035$) in ADG were observed between the calves fed soybean-containing diets (0.38 for Treatment 3 and 0.37 kg per day for Treatment 4) and those that received diets without soybean (0.32 kg per day for treatments 1 and 2). Average daily gain was similar between treatments 1 and 2, or between treatments 3 and 4 ($p > 0.05$). The best feed conversion ratios (FCR) were observed from the calves fed the soybean-containing diets (9.3 and 9.6 in treatments 3 and 4, respectively). These ratios were substantially better ($p < 0.05$) compared to the FCR observed in Treatment 1 (11.5).

Table 3. Live weight gain and feed conversion ratio of experimental animals.

	Treatment 1 (*n* = 4)	Treatment 2 (*n* = 4)	Treatment 3 (*n* = 4)	Treatment 4 (*n* = 4)	SEM	*p*
Initial LW (kg)	130.3	125.8	134.4	132.5	5.1	0.954
Final LW (kg)	157.5	152.5	166.3	163.9	6.6	0.856
Total weight gain (kg)	27.3 ^{a}	26.8 ^{a}	31.9 ^{b}	31.4 ^{b}	0.9	0.035
ADG (kg per day)	0.32 ^{a}	0.32 ^{a}	0.38 ^{b}	0.37 ^{b}	0.01	0.035
Feed conversion ratio (kg DMI per kg ADG)	11.5 ^{a}	10.4 ^{ab}	9.3 ^{c}	9.6 ^{bc}	0.35	0.047

Treatment 1 = 70% urea-treated rice straw (UTRS) + 30% Guinea grass; Treatment 2 = 40% UTRS + 30% Guinea grass + 30% Guinea grass silage; Treatment 3 = 40% UTRS + 30% Guinea grass + 30% Guinea grass–soybean silage; Treatment 4 = 40% UTRS + 30% Guinea grass + 30% maize–soybean silage; LW = live weight; ADG = average daily gain; DMI = dry matter intake; SEM = standard error of the mean. Means bearing distinct superscript letters in the same row significantly differ ($p < 0.05$).

3.3. Body Conformation Traits

The initial and final body conformation traits (chest girth, body length and wither height) among treatments did not significantly differ ($p > 0.05$). Similarly, dietary treatment did not cause any significant change ($p > 0.05$) in body measurements (Table 4).

Table 4. Body conformation measurements.

		Treatment 1	Treatment 2	Treatment 3	Treatment 4	SEM	*p*
		(*n* = 4)	(*n* = 4)	(*n* = 4)	(*n* = 4)		
Initial	Chest girth	120.5	117.0	121.0	122.0	1.8	0.813
	Body length	112.0	110.3	108.0	114.3	3.0	0.917
	Withers height	99.5	101.3	101.8	101.3	1.4	0.954
Final	Chest girth	131.0	129.3	132.3	131.8	2.1	0.971
	Body length	128.3	120.3	129.0	126.5	3.0	0.784
	Withers height	108.8	106.0	107.3	106.5	1.5	0.931
Change	Chest girth	10.5	12.3	11.3	9.8	1.0	0.870
	Body length	16.3	10.0	21.0	12.3	1.7	0.096
	Withers height	9.3	4.8	5.5	5.0	0.8	0.179

Treatment 1 = 70% urea-treated rice straw (UTRS) + 30% Guinea grass; Treatment 2 = 40% UTRS + 30% Guinea grass + 30% Guinea grass silage; Treatment 3 = 40% UTRS + 30% Guinea grass + 30% guinea grass–soybean silage; Treatment 4 = 40% UTRS + 30% Guinea grass + 30% maize–soybean silage; SEM = standard error of the mean.

4. Discussion

The results illustrate that although no significant differences were detected in DM, CP and ME intakes for the native Yellow calves fed the different dietary treatments, the inclusion of vegetative soybean in silages enhanced ADG, resulting in greater overall live weight gain. The higher ADG of the animals fed diets containing Guinea grass–soybean silage or maize–soybean silage might be in part due to the higher rumen degradability of soybean-containing silage, which could stimulate microbial growth and nutrient digestibility. Urea, a non-protein nitrogen source, is widely utilised as a substitute for feed protein in ruminant diets due to its cost effectiveness [30]. In the absence of a rumen-protected coating, however, urea may be inadequately utilised as a nitrogen source for microbial protein synthesis because it is rapidly hydrolysed to ammonia by bacterial ureases in the rumen and it often exceeds the use capacity by rumen bacteria with the surplus being absorbed through the rumen wall into the blood [31]. This can reduce the availability of protein in the rumen, which then negatively affects the protein syntheses of the bacterial community.

While UTRS boosts the overall CP percentage, it does not provide the same nutritional value as true protein sources found in other forages [32]. Dewhurst and Newbold [33] concluded that ammonia nitrogen concentration alone does not fully explain the effects of degradable protein on microbial growth. They emphasised that rapid growth of rumen bacteria is supported by true protein sources, such as peptides and amino acids. Based on the effects on fibre digestion, Calsamiglia et al. [34] highlighted that true proteins are essential for proper rumen function, including enhanced fibre digestion. Furthermore, Weimer [35] pointed out that branched-chain volatile fatty acids, which are the primary products of peptide and amino acid fermentation, serve as an important nutrient source for cellulolytic and hemicellulolytic bacteria. Aquino et al. [36] also affirmed that the digestibility of the protein in UTRS remains lower compared to other forages and this can limit the efficiency of protein utilisation by ruminants. When mixing vegetative soybean with corn stover at various ratios, Cheng et al. [37] concluded that increased vegetative soybean levels resulted in an enhanced nutritive value of the silage, demonstrated by increased CP and ME values. This trend is consistent with the present study when comparing those values between Guinea grass silage and Guinea grass–soybean silage. Similarly, Gandra et al. [38] reported that Jersey heifers fed diets containing maize–soybean silage (either 3:1 or 1:1) had higher digestibility of DM, organic matter, CP, and NDF than those on a diet containing maize silage.

The daily feed intake of the animals in this study were within the range recommended by Kearl [39], who found that a 100–200 kg beef cow would need to consume DMI of approximately 2.3–2.8% of their liveweight to achieve a 0.25–0.50 kg weight gain per day. Our findings align with those reported by Dung et al. [5] whose young male Yellow cattle increased 0.51 kg per day when consuming DMI at 2.7% of their liveweight. The smallest feed conversion ratio (9.3) was observed in the animals fed a diet containing Guinea grass-soybean silage, the observation was comparable to that found by Nguyen et al. [4] and Vu et al. [40] when feeding native Yellow and Sindhi x Yellow crossed cattle, respectively. Therefore, our findings align with previous studies suggesting that the addition of legume and sorghum mixed silage in diets increases growth performance of beef cattle [41,42]. Both Song et al. [21] and Zeng et al. [15] stated the combination of maize and soybeans in silage offers a good balance of carbohydrates and proteins, which is beneficial for ruminant production as a reliable source of high-quality roughage during periods when fresh forage is limited or unavailable.

The findings in the present study support the development of management strategies aimed at achieving more sustainable mixed-crop livestock production systems. Incorporating legumes such as soybean in maize production systems could lead to soil benefits including improved nitrogen cycling and fertility, and reduced erosion, especially on steep slopes [14,43]. Furthermore, maize–soybean intercropping could disrupt pest lifecycles and reduce plant pathogens and disease spread, leading to healthier crops and increased yields [44]. Incorporating legumes into agricultural systems offers practical, cost-effective and environmentally sustainable methods for the crop-livestock sustainable agriculture production by reducing reliance on external inputs while enhancing soil nutrient levels, maintaining availability of good forage for ruminant production.

There are some limitations in research and extension services that provide region-specific guidance on best practices for maize–soybean intercropping and silage making [15]. In Vietnam, while numerous agronomic studies have optimised spatial arrangements and planting densities to maximise biomass yield, control weeds, and reduce pesticide use [12,45,46], investigations into silage fermentation characteristics, ruminal digestibility, and the biochemical responses of animals to intercropped maize–soybean silage remain insufficient. Additionally, resource constraints have led to small sample sizes in various studies, including the current one, potentially compromising the validity of their findings [47,48]. Extension services, essential for translating scientific knowledge into practical applications and improving mixed crop-livestock systems, are also underdeveloped in remote areas of Vietnam [49,50]. Consequently, farmers, particularly small-scale producers, often lack training and knowledge in production costs, resource management, pest control, ensiling techniques, and dietary silage inclusion.

To maximise the potential benefits of silage made from maize–soybean intercrop-ping, future research should employ larger sample sizes to validate findings and provide more robust conclusions. In addition, it is important to investigate other factors influencing the characteristics and quality of maize–soybean silage, and the biochemical responses, health and productivity of animals fed the silage. Substantially expanding the extension network and enhancing extension agents' knowledge and skills are essential to making extension services more effective and responsive, ultimately benefiting farmers and communities. Finally, comprehensive economic analyses are required to evaluate the cost–benefit ratio of maize–soybean silage systems and their impact on farm profitability.

5. Conclusions

The feed conversion ratio, total weight gain and average daily gain of growing Yellow calves were significantly improved by substituting 30% urea-treated rice straw with Guinea grass–soybean silage or maize–soybean silage; however, the silage substitution had no considerable effect on feed intake and body conformation traits. It is therefore recommended that replacing 30% Guinea grass–soybean silage or 30% maize–soybean silage in urea-treated rice straw-based diets of calves over an 84-day feeding period could represent an effective strategy to enhance average daily gain and feed conversion ratio. Furthermore, maize–soybean intercropping for silage could deliver environmental co-benefits such as reducing fertiliser requirements and improving soil health. Future research should include a larger number of animals to provide more robust findings and focus on improving intercropped silage quality, ruminant responses, extension services and conducting economic analyses to assess profitability.

Author Contributions: Conceptualization, D.V.N., N.B.T.T., H.T.T.L. and S.I.; methodology, D.V.N., N.B.T.T., H.T.T.L., H.T.T. and S.I.; formal analysis, D.V.N., B.P. and S.I.; investigation, D.V.N., N.B.T.T. and H.T.T.; writing—original draft preparation, D.V.N.; writing—review and editing, D.V.N., B.P., N.B.T.T., H.T.T.L. and S.I.; visualization, D.V.N.; supervision, H.T.T.L., B.P. and S.I.; project administration, H.T.T.L. and S.I.; funding acquisition, H.T.T.L. and S.I. All authors have read and agreed to the published version of the manuscript.

Funding: The research was financially supported by the Australian Centre for International Agricultural Research (ACIAR, http://aciar.gov.au/, accessed on 18 November 2024) through LPS/2015/037: "Intensification of beef cattle production in upland cropping systems in Northwest Vietnam" and SMCN/2014/049: "Improving maize-based farming systems on sloping lands in Vietnam and Laos".

Institutional Review Board Statement: The welfare of all experimental animals was maintained in compliance with the 2013 Australian Code of Practice for the Care and Use of Animals for Scientific Purposes. All procedures were approved by the University of Tasmania Animal Ethics Committee (Permit No. A0017801).

Informed Consent Statement: Not applicable.

Data Availability Statement: Dataset available on request from the authors.

Acknowledgments: We are grateful to Provincial Extension Centre for Livestock Breeds and Crop Varieties, Dien Bien, Vietnam for their cooperation, experimental forage and facility supplies. We acknowledge Fiona Brodribb from University of Tasmania, Australia for her assistance in English editing and proofreading.

Conflicts of Interest: The authors declare no conflicts of interest.

References

1. Le, T.N.; Vu, H.V.; Okuda, Y.; Duong, H.T.; Nguyen, T.B.; Nguyen, V.H.; Le, P.D.; Kunieda, T. Genetic characterization of Vietnamese Yellow cattle using mitochondrial DNA and Y-chromosomal haplotypes and genes associated with economical traits. *Anim. Sci. J.* **2018**, *89*, 1641–1647. [CrossRef] [PubMed]
2. Uyen, N.T.; Cuong, D.V.; Thuy, P.D.; Son, L.H.; Ngan, N.T.; Quang, N.H.; Tuan, N.D.; Hwang, I.H. A Comparative Study on the Adipogenic and Myogenic Capacity of Muscle Satellite Cells, and Meat Quality Characteristics between Hanwoo and Vietnamese Yellow Steers. *Food Sci. Anim. Resour.* **2023**, *43*, 563–579. [CrossRef] [PubMed]
3. Gioi, P.V.; Dung, D.V.; Son, P.V.; Lap, D.V.; Tiem, P.V. Effects of Brahman genetic resource proportion on growth performance traits of beef crossbreds in Western Highlands of Vietnam. *Agric. Nat. Resour.* **2023**, *57*, 677–688.
4. Nguyen, D.V.; Tran, N.B.T.; Vang, M.A.; Le, H.T.T.; Nguyen, G.T.T.; Nguyen, Q.H.; Blanchard, M.; Bailey, A.; Ives, S. Live weight and body conformation responses of culled local Yellow cows fed maize silage and urea-treated rice straw in an intensive feedlot system in Northwest Vietnam. *Adv. Anim. Vet. Sci.* **2021**, *9*, 1283–1291.
5. Dung, D.; Phung, L.; Roubik, H. Performance and estimation of enteric methane emission from fattening vietnamese yellow cattle fed different crude protein and concentrate levels in the diet. *Adv. Anim. Vet. Sci.* **2019**, *7*, 962–968. [CrossRef]
6. Hue, P.T. Growth and development of VA06 grass and Ghine TD58 in Eakar district, Dak Lak province. *CTU J. Sci.* **2017**, *51*, 1–6.
7. Nguyen, D.V.; Vu, C.C.; Nguyen, T.V. Current Possibilities and Treating Strategies of Rice Straw as a Ruminant Feed Source in Vietnam: A Review. *Pak. J. Nutr.* **2020**, *19*, 94–104. [CrossRef]
8. Nguyen, D.V.; Dang, L.H. Using treated fresh rice straw partly replacing grass in growing beef cattle diets. *Asian J. Anim. Sci.* **2020**, *14*, 6–24.
9. Thiet, N.; Ngu, N.T. Effect of urea treatment and preservation duration on chemical composition of rice straw offer for growing Sind crossbred cattle. *Livest. Res. Rural Dev.* **2023**, *35*, 5.
10. GSO. *Statistical Yearbook of Vietnam*; Statistical Publishing House: Hanoi, Vietnam, 2022.
11. Belete, T.; Yadete, E. Effect of Mono Cropping on Soil Health and Fertility Management for Sustainable Agriculture Practices: A Review. *J. Plant Sci.* **2023**, *11*, 192–197. [CrossRef]

12. Ha, T.M.; Voe, P.; Boulom, S.; Le, T.T.L.; Dao, C.D.; Yang, F.; Dang, X.P.; Hoang, T.T.H.; Abu Hatab, A.; Hansson, H. Factors associated with smallholders' uptake of intercropping in Southeast Asia: A cross-country analysis of Vietnam, Laos, and Cambodia. *Clim. Risk Manag.* **2024**, *45*, 100646. [CrossRef]
13. Raza, M.A.; Yasin, H.S.; Gul, H.; Qin, R.; Mohi Ud Din, A.; Khalid, M.H.B.; Hussain, S.; Gitari, H.; Saeed, A.; Wang, J. Maize/soybean strip intercropping produces higher crop yields and saves water under semi-arid conditions. *Front. Plant Sci.* **2022**, *13*, 1006720. [CrossRef] [PubMed]
14. Feng, C.; Sun, Z.; Zhang, L.; Feng, L.; Zheng, J.; Bai, W.; Gu, C.; Wang, Q.; Xu, Z.; van der Werf, W. Maize/peanut intercropping increases land productivity: A meta-analysis. *Field Crops Res.* **2021**, *270*, 108208. [CrossRef]
15. Zeng, T.; Wu, Y.; Xin, Y.; Chen, C.; Du, Z.; Li, X.; Zhong, J.; Tahir, M.; Kang, B.; Jiang, D. Silage quality and output of different maize–soybean strip intercropping patterns. *Fermentation* **2022**, *8*, 174. [CrossRef]
16. Garay-Martínez, J.R.; Godina-Rodríguez, J.E.; Maldonado-Jáquez, J.A.; Lucio-Ruíz, F.; Joaquín-Cancino, S.; Bautista-Martínez, Y.; Granados-Rivera, L.D. Nutritive value of maize and soybean silages at different ratio in a subtropical climate condition. *Chil. J. Agric. Res.* **2024**, *84*, 540–547.
17. Bolson, D.C.; Jacovaci, F.A.; Gritti, V.C.; Bueno, A.V.I.; Daniel, J.L.P.; Nussio, L.G.; Jobim, C.C. Intercropped maize-soybean silage: Effects on forage yield, fermentation pattern and nutritional composition. *Grassl. Sci.* **2022**, *68*, 3–12. [CrossRef]
18. Carpici, E.B. Nutritive values of soybean silages ensiled with maize at different rates. *Legume Res. Int. J.* **2016**, *39*, 810–813. [CrossRef]
19. Parra, C.S.; Bolson, D.C.; Jacovaci, F.A.; Nussio, L.G.; Jobim, C.C.; Daniel, J.L.P. Influence of soybean-crop proportion on the conservation of maize-soybean bi-crop silage. *Anim. Feed Sci. Technol.* **2019**, *257*, 114295. [CrossRef]
20. Li, L.; Zou, Y.; Wang, Y.; Chen, F.; Xing, G. Effects of corn intercropping with soybean/peanut/millet on the biomass and yield of corn under fertilizer reduction. *Agriculture* **2022**, *12*, 151. [CrossRef]
21. Song, Y.; Lee, S.-H.; Woo, J.H.; Lee, K.-W. Evaluation of the growth characteristics, forage productivity, and feed value of the maize–soybean intercropping system under different fertilization levels. *J. Crop Sci. Biotechnol.* **2023**, *26*, 107–118. [CrossRef]
22. Zaeem, M.; Nadeem, M.; Pham, T.H.; Ashiq, W.; Ali, W.; Gilani, S.S.M.; Elavarthi, S.; Kavanagh, V.; Cheema, M.; Galagedara, L. The potential of corn-soybean intercropping to improve the soil health status and biomass production in cool climate boreal ecosystems. *Sci. Rep.* **2019**, *9*, 13148. [CrossRef] [PubMed]
23. Bolson, D.; Jobim, C.; Daniel, J.; Jacovaci, F.; Bueno, A.; Ribeiro, M.; Gritti, V. Performance of lambs fed maize, soybean or maize-soybean intercrop silages. *Grassl. Resour. Extensive Farming Syst. Marg. Lands Major Driv. Future Scenar.* **2017**, *22*, 112.
24. Kang, J.; Song, J.; Marbun, T.D.; Kwon, C.H.; Kim, E.J. Effect of intercropped corn and soybean silage on nutritive values, in vitro ruminal fermentation, and milk production of Holstein dairy cows. *J. Kor. Grassl. Forage. Sci.* **2017**, *37*, 216–222. [CrossRef]
25. NRC. *Nutrient Requirements of Beef Cattle, Eighth Revised Edition*; National Academies Press: Washington, DC, USA, 2016.
26. AOAC. *Official Methods of Analysis of AOAC International*, 15th ed.; Association of Official Analytical Chemists: Arlington, VA, USA, 1990; Volume 489.
27. Mertens, D.R. Gravimetric determination of amylase-treated neutral detergent fiber in feeds with refluxing in beakers or crucibles: Collaborative study. *J. AOAC Int.* **2002**, *85*, 1217–1240. [PubMed]
28. Jayanegara, A.; Ridla, M.; Laconi, E. Estimation and validation of total digestible nutrient values of forage and concentrate feedstuffs. *IOP Conf. Ser. Mater. Sci. Eng.* **2019**, *546*, 042016. [CrossRef]
29. Minitab. *Minitab 16 Statistical Software*; Minitab, LLC: State College, PA, USA, 2010.
30. Zhou, Z.; Meng, Q.; Li, S.; Jiang, L.; Wu, H. Effect of urea-supplemented diets on the ruminal bacterial and archaeal community composition of finishing bulls. *Appl. Microbiol. Biotechnol.* **2017**, *101*, 6205–6216. [CrossRef]
31. Guo, Y.; Xiao, L.; Jin, L.; Yan, S.; Niu, D.; Yang, W. Effect of commercial slow-release urea product on in vitro rumen fermentation and ruminal microbial community using RUSITEC technique. *J. Anim. Sci. Biotechnol.* **2022**, *13*, 56. [CrossRef]
32. Katoch, R. Forage Nutritional Quality Management. In *Techniques in Forage Quality Analysis*; Katoch, R., Ed.; Springer Nature Singapore: Singapore, 2023; pp. 211–221.
33. Dewhurst, R.J.; Newbold, J.R. Effect of ammonia concentration on rumen microbial protein production in vitro. *Br. J. Nutr.* **2022**, *127*, 847–849. [CrossRef]
34. Calsamiglia, S.; Ferret, A.; Reynolds, C.K.; Kristensen, N.B.; van Vuuren, A.M. Strategies for optimizing nitrogen use by ruminants. *Animal* **2010**, *4*, 1184–1196. [CrossRef]
35. Weimer, P.J. Degradation of cellulose and hemicellulose by ruminal microorganisms. *Microorganisms* **2022**, *10*, 2345. [CrossRef]
36. Aquino, D.; Del Barrio, A.; Trach, N.X.; Hai, N.T.; Khang, D.N.; Toan, N.T.; Van Hung, N. Rice Straw-Based Fodder for Ruminants. In *Sustainable Rice Straw Management*; Gummert, M., Hung, N.V., Chivenge, P., Douthwaite, B., Eds.; Springer International Publishing: Cham, Switzerland, 2020; pp. 111–129.
37. Cheng, Q.; Li, P.; Xiao, B.; Yang, F.; Li, D.; Ge, G.; Jia, Y.; Bai, S. Effects of LAB inoculant and cellulase on the fermentation quality and chemical composition of forage soybean silage prepared with corn stover. *Grassl. Sci.* **2021**, *67*, 83–90. [CrossRef]
38. Gandra, J.R.; Del Valle, T.A.; Takiya, C.S.; Oliveira, E.R.; Goes, R.H.; Batista, J.D.; Acosta, A.P.; Noia, I.Z.; Antônio, G.; Urio, G.S. Soybean silage in dairy heifers' diets: Ruminal fermentation, intake and digestibility of nutrients. *N. Z. J. Agric. Res.* **2022**, *65*, 202–212. [CrossRef]
39. Kearl, L.C. *Nutrient Requirements of Ruminants in Developing Countries*; International Feedstuffs Institute, Utah State University: Logan, UT, USA, 1982.

40. Da Silva, L.; Pereira, O.; Da Silva, T.; Valadares Filho, S.; Ribeiro, K. Effects of silage crop and dietary crude protein levels on digestibility, ruminal fermentation, nitrogen use efficiency, and performance of finishing beef cattle. *Anim. Feed Sci. Technol.* **2016**, *220*, 22–33. [CrossRef]
41. Zhang, H.; Zhang, L.; Xue, X.; Zhang, X.; Wang, H.; Gao, T.; Phillips, C. Effect of feeding a diet comprised of various corn silages inclusion with peanut vine or wheat straw on performance, digestion, serum parameters and meat nutrients in finishing beef cattle. *Anim. Biosci.* **2022**, *35*, 29–38. [CrossRef] [PubMed]
42. Li, C.; Hoffland, E.; Kuyper, T.W.; Yu, Y.; Zhang, C.; Li, H.; Zhang, F.; van der Werf, W. Syndromes of production in intercropping impact yield gains. *Nat. Plants* **2020**, *6*, 653–660. [CrossRef]
43. Liu, M.; Zhao, H. Maize-soybean intercropping improved maize growth traits by increasing soil nutrients and reducing plant pathogen abundance. *Front. Microbiol.* **2023**, *14*, 1290825. [CrossRef]
44. Kebede, E. Contribution, Utilization, and Improvement of Legumes-Driven Biological Nitrogen Fixation in Agricultural Systems. *Front. Sustain. Food Syst.* **2021**, *5*, 767998. [CrossRef]
45. Thu, T.T.P.; Thiem, T.T.; Loan, N.T.; Van An, P. Effects of Baby Corn Density on the Crop and Weed Performance under Different Maize-Soybean Intercropping Systems: Effects of baby corn density on crop and weed performance under different maize-soybean intercropping systems. *Vietnam J. Sci. Technol.* **2021**, *4*, 1007–1020.
46. Hoang, V.D.; Binh, H.T.T. Effect of maize-soybean intercropping and hand weeding on weed control. *Vietnam J. Sci.* **2015**, *13*, 354–363.
47. Slayi, M.; Zhou, L.; Jaja, I.F. Exploring farmers' perceptions and willingness to tackle drought-related issues in small-holder cattle production systems: A case of rural communities in the eastern cape, South Africa. *Appl. Sci.* **2023**, *13*, 7524. [CrossRef]
48. Gonzales-Malca, J.A.; Tirado-Kulieva, V.A.; Abanto-López, M.S.; Aldana-Juárez, W.L.; Palacios-Zapata, C.M. Worldwide research on the health effects of bovine milk containing A1 and A2 β-casein: Unraveling the current scenario and future trends through bibliometrics and text mining. *Curr. Res. Food Sci.* **2023**, *7*, 100602. [CrossRef] [PubMed]
49. Nguyen, V.D.; Nguyen, C.O.; Chau, T.M.L.; Nguyen, D.Q.D.; Han, A.T.; Le, T.T.H. Goat production, supply chains, challenges, and opportunities for development in Vietnam: A Review. *Animals* **2023**, *13*, 2546. [CrossRef] [PubMed]
50. Olmo, L.; Nguyen, H.V.; Nguyen, X.B.; Bui, T.N.; Ngo, C.T.K.; Nguyen, V.D.; Hoang, N.; Morales, L.E.; Walkden-Brown, S. Goat meat supply and demand in Vietnam: Global context and opportunities and risks for smallholder producers. *Anim. Prod. Sci.* **2024**, *64*, AN23416. [CrossRef]

ruminants

MDPI

Article

Behaviour of Cows with Johne's Disease (Paratuberculosis)

Gemma L. Charlton [1,*], Jeanette Churches [1], Emma C. L. Bleach [2] and Vivi M. Thorup [3]

[1] Animal Behaviour and Welfare Research Group, Animal Science Research Centre, Harper Adams University, Edgmond, Shropshire TF10 8NB, UK; jechurches00@gmail.com
[2] The Dairy Cow, Heifer and Calf Group, Animal Science Research Centre, Harper Adams University, Edgmond, Shropshire TF10 8NB, UK; ebleach@harper-adams.ac.uk
[3] Department of Animal and Veterinary Sciences, Aarhus University, 8830 Tjele, Denmark; vivim.thorup@anivet.au.dk
* Correspondence: gcharlton@harper-adams.ac.uk

Simple Summary: Johne's disease (JD) has detrimental effects on production and health and significantly reduces animal welfare. A recent study during peak lactation revealed that cows with JD (JD5) reduced their lying time compared to cows without JD (JD0). However, their step count was unaffected by JD, suggesting that the cows were standing idle, ruminating, or possibly standing eating. The objective of our study was to compare feeding behaviours and activity of JD5 cows to JD0 cows. The results support previous findings, with JD5 cows having lower lying times but no difference in step count. In addition, around week 8 of lactation, JD5 cows spent over 1 h/d longer ruminating, but there was no difference in eating times. Although it is still unclear what the cows were doing during these periods of reduced lying, this study does confirm behavioural difference between JD5 and JD0. Future research to further explore the difference in behaviour between JD5 and JD0 cows is recommended.

Abstract: Johne's disease (JD) significantly reduces the welfare of cattle worldwide. As changes in lying and feeding behaviours are considered important tools for assessing health and early detection of diseases, the aim of this study was to compare lying and feeding behaviours of JD-positive (JD5) and JD-negative (JD0) cows around peak lactation. The cows were fitted with an accelerometer-based sensor to record step counts and lying behaviour. They were also fitted with a pressure-based halter from approximately 56 d post-partum to collect feeding and rumination data. Every 3 months, the cows were milk sampled to test for naturally occurring JD using an ELISA. JD5 cows [n = 14 (two positive results in any four consecutive ELISAs)] were matched to JD0 cows [n = 14 (consecutive negative ELISAs)] based on lactation stage, parity, age, and milk yield. Of the 28 cows, 9 JD5 and 9 JD0 cows provided sensor data for analysis. JD5 cows spent 1.7 h/d less lying compared to JD0 cows. No differences in time spent eating were found; however, JD5 cows spent 1.1 h/d longer ruminating and produced 80 more feed boluses/d than JD0 cows around week 8 of lactation. The reason JD5 and JD0 cows behave differently around peak lactation is unclear and therefore warrants further investigation focusing on behaviour, milk yield, and feed intake among cows with JD.

Keywords: Johne's disease; paratuberculosis; MAP; lying behaviour; feeding behaviour; rumination

1. Introduction

Johne's disease (JD, also known as paratuberculosis) is a chronic wasting disease in ruminants caused by *Mycobacterium avium* subspecies paratuberculosis (MAP; [1]). JD is a complex, global health issue, and it is considered endemic in countries with large dairy industries, including Australia, Canada, Denmark, the Netherlands, the United Kingdom, and the United States of America [2,3]. Annual global losses due to JD are estimated at USD 4B [4], with economic losses caused by decreased milk yield, fertility, and feed conversion efficiency, as well as increased culling rates, replacement costs, and disease susceptibility [1,5,6]. Johne's disease in dairy cattle has an environmental impact, increasing greenhouse gas emissions per unit of milk by up to 24% [7]. Additionally, associations have been made between MAP and Crohn's disease in humans [8].

Detection of JD is difficult, as MAP has a long incubation period of between two and seven years [9]; therefore, the majority of infected cattle are subclinically ill. Despite appearing healthy, subclinical cows can still shed bacteria in faeces, thus contaminating the environment and exposing other animals to MAP [10]. Primarily, clinical signs are loose faeces and gradual loss of body condition, which progresses to profuse scour, rapid weight loss, and lethargy in the terminal stage [10,11]. The delay in onset of these non-specific clinical signs makes JD difficult to diagnose [12].

Behavioural changes resulting from the disease may occur before clinical signs develop [13]. Hence, recording and interpreting animal behaviour can aid in the diagnosis of health issues [14]. For example, in the 2 weeks before calving, cows later diagnosed with metritis spent around 40 min/d less lying and had fewer lying bouts compared with healthy cows [15]. In addition, during the 3 d before metritis diagnosis, metritic cows ate approximately 1 kg/d less and consumed fewer meals compared with healthy cows. Furthermore, on the day that lame cows were diagnosed, they spent half as long ruminating compared to healthy cows [16]. Being able to diagnose the disease earlier could improve animal welfare and reduce economic losses [13]. Although there is no effective treatment for cattle with JD, identifying infected cows earlier would allow control measures to be more effectively implemented, thus helping to prevent further disease transmission [1,17].

A study by Charlton et al. [18] was the first to investigate the activity of JD-positive (at least two positive ELISA results; JD5) cows during the preclinical phase. The study found that during peak lactation (week 8), JD5 cows had reduced lying time, spending an average of two hours per day less lying than JD-negative (consecutive negative ELISA results; JD0) cows. However, there was no difference in the number of steps per day of JD5 and JD0 cows, indicating that cows spent more time standing still during this period of reduced lying. As no feeding behaviours were recorded by Charlton et al. [18], it is unclear what the cows were doing during this time.

As a gastrointestinal disease, JD causes malabsorption of nutrients. Consequently, JD5 cows may have spent more time standing and eating each day to meet the energy demands of peak lactation [19]. Therefore, the aim of this study was to determine whether JD5 dairy cows reduce their lying time and spend more time eating and ruminating around peak lactation compared to JD0 cows.

2. Materials and Methods

2.1. Animals and Management

Data collection was carried out in 2018, 2019, and 2021 at Harper Adams University, UK, using 28 lactating, multiparous (range: 2–9 lactations) Holstein Friesian cows. The cows were milked twice per day in a 40-point Westfalia internal rotary parlour, starting at 5:00 a.m. and 3:00 p.m. The cows were housed in freestall yards with others from the milking group of approximately 180 cows with 1.3 m × 2.5 m cubicles (approximately 110 cubicles per 100 cows) equipped with 3 cm deep latex mattresses with a rubber top-cover. The cubicles were bedded with fresh sawdust twice weekly. Passageways were scraped five times a day by automatic scrapers.

Fresh total mixed ration (dry matter (DM) 41.4%, metabolizable energy 12.1 MJ per kg DM, crude protein 16.9% of DM, neutral detergent fibre 35.6% of DM) was given at approximately 8:00 a.m. daily, which is sufficient for ad libitum intake, and it was pushed up to the feed barrier a minimum of five times a day by an automated feed pusher. The cows had ad libitum access to water from tipping water troughs. Ethical approval for the study was given by Harper Adams University Research Ethics Committee (protocol-5646-201712).

2.2. Selection of the Study Cows

Milk samples were collected from all lactating cows in the dairy herd and tested for naturally occurring JD every three months (National Milk Records (NMR; Chippenham,

UK)) using the commercial milk Mycobacterium Paratuberculosis Antibody Screening ELISA (Idexx Laboratories Inc., Westbrook, ME, USA; [20]). According to NMR [21], when this ELISA is conducted on milk samples from individual cows, it has a sensitivity of 40–80% and specificity > 99%. Table 1 shows how the JD status of cows was classified based on these milk ELISA results. Cows who had repeated negative test results (minimum two tests) were classed as JD-negative (JD0), and cows who had two or more positive results in the previous four tests, at any time in their testing history, were classed as JD-positive (JD5). The cows' JD status was monitored throughout and after the end of the study to ensure it did not change. Additionally, their health was checked, and apart from half of them testing positive for JD, all cows were healthy upon selection and remained healthy. The JD5 cows were in the subclinical phase of JD throughout the study, exhibiting no apparent clinical signs. Prior to the start of the study, JD5 cows (n = 14) were paired with JD0 cows (n = 14) based on the stage of lactation, parity, age, and milk yield. The data from each pair were collected within a 4-week period to balance for the season and year. The number of JD5 cows was limited due to a JD control program on the research farm.

Table 1. Classification and definition of Johne's disease (JD) infection groups from National Milk Records (NMR), UK.

Risk Level	Classification	JD Infection Group	Definition
Low	Green	JD0	Repeat ELISA negative (minimum 2 tests)
Low	Green	JD1	ELISA negative but only one test
High	Amber	JD2	ELISA negative but 1 positive within 3 tests
High	Amber	JD3	ELISA negative but positive on previous test
High	Amber	JD4	ELISA positive, all previous tests negative
High	Red	JD5	Repeat ELISA positive (minimum 2 tests)

[21].

2.3. Behaviour Recordings

From after parturition and throughout the study, all cows wore an IceQube accelerometer-based sensor (Peacock Technology (formerly IceRobotics), Stirling, UK), attached using a Velcro strap on the back-left leg. IceQube sensors were previously validated for lying time, standing time, frequency of lying bouts, and step count in cubicle housing and straw yards [22]. The IceQubes use a 3-axis accelerometer to record the lying time, standing time, frequency of lying bouts, and step count with a granularity of 15 min (Table 2). These activity data were stored within the sensor before automatic downloading onto the CowAlert system (Peacock Technology (formerly IceRobotics), Stirling, UK) when the cow passed a reader in the entrance to the milking parlour twice daily. Daily sums were calculated for all variables (see Table 2). Due to missing data, accelerometer data were available for n = 9 pairs of cows (n = 9 JD0 cows and n = 9 JD5 cows).

From approximately 56 d post-partum (when the largest behavioural difference was previously observed [18]), the cows were each fitted with a RumiWatch halter (Itin and Hoch GmbH, Liestal, Switzerland) to collect seven consecutive days of data. The RumiWatch halter collects data via a pressure sensor and a triaxial accelerometer, whereby it measures feeding behaviour and rumination from jaw activity, which was previously validated [23]. The feeding-related activities analysed in the present study are described in Table 2. For this study, eating time with head up and eating time with head down ('Eat1 time' and 'Eat2 time', respectively) and chewing feed before swallowing with head up and chewing feed before swallowing with head down ('Eat1 chew' and 'Eat2 chew', respectively) were combined to generate the total eating time and the total number of eating chews, respectively. The halter was fitted so that the band around the lower jaw and the nose was 11–16 cm above the tip of the nose and it had 3–5 cm of movement space between it and the nose bridge [23]. Data were downloaded using the RumiWatch Manager software (Version 1.16; Itin and Hoch GmbH, Liestal, Switzerland) twice per week (at three- to four-day intervals). The first 48 h of recorded data were not analysed to allow the cows an adjustment period to the halter. RumiWatch data were available for n = 9 pairs of cows (n = 9 JD0 and n = 9 JD5 cows), with some data from the original dataset of 28 cows excluded due to device malfunctions and the analysis being limited to lactation weeks 8 and 9.

Table 2. Descriptions of behaviours recorded by the IceQube and RumiWatch sensors.

Behaviour	Description
IceQube variables	
Lying time	When the sensor is horizontal and the cow lies on the sternum or side.
Lying bout	The number of times the cow lies down.
Standing time	When the sensor is vertical and the body is supported by all four legs.
Step count	The number of times the cow lifts a foot and sets it down to walk or run to a new position.
RumiWatch variables	
Eating time	Ingesting and masticating food with head positioned down or up.
Eating chews	Opening and closing the jaw to masticate fresh food with head positioned down or up.
Bolus count	A regurgitated mass of cud, which is swallowed again after chewing, counted when a mass of cud is regurgitated.
Rumination time	Time spent regurgitating and rechewing boluses of food.
Rumination chews	Opening and closing the jaw to chew regurgitated food.
Chews per bolus	Chews between the regurgitation and swallowing of one bolus.

(Source: adapted from Peacock Technology, formerly IceRobotics, not dated; [23]).

2.4. Statistical Analysis

Weekly means per cow for weeks 8 (days in milk (DIM) 50–56) and 9 (DIM 57–63) of lactation were calculated from the daily sums for all variables measured by the RumiWatch halters and the IceQubes, and they were analysed separately through one-way ANOVA to enable comparison with a previous study [18].

Data were checked for normality. Step counts and lying bouts were log-transformed for normal distribution. The leg sensor variables were tested separately in the same non-linear mixed effects model using the lme function in R-package nlme (version 3.1-163). Longitudinal studies of loose-housed cows suggest that the lying time and the step count decreased to a nadir around 4–5 weeks post-partum and then increased [24–26]. Considering these indications of non-linear trends in the behaviour of cows during early lactation and the current study using frequent observations during DIM 1–70, cow behaviour was approximated with a polynomial function for DIM in model M1:

$$[M1]\ Y \sim \alpha + \text{JD-status} + (\beta + \text{JD-status}) \times \text{DIM} + (\gamma + \text{JD-status}) \times \text{DIM}^2 + \text{calving year} + \text{calving season} + \text{cow}$$

where JD-status (JD0; JD5), calving year (2018, 2019, 2021), and calving season (early = January to April; late = June to September) were fixed factors, DIM and DIM^2 were fixed, continuous effects, two-way interactions between JD-status and DIM and JD-status and DIM^2 were included as fixed effects, and the cow was a random effect. Temporal autocorrelations in covariance were accounted for. More specifically, the AR1 structure was used, which specifies that the correlations between the repeated measurements of each cow decrease with the time lag, and DIM was used to determine the correct order and define the time lag. Variance heterogeneity was accounted for by allowing different variance for each level of JD-status.

The effect of JD-status was tested using a χ^2 likelihood ratio test (LRT) comparing M1 with the model with the same polynomial coefficients for the two levels, that is, the reduced model M1, where JD-status (and interactions with it) were removed. When the polynomials differed significantly, the coefficients for each level of JD-status were reported. When the polynomials did not differ, the model was reduced, and the LS means (S.E.) reported for each level of the JD-status (JD0 or JD5) were averaged over the levels of the calving year (2018, 2019, or 2021) and the calving season (early or late).

Variance heterogeneity was tested by comparing models with and without heterogeneity components using an LRT after REML (restricted maximum likelihood) estimation. Significance was declared at $p \leq 0.05$.

3. Results

The comparison of variables for weeks 8 and 9 of lactation is provided in Table 3. During week 8, the time spent ruminating and the number of boluses regurgitated were significantly higher in JD5 cows compared with JD0 cows (Table 3). JD5 cows ruminated for 1.1 h/d longer and regurgitated 80 more boluses of feed/d compared to JD0 cows. There was no difference in the time spent eating or the other variables measured using the RumiWatch halters with respect to JD-status during week 8 or during week 9.

Table 3. Means (s.d.) of weekly averages of sensor variables of Johne's disease negative (JD0) and positive (JD5) cows during weeks 8 and 9 of lactation. IceQube sensors: 18 cows (n = 9 JD0 and n = 9 JD5) contributed 234 days of data, n = 35 weekly averages. RumiWatch sensors: 18 cows contributed 112 days, n = 30 weekly averages.

	Week 8			Week 9		
	JD0	**JD5**	***p* Value**	**JD0**	**JD5**	***p* Value**
		IceQube variables				
Lying Time, h/d	12.5 (2.73)	11.1 (1.08)	0.19	12.2 (2.64)	10.9 (1.37)	0.21
Lying Bouts, no/d	12.0 (4.04)	10.6 (4.31)	0.50	11.1 (3.92)	9.5 (2.99)	0.34
Standing Time, h/d	11.5 (2.73)	12.9 (1.08)	0.19	11.8 (2.64)	13.1 (1.37)	0.21
Step Count, no/d	778 (321)	1061 (534)	0.20	801 (378)	1192 (684)	0.15
		RumiWatch variables				
Eating Time, h/d	5.5 (0.67)	5.9 (0.78)	0.34	5.3 (0.96)	5.1 (0.52)	0.56
Eating Chews, no/d	21,998 (2374)	23,576 (4387)	0.32	21,239 (3489)	19,826 (4379)	0.50
Bolus, no/d	553 (46.1)	633 (93.0)	0.04	562 (55.0)	614 (124.1)	0.28
Rumination Time, h/d	9.1 (0.70)	10.2 (0.82)	0.02	9.2 (0.76)	9.8 (1.6)	0.38
Rumination Chews, no/d	35,744 (2758)	38,456 (3577)	0.13	35,441 (2481)	37,294 (7566)	0.50
Chews per Bolus, no/d	51.1 (4.48)	54.6 (4.53)	0.18	51.6 (5.24)	49.2 (8.02)	0.49

The mixed modelling showed that the lying time curves for JD0 (n = 9) and JD5 (n = 9) differed significantly (LRT = 14.9; df = 3; $p = 0.002$), that is, JD5 cows laid down less (−1.7 h/d) during days 1 to 70 compared to JD0 cows (Figure 1A). Equations (1) and (2) enabled us to plot the estimated curves for the lying time of JD0 cows and JD5 cows (see Figure 1A).

$$Y_{JD0} = 14.14 - 0.012 \times DIM - 0.0001 \times DIM^2 \text{ (h/d)} \quad (1)$$

$$Y_{JD5} = 12.36 - 0.078 \times DIM + 0.0009 \times DIM^2 \text{ (h/d)} \quad (2)$$

The JD0 and JD5 curves for Log(lying bouts) did not differ (LRT = 7.66; df = 3; $p = 0.054$); see Equations (3) and (4), with the estimated curves shown in Figure 1B. The back-transformed intercepts of the JD0 and JD5 curves were 12.3 and 10.6 lying bouts/d, respectively.

$$\text{Lying bouts}_{JD0} = 1.091 - 0.00005 \times DIM - 0.00002 \times DIM^2 \text{ (h/d)} \quad (3)$$

$$\text{Lying bouts}_{JD5} = 1.024 - 0.00223 \times DIM + 0.00003 \times DIM^2 \text{ (h/d)} \quad (4)$$

The curves of the two JD statuses did not differ significantly (LRT = 1.49; df = 3; p = 0.684) for log(Step count); thus, JD0 = 3.04 (0.071) and JD5 = 3.09 (0.054), which corresponds to 1096 and 1230 steps/d when back-transformed.

The variance was not heterogenous for any of the features (lying time, lying bouts, and step count), meaning the JD status did not cause more or less variation in these features.

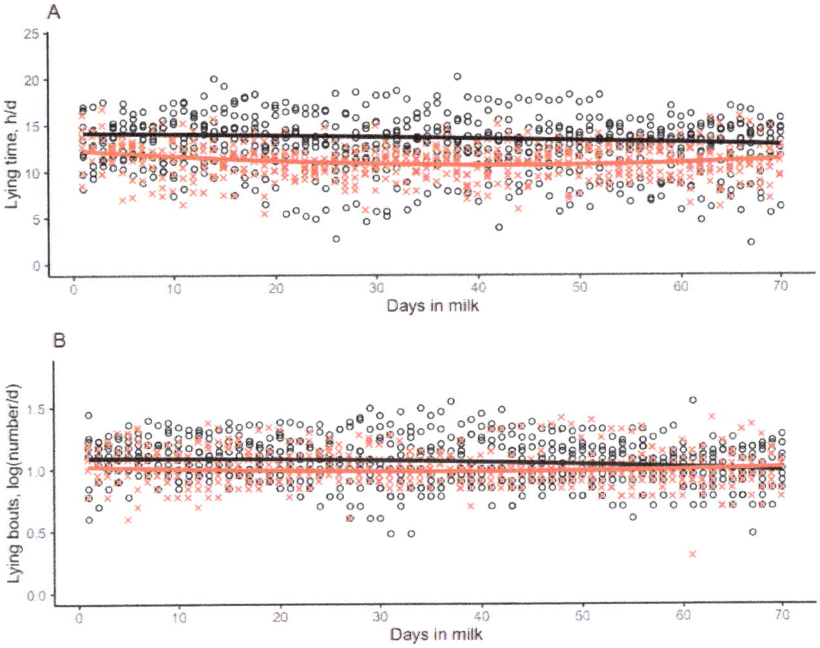

Figure 1. Lying behaviour during 1 to 70 days in milk in Johne's disease positive (JD5; n = 9) and negative (JD0; n = 9) cows under model M1. (**A**) Lying time (h/d). (**B**) Lying bouts (log (number/d)). JD0 estimate: black line; JD5 estimate: red line. Individual observations shown as black hollow circles (JD0) or red crosses (JD5).

4. Discussion

JD5 cows spent less time lying down than JD0 cows in early lactation, which confirms previous findings [18]. It was speculated by Charlton et al. [18] that the JD5 cows may have spent their increased standing time at the feed fence eating. The current study does not support this supposition, as eating times during week 8 and 9 of lactation were similar in JD5 and JD0 cows. However, during week 8 of lactation, in the present study, JD5 cows increased their rumination time and the number of boluses produced, so it is possible that the JD5 cows were standing and ruminating, although cows have been shown to more commonly ruminate while lying down [27].

Lying is a high-priority activity for dairy cattle, and lying time must be sufficient to promote milk production, welfare, and foetal development [28–30]. During the current study, the cows laid down for an average of 11.8 h/d. This is within the expected range of 9.5 to 12.9 h/d for lactating cows housed in freestalls [31], but lying times did vary, ranging from 6.9 to 14.9 h/d. A study using 3649 cows from 100 dairy herds also found a similarly wide variation in individual daily lying time, ranging from 6.3 to 17.9 h/d [32]. In the present study, there was a significant difference in the lying time curves of JD5 and JD0 cows, with JD5 cows lying on average 1.7 h/d less than JD0 cows. Similarly, during our previous study on the effects of JD on cow activity, JD5 cows spent up to 2 h/d less lying compared to JD0 cows [18]. The stocking rate of cubicles and at the feed fence can influence the lying time, the time spent standing idle, and the eating time [33,34]. In the current study, sufficient cubicles and feed space were provided to allow all cows to lie down and to eat simultaneously; therefore, the stocking rate should not have affected activity like lying times or feeding behaviours. However, paratuberculosis may cause gastrointestinal discomfort [9], which may have caused the JD5 cows to reduce their lying time and spend more time standing.

During weeks 8 and 9 of lactation, the average lying bout frequency was 10.8 times per day, which is within the expected range of 9 to 11 bouts per day [35,36]. Lying bout frequency curves did not differ between JD0 and JD5 cows. In comparison, Charlton et al. [18] recorded significant differences, with JD5 cows having 2.7 and 2.5 fewer lying bouts/d at weeks 8 and 9, respectively, of lactation, with this further decreasing to 3.6 fewer lying bouts/d at week 11. Other conditions affect lying bout frequency. Cows with mastitis have shown an increase in lying bouts, thought to be due to udder discomfort [37]. The frequency of lying bouts also increased in the 6 h prior to parturition [38,39], possibly due to pain and restlessness. Throughout the current study, apart from half of the cows having JD, they were all otherwise healthy, with no signs of pain, distress, or disease.

Similarly to previous findings [18], there was no difference in step count between JD0 and JD5 cows in the present study. Other diseases have caused cows to increase or decrease their step count. For example, cows suffering from mastitis take more steps compared to healthy cows [37], whereas lame cows decrease their stepping behaviour, possibly due to pain and discomfort caused by hoof and leg disorders [40–42]. Our results suggest that during periods of reduced lying, the cows were standing still. It remains unclear what they were doing during this time, which warrants further investigation. However, it is possible that the JD5 cows spent more time standing and ruminating. In support of this, cows with digital dermatitis reduced their lying time and spent more time ruminating while standing compared to healthy control cows [43].

Rumination is a vital part of the feeding process for ruminants, playing a key role in maintaining the optimum rumen pH, minimising the risk of rumen acidosis, enhancing fibre digestion, and breaking down particles for their movement from the rumen to the lower gastrointestinal tract [44]. In a review, a large variability in rumination time was reported, with mean rumination of 7.3 h/d, ranging from 3.9 to 10.2 h/d [45]. In the current study, average rumination times were at the higher end of this range at 9.2 and 10.0 h/d for JD0 and JD5 cows, respectively, during weeks 8 and 9 of lactation, possibly due to the chemical and physical composition of the diet [44].

During week 8 of lactation, JD5 cows spent 1.1 h/d longer ruminating and produced 80 more boluses per day compared to JD0 cows. This finding contrasts with the decrease in eating and rumination behaviours often seen in lactating dairy cows in response to disease or illness [46]. For example, lame cows reduced their rumination time and number of boluses up to 14 days before the onset of the clinical signs of lameness [16]. In addition, they reduced their eating time, drinking time, and eating and rumination chews. Fogsgaard et al. [47] found that cows with acute, clinical mastitis spent less time ruminating and eating following diagnosis. In addition, cows with subclinical ketosis have been found to reduce their time spent ruminating compared to cows with no recorded health problems [48]. Rumination time is considered an important tool for assessing dairy cow health and the early detection of diseases [49]; therefore, it may be a useful addition to a multifaceted diagnostic tool for JD.

The reason that JD5 cows produced more boluses and spent longer time ruminating is unclear. Animal variability, physical and chemical composition of the diet, and accuracy of measurement techniques have been discussed as reasons for variability in rumination time [44]. As the JD5 and JD0 cows were housed together and offered the same diet, and the same halters were used to collect data, neither the diet nor the measurement technique should have affected the rumination time in the present study. JD causes malfunction of the intestinal tract, particularly whilst nutrient demand is at its highest during peak lactation [19,50], and this may have affected rumination times. However, rumination is a complex phenomenon, and further investigation, including a much larger sample size, is required to understand the complexities of rumination and the effect of JD on rumination times.

Eating times and eating chews were similar for JD0 and JD5 cows at week 8 and week 9 of lactation. Unfortunately, information on eating behaviour was only available from 18 cows due to a limited number of JD5 cows in the herd and because of technological failures, resulting in missing data, which is a limitation of the current study. Research has shown large variability in eating times between cows fed TMR [44,45]; therefore, a greater sample size may have been required to find significant differences in eating times between JD0 and JD5 cows. Future research should aim to include a much larger sample size to further investigate eating behaviours and explore the DMI of dairy cows in the subclinical and clinical stages of JD. With more focus on sustainable dairy herds, it would be useful, alongside milk production, to determine the feed efficiency of JD cows, so that farmers can keep the most profitable cows in the herd and make strategic culling decisions based on lifetime cumulative costs and revenues [51] and to improve herd health. Despite the low sample size in the current study, the behaviour of cows with JD is an under-investigated research area, and the findings of this study and our previous study [18] confirm behavioural differences between JD5 and JD0 cows.

With dairy farmers increasingly using wearable sensors to constantly record activities, such as the lying times and step counts of cows [52], the present study suggests that there are opportunities for future research to further focus on the behaviour of cows with JD under differing management systems and environments. Knowing how infected individuals deviate from those that are healthy will help farmers identify subclinically ill cows and allow us to understand how they cope in these systems.

5. Conclusions

Cows with subclinical JD showed reduced lying time during early lactation, but their step count was not different to JD0 cows. There were no differences in eating times, but JD5 cows spent longer ruminating during week 8 of lactation. This study indicates that there are some differences in the feeding behaviours of JD0 and JD5 cows that warrant further investigation. Future research should also investigate the DMI and feed efficiency of cows with JD.

Author Contributions: Conceptualisation, G.L.C., V.M.T., and E.C.L.B.; methodology, G.L.C.; software, V.M.T. and G.L.C.; validation, G.L.C.; formal analysis, V.M.T.; investigation, G.L.C. and J.C.; resources, G.L.C.; data curation, V.M.T.; writing—original draft preparation, G.L.C.; writing—review and editing, G.L.C., V.M.T., E.C.L.B., and J.C.; visualisation, G.L.C.; supervision, G.L.C.; project administration, G.L.C.; funding acquisition, G.L.C. and V.M.T. All authors have read and agreed to the published version of the manuscript.

Funding: The RumiWatch halters were funded by Innovate UK.

Institutional Review Board Statement: The animal study protocol was approved by the Ethics Committee of Harper Adams University (protocol code 5646-201712 and date of approval 9 December 2017).

Informed Consent Statement: Not applicable.

Data Availability Statement: The datasets used and analysed during the current study are available from the corresponding author upon reasonable request.

Acknowledgments: The authors thank Carrie Gauld and Sarah Williams at Harper Adams University for their assistance in collecting the data.

Conflicts of Interest: The authors declare no conflicts of interest. The funders had no role in the design of the study; in the collection, analyses, or interpretation of data; in the writing of the manuscript; or in the decision to publish the results.

References

1. Garcia, A.B.; Shalloo, L. Invited review: The economic impact and control of paratuberculosis in cattle. *J. Dairy Sci.* **2015**, *98*, 5019–5039. [CrossRef] [PubMed]
2. Kennedy, D.; Holmstrom, A.; Plym Forshell, K.; Vmdel, E.; Suarez, G. On-farm management of paratuberculosis (Johne's disease) in dairy herds. In Mycobacterium paratuberculosis. *Bull. Int. Dairy Fed.* **2001**, *362*, 18–31.
3. Geraghty, T.; Graham, D.A.; Mullowney, P.; More, S.J. A review of bovine Johne's disease control activities in 6 endemically infected countries. *Prev. Vet. Med.* **2014**, *116*, 1–11. [CrossRef] [PubMed]
4. Rasmussen, P.; Barkema, H.W.; Osei, P.P.; Taylor, J.; Shaw, A.P.; Conrady, B.; Chaters, G.; Muñoz, V.; Hall, D.C.; Apenteng, O.O.; et al. Global losses due to dairy cattle diseases: A comorbidity-adjusted economic analysis. *J. Dairy Sci.* **2024**, *107*, 6945–6970. [CrossRef] [PubMed]
5. Ott, S.L.; Wells, S.J.; Wagner, B.A. Herd-level economic losses associated with Johne's disease on US dairy operations. *Prev. Vet. Med.* **1999**, *40*, 179–192. [CrossRef]
6. Rossi, G.; Grohn, Y.T.; Schukken, Y.H.; Smith, R.L. The effect of Mycobacterium avium ssp. paratuberculosis infection on clinical mastitis occurrence in dairy cows. *J. Dairy Sci.* **2017**, *100*, 7446–7454. [CrossRef]
7. Williams, A.; Chatterton, J.; Hateley, G.; Curwen, A.; Elliott, J.A. A systems-life cycle assessment approach to modelling the impact of improvements in cattle health on greenhouse gas emissions. *Adv. Anim. Biosci.* **2015**, *6*, 29–31. [CrossRef]
8. Behr, M.A.; Kapur, V. The evidence for Mycobacterium paratuberculosis in Crohn's disease. *Curr. Opin. Gastroenterol.* **2008**, *24*, 17–21. [CrossRef]
9. Fecteau, M.-E. Paratuberculosis in Cattle. *Vet. Clin. Food Anim.* **2018**, *34*, 209–222. [CrossRef]
10. Grant, I.R. Zoonotic potential of Mycobacterium avium ssp. paratuberculosis: The current position. *J. Appl. Microbiol.* **2005**, *98*, 1282–1293. [CrossRef]
11. Stabel, J.R. Johne's disease: A hidden threat. *J. Dairy Sci.* **1998**, *81*, 283–288. [CrossRef] [PubMed]
12. Meyer, A.; Bond, K.; Van Winden, S.; Green, M.; Guitian, J. A probabilistic approach to the interpretation of milk antibody results for diagnosis of Johne's disease in dairy cattle. *Prev. Vet. Med.* **2018**, *150*, 30–37. [CrossRef] [PubMed]
13. González, L.A.; Tolkamp, B.J.; Coffey, M.P.; Ferret, A.; Kyriazakis, I. Changes in feeding behavior as possible indicators for the automatic monitoring of health disorders in dairy cows. *J. Dairy Sci.* **2008**, *91*, 1017–1028. [CrossRef] [PubMed]
14. Weary, D.M.; Huzzey, J.M.; von Keyserlingk, M.A.G. Board-Invited Review: Using behavior to predict and identify ill health in animals. *J. Anim. Sci.* **2009**, *87*, 770–777. [CrossRef] [PubMed]
15. Neave, H.W.; Lomb, J.; Weary, D.M.; LeBlanc, S.J.; Huzzey, J.M.; von Keyserlingk, M.A.G. Behavioral changes before metritis diagnosis in dairy cows. *J. Dairy Sci.* **2018**, *101*, 4388–4399. [CrossRef]
16. Antanaitis, R.; Juozaitienė, V.; Urbonavičius, G.; Malašauskienė, D.; Televičius, M.; Urbutis, M.; Baumgartner, W. Impact of Lameness on Attributes of Feeding Registered with Noseband Sensor in Fresh Dairy Cows. *Agric.* **2021**, *11*, 851. [CrossRef]
17. NADIS (National Animal Disease Information Service). Johne's (Paratuberculosis). Available online: https://www.nadis.org.uk/disease-a-z/cattle/johnes-paratuberculosis/ (accessed on 1 May 2022).
18. Charlton, G.L.; Bleach, E.C.; Rutter, S.M. Cows with paratuberculosis (Johne's disease) alter their lying behavior around peak lactation. *J. Dairy Sci.* **2019**, *102*, 11328–11336. [CrossRef]
19. Knight, C.H. Lactation and gestation in dairy cows: Flexibility avoids nutritional extremes. *Proc. Nutr. Soc.* **2001**, *60*, 527–537. [CrossRef]
20. Bartlett, B.; Pearse, H.L. Integrating milk recording data and disease test results to provide a system for the management of paratuberculosis in UK dairy herds. In Proceedings of the ICAR, Cork, Ireland, 28 May 2012.
21. NMR (National Milk Records). HerdWise: A Step by Step Guide on How to Read Your HerdWise Results and Reports. Available online: https://craft-web.files.svdcdn.com/production/documents/J086010-NMR-Herdwise-Results-and-Reports-Booklet-FINAL.pdf?dm=1725884613 (accessed on 16 September 2024).
22. Charlton, G.; Gauld, C.; Veronesi, F.; Rutter, S.M.; Bleach, E. Assessing the accuracy of leg mounted sensors for recording dairy cow behavioural activity at pasture, in cubicle housing and a straw yard. *Animals* **2022**, *12*, 638. [CrossRef]
23. Rombach, M.; Münger, A.; Niederhauser, J.; Südekum, K.H.; Schori, F. Evaluation and validation of an automatic jaw movement recorder (RumiWatch) for ingestive and rumination behaviors of dairy cows during grazing and supplementation. *J. Dairy Sci.* **2018**, *101*, 2463–2475. [CrossRef]
24. Thorup, T.; Charlton, G.; Gauld, C.; Bleach, E.; Rutter, M. Towards life-long sensing of dairy cow behaviour using accelerometers. In Proceedings of the Fourth DairyCare Conference, Lisbon, Portugal, 13 October 2016.
25. Maselyne, J.; Pastell, M.; Thomsen, P.; Thorup, V.; Hänninen, L.; Vangeyte, J.; Van Nuffel, A.; Munksgaard, L. Daily lying time, motion index and step frequency in dairy cows change throughout lactation. *Res. Vet. Sci.* **2017**, *110*, 1–3. [CrossRef] [PubMed]
26. Munksgaard, L.; Weisbjerg, M.R.; Henriksen, J.C.S.; Løvendahl, P. Changes to steps, lying, and eating behavior during lactation in Jersey and Holstein cows and the relationship to feed intake, yield, and weight. *J. Dairy Sci.* **2020**, *103*, 4643–4653. [CrossRef]
27. Schirmann, K.; Chapinal, N.; Weary, D.M.; Heuwieser, W.; von Keyserlingk, M.A.G. Rumination and its relationship to feeding and lying behavior in Holstein dairy cows. *J. Dairy Sci.* **2012**, *95*, 3212–3217. [CrossRef] [PubMed]
28. Munksgaard, L.; Jensen, M.B.; Pedersen, L.J.; Hansen, S.W.; Matthews, L. Quantifying behavioural priorities—Effects of time constraints on behaviour of dairy cows, Bos taurus. *Appl. Anim. Behav. Sci.* **2005**, *92*, 3–14. [CrossRef]

29. Jensen, M.B.; Pedersen, L.J.; Munksgaard, L. The effect of reward duration on demand functions for rest in dairy heifers and lying requirements as measured by demand functions. *Appl. Anim. Behav. Sci.* **2005**, *90*, 207–217. [CrossRef]
30. Blackie, N.; Scaife, J.R.; Bleach, E.C.L. Lying behaviour and activity of early lactation Holstein dairy cattle measured using an activity monitor. *Cattle Pract.* **2006**, *14*, 139–142. [CrossRef]
31. Tucker, C.B.; Jensen, M.B.; de Passillé, A.M.; Hänninen, L.; Rushen, J. Invited review: Lying time and the welfare of dairy cows. *J. Dairy Sci.* **2021**, *104*, 20–46. [CrossRef]
32. Charlton, G.L.; Bouffard, V.; Gibbons, J.; Vasseur, E.; Haley, D.B.; Pellerin, D.; Rushen, J.; de Passillé, A.M. Can automated measures of lying time help assess lameness and leg lesions on tiestall dairy farms? *Appl. Anim. Behav. Sci.* **2016**, *175*, 14–22. [CrossRef]
33. Huzzey, J.M.; DeVries, T.J.; Valois, P.; von Keyserlingk, M.A.G. Stocking density and feed barrier design affect the feeding and social behavior of dairy cattle. *J. Dairy Sci.* **2006**, *89*, 126–133. [CrossRef]
34. Hill, C.T.; Krawczel, P.D.; Dann, H.M.; Ballard, C.S.; Hovey, R.C.; Falls, W.A.; Grant, R.J. Effect of stocking density on the short-term behavioural responses of dairy cows. *Appl. Anim. Behav. Sci.* **2009**, *117*, 144–149. [CrossRef]
35. Ito, K.; Weary, D.M.; von Keyserlingk, M.A.G. Lying behavior: Assessing within- and between-herd variation in free-stallhoused dairy cows. *J. Dairy Sci.* **2009**, *92*, 4412–4420. [CrossRef] [PubMed]
36. Charlton, G.L.; Haley, D.B.; Rushen, J.; de Passillé, A.M. Stocking density, milking duration, and lying times of lactating cows on Canadian freestall dairy farms. *J. Dairy Sci.* **2014**, *97*, 2694–2700. [CrossRef] [PubMed]
37. Fogsgaard, K.K.; Bennedsgaard, T.W.; Herskin, M.S. Behavioral changes in freestall-housed dairy cows with naturally occurring clinical mastitis. *J. Dairy Sci.* **2015**, *98*, 1730–1738. [CrossRef] [PubMed]
38. Jensen, M.B. Behaviour around the time of calving in dairy cows. *Appl. Anim. Behav. Sci.* **2012**, *139*, 195–202. [CrossRef]
39. Borchers, M.R.; Chang, Y.M.; Proudfoot, K.L.; Wadsworth, B.A.; Stone, A.E.; Bewley, J.M. Machine-learning-based calving prediction from activity, lying, and ruminating behaviors in dairy cattle. *J. Dairy Sci.* **2017**, *100*, 5664–5674. [CrossRef]
40. Chapinal, N.; Tucker, C.B. Validation of an automated method to count steps while cows stand on a weighing platform and its application as a measure to detect lameness. *J. Dairy Sci.* **2012**, *95*, 6523–6528. [CrossRef]
41. Thorup, V.M.; Munksgaard, L.; Robert, P.E.; Erhard, H.W.; Thomsen, P.T.; Friggens, N.C. Lameness detection via leg-mounted accelerometers on dairy cows on four commercial farms. *Animal* **2015**, *9*, 1704–1712. [CrossRef]
42. Hut, P.R.; Hostens, M.M.; Beijaard, M.J.; van Eerdenburg, F.J.C.M.; Hulsen, J.H.J.L.; Hooijer, G.A.; Stassen, E.N.; Nielen, M. Associations between body condition score, locomotion score, and sensor-based time budgets of dairy cattle during the dry period and early lactation. *J. Dairy Sci.* **2021**, *104*, 4746–4763. [CrossRef]
43. Pavlenko, A.; Bergsten, C.; Ekesbo, I.; Kaart, T.; Aland, A.; Lidfors, L. Influence of digital dermatitis and sole ulcer on dairy cow behaviour and milk production. *Animal* **2011**, *5*, 1259–1269. [CrossRef]
44. Beauchemin, K.A. Invited review: Current perspectives on eating and rumination activity in dairy cows. *J. Dairy Sci.* **2018**, *101*, 4762–4784. [CrossRef]
45. White, R.R.; Hall, M.B.; Firkins, J.L.; Kononoff, P.J. Physically adjusted neutral detergent fiber system for lactating dairy cow rations. I: Deriving equations that identify factors that influence effectiveness of fiber. *J. Dairy Sci.* **2017**, *100*, 9551–9568. [CrossRef] [PubMed]
46. Paudyal, S. Using rumination time to manage health and reproduction in dairy cattle: A review. *Vet. Q.* **2021**, *41*, 292–300. [CrossRef] [PubMed]
47. Fogsgaard, K.K.; Røntved, C.M.; Sørensen, P.; Herskin, M.S. Sickness behavior in dairy cows during Escherichia coli mastitis. *J. Dairy Sci.* **2012**, *95*, 630–638. [CrossRef] [PubMed]
48. Kaufman, E.I.; LeBlanc, S.J.; McBride, B.W.; Duffield, T.F.; DeVries, T.J. Association of rumination time with subclinical ketosis in transition dairy cows. *J. Dairy Sci.* **2016**, *99*, 5604–5618. [CrossRef]
49. Cocco, R.; Canozzi, M.E.A.; Fischer, V. Rumination time as an early predictor of metritis and subclinical ketosis in dairy cows at the beginning of lactation: Systematic review-meta-analysis. *Prev. Vet. Med.* **2021**, *189*, 105309. [CrossRef]
50. Henderson, D.C.; Caldow, G.; Low, C.J. Paratuberculosis in cattle: Pathology and clinical disease (chapter 3). In *Assessment of Surveillance and Control of Johne's Disease in Farm Animals in GB*, 1st ed.; Caldow, G., Greig, A., Gunn, G.J., Humphry, R., Low, C.J., Ashworth, S.W., Jones, G.M., Stott, A.W., Cranwell, M.V., Sharp, J.M., Eds.; Veterinary Science Division, Scottish Agricultural College: Perth, UK, 2001; pp. 15–19.
51. Warner, D.; Dallago, G.M.; Dovoedo, O.W.; Lacroix, R.; Delgado, H.A.; Cue, R.I.; Wade, K.M.; Dubuc, J.; Pellerin, D.; Vasseur, E. Keeping profitable cows in the herd: A lifetime cost-benefit assessment to support culling decisions. *Animal* **2022**, *16*, 100628. [CrossRef]
52. Kok, A.; van Knegsel, A.; Bokkers, E.A.M.; Kemp, B.; Thorup, V. Exploring synchrony of lying on commercial dairy farms in relation to management. *Appl. Anim. Behav. Sci.* **2023**, *262*, 105906. [CrossRef]

ruminants

MDPI

Article

Changes in the Fatty Acid Composition of Vegetable Oils Affect the Feeding Behavior, Feed Preference, and Thermoregulatory Responses of Sheep

Évyla Layssa G. Andrade [1], José M. Pereira Filho [1], Kevily Henrique de O. S. de Lucena [1], Yuri C. S. Barreto [1], Ronaldo L. Oliveira [2], Bonifácio B. de Sousa [1], Antônio Fernando de M. Vaz [1], Juliana Paula F. de Oliveira [3], Mozart A. Fonseca [4,5] and Leilson R. Bezerra [1,4,5,*]

[1] Department of Animal Science, Federal University of Campina Grande, Patos 58708-110, Paraíba, Brazil; evylalayssa@hotmail.com (É.L.G.A.); jmpfpiaui@gmail.com (J.M.P.F.); yuri.caie@estudante.ufcg.edu.br (Y.C.S.B.); bonifacio.ufcg@gmail.com (B.B.d.S.); antonio.melo@ufcg.edu.br (A.F.d.M.V.)

[2] Animal Science Department, Federal University of Bahia, Salvador 40170-155, Bahia, Brazil; ronaldooliveira@ufba.br

[3] Animal Science Department, Federal University of Sergipe, Nova Esperança, Nossa Senhora da Glória 49680-000, Sergipe, Brazil; jupaula.oliv@academico.ufs.br

[4] Department of Agriculture, Veterinary, and Rangeland Sciences, University of Nevada, Reno, NV 89557, USA; mozart@nmsu.edu

[5] Clayton Livestock Research Center, Department of Animal and Range Sciences, New Mexico State University, Clayton, NM 88415, USA

* Correspondence: leilson@ufpi.edu.br

Simple Summary: Different vegetable oils have been tested in ruminant diets to increase the energy concentration of the diet and improve the lipid quality of milk and meat. However, when we add vegetable oil to the diet, we change the ruminal fermentation pattern, which directly interferes with sheep's digestive behavior and selectivity. Furthermore, the increase in calories in the diet can change animals' thermoregulation, especially in tropical environments. In our study, it was evidenced that sheep showed better diet selectivity in flocks in which the oil used was canola oil with higher monounsaturated fatty acid content and that the canola oil + cashew nut shell liquid blend increased dietary energy content, improving selectivity due to a lower concentration of polyunsaturated fatty acids intake.

Abstract: This research evaluated the effects of energy supplementation on sheep's feeding behavior, feed preference, and thermoregulatory responses using technical cashew nutshell liquid (CNSL) and different vegetable oils with different unsaturated fatty acid (UFA) compositions. The experiment was completely randomized with five treatments: a mixture of CNSL (0.5%) + vegetable oils [canola (high in monounsaturated fatty acids—MUFA), and corn, soybean, sunflower, or cottonseed oil (high in polyunsaturated fatty acids-PUFA) at 1.5%] based on total diet dry matter, with eight replications. Forty uncastrated male sheep, with an average initial BW of 24.44 ± 1.5 kg, were evaluated for 70 days. The CNSL + vegetable oil blend did not affect DM and neutral detergent fiber (aNDF) intake ($p > 0.05$). However, diets with canola oil resulted in higher SFA intake ($p < 0.05$) than other oils. The canola oil + CNSL blend led to a higher intake of UFA and MUFA and lower PUFA intake than other oil blends ($p < 0.05$). Sheep fed canola oil ruminated fewer boli per day than those fed soybean and sunflower oils. Using three sieves ($pef_{1.18}$) reflected in higher sheep aNDF intake. Respiratory frequency and surface temperature of sheep were lower before feeding than 3 h after, without effects of the type of oil. Higher serum creatinine and cholesterol levels were observed in sheep fed CNSL with corn and canola oils compared to other oils. Serum calcium was lower in sheep fed CNSL with soybean and canola compared to sunflower and corn. Including CNSL with vegetable oils with different FA compositions did not affect physiological and thermographic variables. However, sheep showed better diet selectivity and lower bolus rumination with higher MUFA (canola oil) content.

Including CNSL with canola oil in sheep diets is recommended, as it increases dietary energy content, enhances diet selectivity, reduces PUFA intake, and does not impact animal health.

Keywords: caloric increment; fatty acids; intake; ruminating; thermal comfort

1. Introduction

Growing sheep is a category with high nutrient requirements, particularly protein and energy. The efficiency of the digestive process is related to rumen fermentation and microbial activity, which, depending on the diet, can represent a loss of gross energy from the feed ingested (which can vary between 6% and 12%) and worsen the animal's performance [1]. A widely used strategy to increase dietary energy is using vegetable oils, which improve nutrient use efficiency, especially energy nutrients [2].

In semi-arid regions, environmental conditions such as high temperatures, relative humidity, and solar radiation can lead to discomfort and stress in animals, significantly affecting livestock production. The complexity of the thermal environment surrounding animals means that the exchange of thermal energy between them and their surroundings is influenced by a range of physiological and environmental factors, which are interconnected in various ways [3–6]. Consequently, heat stress is one of the primary constraints on animal production, underscoring the importance of maintaining production activities within a thermal comfort zone to optimize animal performance [4]. To help animals in this stressful situation, lipid supplementation has been highlighted as an efficient nutritional practice since lipids are 2.25 times more energetic than carbohydrates. In addition, they have a lower caloric increase when compared to carbohydrates and proteins, making them an excellent dietary option, especially during the hottest periods of the year [7]. Increasing fatty acid intake can increase energy efficiency, as the direct deposition of dietary fatty acids in animal tissues replaces the metabolic steps of converting carbohydrates or short-chain fatty acids (SCFA) [8,9]. However, a very important aspect of fatty acid metabolism in the rumen is that it does not contribute to the growth of rumen microbial protein. This must, therefore, be considered when adjusting energy and protein [10].

Fat is a concentrated energy source, and energy production is linked to the length of the carbon chain of each fatty acid, as well as the presence of double bonds in unsaturated fatty acids. When sheep ingest fat, it is metabolized and generates energy, part of which is released as heat [11]. Animals that eat diets with higher energy levels result in reduced feed intake and fermentation heat, consequently losing less energy and improving feed conversion of the animals, and this is even more important in tropical environments where temperatures are quite high [12]. Thus, adding fat to the diets of growing and finishing sheep generally improves feed efficiency, increasing weight gain and influencing carcass measurements [11,13].

Several sources of lipids can be added to ruminant diets, including vegetable oils that influence the rumen biohydrogenation process and reduce the caloric increase in the diet, improving the animal's ability to withstand high daily temperatures, especially in semi-arid zones. To maximize these benefits, the liquid from the cashew nutshell (*Anacardium occidentale*), known as CNSL, a byproduct and waste product, can be used to feed ruminants.

The CNSL, a phenolic lipid, primarily consists of anacardic acid, cardanol, and cardol, long-chain unsaturated phenols [14]. Including technical CNSL in ruminant diets has proven antioxidant and bactericidal activity, offering potential applicability in animal production [15,16]. This addition can result in changes to the fermentation profile in the rumen environment, including the mitigation of enteric methane, leading to improvements in production performance and nutrient metabolism [17]. In vitro studies conducted by Watanabe et al. [18] demonstrated that the application of CNSL reduced methane production by 70% and increased propionate production by 44%. This could also contribute to thermal comfort, as metabolic production is a factor in thermal stress. Shinkai et al. [19]

observed a rise in rumen propionate production and a 38% reduction in methane emissions in cows fed a diet of hay and concentrate (60:40 ratio) with the inclusion of 4 g of CNSL per 100 kg of body weight. However, its composition can cause astringency due to phenolic compounds and alter the selectivity of animals, making it necessary to evaluate it. Both Ramos et al. [16] and Araújo [17] recommended a maximum level of 0.75% CNSL in sheep diets, given the risk of toxicity and astringency negatively affecting nutrient intake and digestibility. Compton [20] evaluated levels up to 300 ppm of CNSL in cows' diets and found no impact on the rumen fermentation profile and apparent nutrient digestibility. However, studies have reported positive effects from including of vegetable oils [21] and CNSL in sheep diets, such as no change in intake and no interference in physiological and thermographic variables [17,20].

Thus, this study tested the hypothesis that energy supplementation with CNSL associated with vegetable oils of the UFA compositions will change the feeding behavior and feed preference and contribute to the thermoregulation of sheep. Therefore, intake and factors related to ingestive behavior and selectivity were evaluated, as well as serum and physiological parameters of sheep raised in a confinement system in a semi-arid region.

2. Materials and Methods

The study took place in the semi-arid region of Brazil, classified as having a Bsh (hot semi-arid) climate according to the Köppen climate classification, with an average temperature of 28 °C, ranging from a minimum of 22.4 °C to a maximum of 33.5 °C. The historical average annual rainfall is 996.6 ± 300.0 mm, with precipitation primarily occurring between December and May, and 43% falling between March and April [22].

2.1. Animals, Experimental Design, Diets, and Facilities

Forty uncastrated male Santa Ines × Dorper crossbred sheep, approximately three months old with an initial average weight of 24.44 ± 1.5 kg, were utilized in this study. The animals were allocated in a randomized complete design with five treatments, each combining 0.5% CNSL of the total diet dry matter (DM) with 1.5% DM of different vegetable oils (soybean, corn, cottonseed, sunflower, and canola) across eight experimental units.

At the start of the experiment, all sheep were weighed, identified, vaccinated against clostridium and rabies, dewormed, and housed individually in wooden stalls elevated 0.5 m above the ground. These stalls, measuring 1.3 × 1.5 m^2 and equipped with feeders and drinkers, were in an open shed with an asbestos cement roof, concrete central floor, and wooden slats measuring 16 m long and 6 m wide. The shed had a central aisle of 1.8 m wide and a ceiling height of 2.5 m. The experimental period was 70 days, with the first 15 days dedicated to acclimating the sheep to their new environment, management practices, and diets, followed by 55 days of sampling and data collection.

The experimental diets had a roughage-to-concentrate ratio of 40:60 in a total mixed ration, formulated to meet requirements for an average daily weight gain of 250 g [1]. Sorghum silage served as the roughage, while the concentrate consisted of soybean meal, ground corn, mineral salt, and the specified treatments combining 0.5% CNSL and 1.5% of various vegetable oils (canola, corn, soybean, sunflower, or cottonseed) based on total diet DM (Table 1). The animals were fed twice daily at 8:00 and 15:00, with intake measured daily to maintain 10% leftovers. Water was provided ad libitum.

To extract the cashew nutshell liquid (CNSL), the shells were steam-heated to 80 °C and then pressed, yielding CNSL and a residual cake [23]. A thermomechanical process was used for solvent extraction, heating the CNSL to warm the raw nuts to around 190 °C. At this temperature, the outer shell breaks, releasing alkylphenols from the porous mesocarp, and the inner shell is removed, allowing the recovery of the nuts [24].

The extracted CNSL underwent decarboxylation to eliminate CO_2 and moisture by heating it to 140 °C with continuous stirring [23]. This heat treatment decarboxylates anacardic acid to form cardanol, resulting in technical CNSL [24], used in this study. The decarboxylated CNSL was then filtered through a filter press and stored.

Table 1. Proportion of ingredients and chemical composition of experimental diets.

Item	Oils Mixed with CNSL [1]				
	Soybean	Cottonseed	Sunflower	Corn	Canola
Sorghum silage	40.0	40.0	40.0	40.0	40.0
Ground corn	40.0	40.0	40.0	40.0	40.0
Soybean meal	16.5	16.5	16.5	16.5	16.5
CNSL [1]	0.50	0.50	0.50	0.50	0.50
Oils	1.50	1.50	1.50	1.50	1.50
Mineral mixture	1.50	1.50	1.50	1.50	1.50
Chemical composition of the diet (g/kg)					
Dry matter	694	694	694	694	694
Ashes	51.9	51.9	51.9	51.9	51.9
Crude protein	135	135	135	135	135
Ether extract	58.0	58.0	58.0	58.0	58.0
aNeutral detergent fiber [2]	371	371	371	371	371
Acid detergent fiber	165	165	165	165	165
Non-fiber carbohydrates	384	384	384	384	384
Total digestible nutrients	782	782	782	782	782

[1] Cashew nut shell liquid. [2] Neutral detergent fiber assayed with a heat stable amylase and expressed exclusive of residual ash.

2.2. Determination of Chemical Composition of Diets and Ingredients

Samples of ingredients, diet, and leftovers were pre-dried in a forced air circulation oven at 55 °C for 72 h before being ground into 1.0 mm particles using a Willey-type knife mill. The chemical composition was analyzed according to the standards of [25], determining dry matter (method 934.01), ash (method 942.05), crude protein (method 968.06), and ether extract (method 920.39).

Neutral detergent fiber (NDF) and acid detergent fiber (ADF) were measured following the procedure by Van Soest [26], with modifications for non-woven fabric [27], using thermostable amylase (Sigma A3306; Sigma-Aldrich, Steinheim, Germany) and expressed exclusive of residual ash (aNDF). Lignin was determined using method 973.18 [25], involving treatment of the ADF residue with 72% sulfuric acid. Neutral and acid detergent insoluble protein levels were determined according to methodology of Licitra et al. [28]. Non-fiber carbohydrates (NFC) were calculated based on Mertens [29], considering aNDF corrected for ash and protein. Metabolizable energy (ME) was calculated using Weiss [30] methodology, based on total digestible nutrients (TDN) contents, using the following formula: TDN = %DCP + %DaNDF + %DEE $\times$ 2.25 + %DNFC; 1 kg TDN = 4409 kcal DE; ME = DE $\times$ 0.82.

2.3. Determination of Fatty Acid Composition in the Oils of the Diets

A quick and efficient method for extracting total lipids was employed [31]. The fatty acid profile of the oil samples (Table 2) was analyzed via gas chromatography following the preparation of fatty acid esters according to Hartman and Lago's [32] procedure. The analysis utilized a Thermo Scientific Focus gas chromatograph (Thermo Electron SpA®, Milan, Italy) equipped with a Supelco® Analytical SPTM-2560 capillary column, measuring 100 m $\times$ 0.25 mm $\times$ 0.20 μm (Supelco® Inc., Bellefonte, PA, USA). Both the injector and detector were set at 250 °C. The column temperature started at 140 °C for 5 min, then increased to 220 °C at a rate of 4 °C/min and was held at this final temperature for 10 min. The split ratio was set to 1:50, with an injection volume of 1.2 μL. The fatty acid results are presented as a relative percentage for each, derived by internal normalization of the chromatographic peak areas. The fatty acids were identified by comparing the relative retention times of sample FAME peaks with those of standards. A mixture of 37 FAMEs (standard 47885-U, TraceCERT®, Sigma Aldrich; Merck KGaA, Darmstadt, Germany) was used as the reference standard.

Table 2. Composition of the main fatty acids identified in the vegetable oils in the diets.

Fatty Acids (g/100 g FAME)		Vegetable Oils					
		Cottonseed	Canola	Sunflower	Corn	Soybean	CNSL [1]
Myristic	C14:0	0.5	nd	0.33	0.13	0.09	1.77
Palmitic	C16:0	21.7	1.91	8.52	11.8	12.2	15.97
Stearic	C18:0	2.12	2.33	7.41	5. 18	3.27	5.76
Oleic	C18:1	17.2	63.7	25.7	20.7	29. 4	14.3
Linoleic	C18:2	50.3	23	54.2	54.0	45.8	10.8
Linolenic	C18:3	2.41	8.82	0.32	5.51	3.78	2.83
SFA	Total	24.3	4.21	16.2	17.1	15.6	23.5
MUFA	Total	17.2	63.7	25.7	20.7	29.4	14.3
PUFA	Total	52.7	31.8	54.5	59.5	49.6	13.7
UFA	Total	69.9	95.5	80.2	80.2	79.0	28.0

[1] Cashew nut shell liquid; unsaturated, saturated, monounsaturated and polyunsaturated fatty acids. [nd] = not detected.

2.4. Intake and Ingestive Behavior

The animals were weighed at the beginning (initial body weight) and end (final body weight) of the experiment and every 15 days to adjust consumption. Nutrient intake was calculated by subtracting the amount of each nutrient found in the leftovers from the total amount offered in the diet, with results expressed in grams per day (g/d). The animals' feeding behavior was observed twice during the experimental period, on the 30th and 50th days, with observations recorded every 10 min over 24 h using the scan sampling method outlined by Martin and Bateson [33]. Trained observers, working in shifts, collected data on the time the animals spent eating, ruminating, or being inactive, positioning themselves strategically to avoid influencing the animals' behavior. Night-time observations were conducted using artificial lighting.

To estimate the average number of boluses ruminated per day, the number of chews per bolus, and the chewing time per bolus, evaluations were carried out in four periods: from 8:00 to 10:00, 12:00 to 14:00, 16:00 to 18:00, and 20:00 to 22:00 over the 24 h ingestive behavior assessment period, using digital chronometers as described by Bürger et al. [34]. Feeding and rumination efficiencies for dry matter (DM) and neutral detergent fiber (aNDF) in kg/h were calculated by dividing the intake of each nutrient by the total feeding and rumination times, respectively. Total chewing time was also recorded. Eating and ruminating efficiency rate and the total number and time of rumen boluses in chewings were calculated according to Polli et al. [35].

2.5. Diet Selectivity

Each animal's feed trough was weighed and sampled on the 41st and 43rd days of the experimental period, at 12 and 24 h post-feeding, respectively, for dry matter (DM) analysis and particle size. A 250 g sample was collected, and particle size was assessed using the sieve particle stratification method with the "Penn State Particle Separator—PSPS", as designed by Lammers et al. [36] and modified by Kononoff et al. [37]. This separator comprises three sieves (1.18 mm, 8 mm, and 19 mm) that divide the samples into four fractions according to particle size. The physical effectiveness factor of fiber (pef) was calculated in two ways: pef8, based on the sum of the percentages of particles larger than 8 mm [36], and $pef_{1.18}$, based on the sum of the percentages of particles larger than 1.18 mm [29,37]. The physically effective neutral detergent fiber (NDFpe) was determined by multiplying the sample's neutral detergent fiber content by the previously calculated pef. The chemical composition of the effectively consumed diet was estimated by dividing the intake of each nutrient by the intake of DM and multiplying by 100.

2.6. Environmental Variables

During the experimental period, the climatological data were recorded using a datalogger (HOBO® brand, model U12-013, Onset, São Paulo, Brazil) attached to two black globes, installed in the sunny and shaded environments, at a similar height to the animals, which was programmed to record the environmental data every hour for 24 h, throughout the experimental period, using the environmental data to calculate the black globe temperature and humidity index (BGHI) through the following formula: BGHI = [Tbg + (0.36 × Tdp) + 41.5], described by Buffington et al. [38], where Tbg is the black globe temperature and Tdp is the dew point temperature (Figure 1).

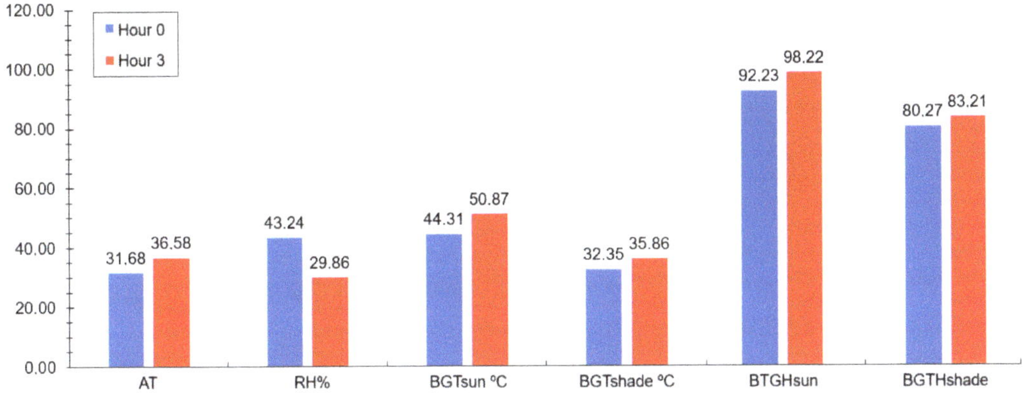

Figure 1. Averages of meteorological data: environmental temperature (AT), relative humidity (RH), black globe temperature in the sun (BGTsun), black globe temperature in the shade (BGTshade), black globe temperature and humidity index in the sun (BTGHsun) and in the shade (BTGHshade), before (0 h or at feeding moment) and 3 h after the feeding. (SEM = 7.86).

2.7. Physiological Variables of the Sheep

Physiological data were gathered on the 15th and 35th days of the experiment, before and three hours after the sheep were offered their diets, denoted as hour 0 (7:00 to 14:00) and hour 3 (10:00 to 17:00) post-feeding.

The respiratory rate was assessed by placing a flexible stethoscope on the right thoracic region, counting the number of respiratory movements for 30 s, and multiplying by two to obtain the rate per minute. Rectal temperature was measured using a veterinary clinical thermometer capable of registering temperatures up to 44 °C, inserted directly into the animal's rectum for two minutes, with results expressed in degrees Celsius (°C).

Surface temperature (ST) was captured using an infrared thermographic camera (Fluke Serie-Ti 25) with automatic calibration and an emissivity of 0.98, as recommended by the manufacturer for biological tissues. Thermograms of both the right and left sides of the animals were taken and analyzed using Smartview software (version 4.3, Copyright© Fluke Corporation, Norwich, UK, 2006–2017) to derive thermographic variables (maximum, minimum, and average surface temperatures) selected from the sheep's right and left flanks (Figure 2).

2.8. Serum Metabolites

To evaluate serum parameters, 5.0 mL of blood was drawn from the animals on days 15 and 35 via jugular venipuncture and placed into pre-labeled tubes with a vacuum system (Becton, Dickson and Co., São Paulo, SP, Brazil). A No. 16 catheter (Medical Supply®, São Paulo, Brazil) was gently inserted into each animal to aid collection and ensure animal comfort. The samples were left at room temperature until clot formation occurred. They were centrifuged (Centrifuge 90-1 model, Coleman®, São Paulo, Brazil) at 2500× g for five minutes to separate the blood serum. For subsequent analysis, serum samples were stored at −20 °C in Eppendorf® tubes (Sigma-Aldrich, São Paulo, Brazil).

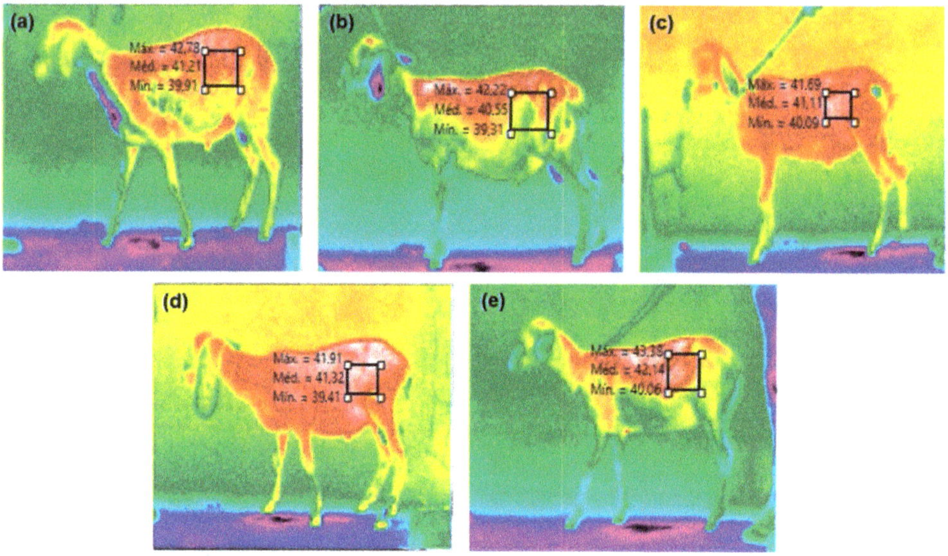

Figure 2. Thermographic variables (maximum, minimum, and medium surface temperatures) selected in the left flank region of the sheep fed diets containing mixtures of 0.5% of cashew nut shell liquid (of total DM) and 1.5% of soybean (**a**), cottonseed (**b**), sunflower (**c**), corn (**d**), or (**e**) canola oil (of total DM).

Biochemical test kits were employed to quantify serum metabolites, including total protein, cholesterol, triglycerides, albumin, aspartate aminotransferase (AST), gamma-glutamyl transferase (GGT), urea, and creatinine using the Cobas® C111 automated biochemical analyzer (Roche, Ludwigsburg, Germany), utilizing enzymatic or colorimetric kinetic assays. Electrolyte levels were analyzed using an automated analyzer (Max Ion-Medmax, Shenzhen, China) via the direction selectivity method for calcium, phosphorus, and magnesium minerals.

2.9. Statistical Analysis

Intake, ingestive behavior, and serum metabolite data underwent analysis following the principles of a randomized complete design, utilizing the MIXED procedure [39]. Collection dates were treated as repeated measurements over time. Each animal served as the experimental unit for all variables examined, as described by the statistical model:

$$Yijk = \mu + Oj + eijk$$

where Yij = value corresponding to the observation of repetition ii of treatment j; μ = overall mean; Oj = effect of treatment j (where j represents CNSL + canola or soybean or corn or cottonseed or sunflower oil); eij = random error associated with the observation.

The physiological variables of sheep were analyzed under a completely randomized design, employing a split-plot arrangement. Treatments (vegetable oil inclusion + CNSL blend) were allocated to plots, while collection times were assigned to subplots, with eight replications, based on the statistical model:

$$Yijk = \mu + Oj + Tk + OTjk + eijk$$

where Yijk = value corresponding to the observation of repetition ii in treatment j and at collection time kk; μ = overall mean; Oj = effect of treatment j (where j represents CNSL + canola or soybean or corn or cottonseed or sunflower oil); Tk = effect of collection time kk (before and 3 h after feeding); OTjk = interaction of the effects of treatment j with collection time kk; eijk = random error associated with the observation

Treatment means were compared using Tukey's test, with significance determined at $p < 0.05$.

3. Results

The CNSL and unsaturated-fatty-acid-rich vegetable oil blend did not affect ($p > 0.05$) the composition of the diet effectively consumed by the sheep, as shown in Table 3. Similarly, the intake of dry matter and aNDF and the duration spent on ruminating, feeding, and idling by the sheep remained unaffected ($p > 0.05$) by incorporating different vegetable oils high in unsaturated fatty acids alongside CNSL.

Table 3. Composition of effectively intake diet of sheep containing mixtures of CNSL (0.5% of total DM) and different vegetable oils rich in unsaturated fatty acids (1.5% of total DM).

Item	Oils Mixed with CNSL [1]					SEM [2]	*p*-Value [3]
	MUFA	PUFA					
	Canola	Cottonseed	Sunflower	Corn	Soybean		
Crude protein	113.85	118.06	118.46	111.93	113.49	0.497	0.845
Ash	49.82	52.81	50.50	54.26	51.96	0.255	0.746
aNeutral detergent fiber [4]	362.49	378.55	391.67	389.80	392.44	1.795	0.732
Acid detergent fiber	208.81	161.20	197.26	157.96	170.05	1.043	0.654
Ether extract	44.74	47.73	45.05	47.45	41.98	0.216	0.341
Non-fibrous carbohydrates	429.08	402.83	394.29	396.53	400.11	1.831	0.670
Total digestible nutrients	768.2	766.87	764.49	764.59	760.38	37.26	0.635
Metabolizable energy	3.797	3.954	4.016	4.00	3.799	0.151	0.728

[1] Cashew nut shell liquid. [2] SEM = standard error of the mean. [3] Means followed by different letters in the line are statistically different according to the Tukey's test at 5% probability. [4] Neutral detergent fiber assayed with a heat stable amylase and expressed exclusive for residual ash.

However, diets containing canola oil (MUFA) promoted a higher intake of SFA ($p < 0.0001$) compared to other vegetable oils. In contrast, sheep fed with cottonseed oil and CNSL blend intake had higher SFA than other vegetable oils ($p < 0.0001$). Sheep fed with canola oil (MUFA) and CNSL blend presented a higher intake of the UFA and MUFA compared to cottonseed oil and CNSL mixture and lower PUFA intake compared to other vegetable oils ($p < 0.0001$). Sheep fed with sunflower or corn oil presented higher PPUFA intake and similar intake of UFA. Cottonseed and corn oil diets promoted similar sheep intake of the MUFA ($p < 0.0001$).

The diets did not affect the intake efficiencies of DM and aNDF, rumination rates of DM and aNDF, amount of DM/bolus, and chewing time ($p > 0.05$). However, the number of boluses ruminated daily was higher ($p = 0.044$) in treatments containing CNSL associated with soybean or sunflower oils (Table 4) compared to CNSL associated with canola oil.

Regarding the selectivity of the diets by the sheep, when the diets were offered 12 h after feeding, the amount of particles retained on the 19 mm sieve ($p = 0.043$) was higher for the diet containing the CNSL and sunflower oil mixture, while in the 8 mm sieve ($p = 0.012$), it was higher for the diet with CNSL and soybean oil mixture, and in the 1.8 mm sieve ($p = 0.035$), it was higher for the diet with CNSL and canola oil (Table 5).

There was an effect of the diets ($p = 0.034$) on the effectiveness, using only two sieves (pef8), reflected in higher NDF_{pe8} intake ($p = 0.055$) for the diet containing the CNSL and sunflower oil compared to the diet with CNSL and canola oil.

There was no effect ($p > 0.05$) for the base sieves, pef1.18, and the intake of $NDF_{pe1.18}$. In the samples taken 24 h after feeding, there was an effect on the number of particles retained only on the 8 mm sieve ($p = 0.036$), with greater retention in the diet with CNSL and soybean oil mixture compared to the diet with CNSL and canola oil, and on the amount of particles retained on the base of the sieve ($p = 0.006$), which was inversely proportional.

Regarding effectiveness, there was an effect ($p = 0.044$) using three sieves (pef1.18), which was reflected in the higher $NDF_{pe1.18}$ intake ($p = 0.045$) for the diet containing the CNSL and soybean oil mixture when compared to the diet with CNSL and canola oil.

Table 4. Intake and ingestive behavior of sheep fed diets containing mixtures of CNSL (0.5% of total DM) and different vegetable oils rich in unsaturated fatty acids (1.5% of total DM).

Item [1]	Oils Mixed with CNSL [1]					SEM [2]	*p*-Value [3]
	MUFA	PUFA					
	Canola	Cottonseed	Sunflower	Corn	Soybean		
Growing							
Initial weight (kg)	23.1	23.9	23.4	24.3	24.4	0.89	0.812
Final weight (kg)	35.8	36.8	36.5	37.0	37.2	1.33	0.959
Intake (g/day)							
SFA	0.057 [d]	0.346 [a]	0.235 [bc]	0.248 [b]	0.215 [c]	0.008	<0.0001
UFA	1.307 [a]	0.997 [c]	1.165 [b]	1.162 [b]	1.091 [bc]	0.476	0.0012
MUFA	0.872 [a]	0.2453 [d]	0.3735 [bc]	0.299 [cd]	0.406 [b]	0.024	<0.0001
PUFA	0.435 [d]	0.751 b[c]	0.792 [ab]	0.862 [a]	0.685 [c]	0.026	<0.0001
Dry matter	1376	1433	1460	1456	1388	60.6	0.440
Dry matter (as %BW)	4.01	4.07	4.24	4.25	3.93	0.15	0.503
Organic matter	131	135	138	137	130	11.65	0.521
aNDF	515	543	572	558	544	29.1	0.405
aNDF (as %BW)	1.58	1.54	1.65	1.66	1.45	0.08	0.848
Acid detergent fiber	289	231	288	230	234	13.42	0.632
Behavioral events (min/day)							
Feeding	183	193	203	198	206	11.7	0.675
Rumination	502	497	481	477	476	17.8	0.762
Idleness	754	748	756	764	757	21.2	0.990
Efficiency rates (g/h)							
DM intake	495	443	457	455	401	15.2	0.626
aNDF intake	175	181	185	181	166	10.7	0.842
DM rumination	160	165	170	175	161	9.74	0.882
aNDF rumination	62.3	68.2	68.6	71.3	66.4	5.39	0.822
Chewing rates							
Quantity (g DM/bolus)	2.19	2.12	1.96	2.20	1.94	0.16	0.700
Boluses ruminated (No./day)	613 [b]	644 [ab]	745 [a]	660 [ab]	741 [a]	14.4	0.044
TCT (min/day)	686	691	683	675	682	19.2	0.990

[1] CNSL is cashew nut shell liquid; SFA, UFA, MUFA, and PUFA are saturated, unsaturated, monounsaturated, and polyunsaturated fatty acids, respectively; DM is dry matter; aNDF is neutral detergent fiber assayed with a heat stable amylase and expressed exclusive of residual ash; and TCT is total chewing time. [2] SEM = standard error of the mean. [3] Means followed by different letters in the line are statistically different according to the Tukey's test at 5% probability.

There was no effect of interaction between the combination of CNSL and different vegetable oils rich in unsaturated fatty acids and the collection time of the sheep's physiological variables ($p > 0.05$).

The combination of CNSL with vegetable oils rich in unsaturated fatty acids did not affect ($p > 0.05$) the rectal temperature and right surface temperature of the sheep before (0 h) and 3 h after feeding (Table 6). There was also no effect of diet on rectal and surface temperatures before (0 h) and 3 h after feeding ($p > 0.05$). However, the average right surface temperature, without considering time, computed by the camera, shows that the fermentation heat was higher ($p < 0.05$) in the animals fed CNSL mixed with sunflower or corn oil (Figure 2) when compared to CNSL mixed with canola oil, with which the animals had the lowest average surface temperature. When comparing times (0 and 3), the respiratory rate was lower in sheep fed CNSL mixed with sunflower or corn oil ($p = 0.0099$) than in animals fed diets with CNSL and soybean, cottonseed, or canola oil.

The diets did not exert any discernible effect ($p > 0.05$) on the serum concentrations of albumin (ALB), total protein (TP), urea (URE), triglycerides (TRI), aspartate aminotransferase (AST), gamma glutamyl transferase (GGT), and the minerals phosphorus (P) and magnesium (Mg), as outlined in Table 7. However, higher serum creatinine concentrations were observed in sheep fed CNSL mixed with corn or canola oil ($p = 0.007$) compared to animals fed CNSL mixed with sunflower oil. Animals fed the CNSL with canola oil had higher serum cholesterol concentrations ($p = 0.050$) than those fed the diet with CNSL and sunflower oil. Serum calcium concentrations were higher in sheep ($p = 0.001$) fed the CNSL with sunflower or corn oil than in sheep fed CNSL and soybean or canola oil.

Table 5. Particle size distribution, physical effectiveness factors (pef), and physically effective neutral detergent fiber (NDFpe) content of the diets containing mixtures of CNSL (0.5% of total DM) and different vegetable oils rich in unsaturated fatty acids (1.5% of total DM).

Variables	Oils Mixed with CNSL [1]					SEM [2]	p-Value [3]
	MUFA	PUFA					
	Canola	Cottonseed	Sunflower	Corn	Soybean		
			Distribution of particles 12 h after feeding [4]				
19 mm	12.5 ^{b}	19.3 ^{ab}	24.7 ^{a}	18.1 ^{ab}	19.9 ^{ab}	2.59	0.043
8 mm	23.3 ^{b}	25.6 ^{ab}	25.3 ^{ab}	28.5 ^{a}	28.7 ^{a}	1.13	0.012
1.18 mm	25.0 ^{a}	20.8 ^{b}	21.2 ^{ab}	23.0 ^{ab}	22.6 ^{ab}	0.93	0.035
Bottom	39.0	34.1	28.5	30.3	28.6	2.73	0.055
[5] pef8	35.8 ^{b}	45.0 ^{ab}	50.1 ^{a}	46.6 ^{ab}	48.7 ^{ab}	3.20	0.034
[6] pef1.18	60.9	65.8	71.4	69.6	71.3	2.73	0.055
[7] NDF_{pe8}	13.3 ^{b}	16.7 ^{ab}	18.6 ^{a}	17.3 ^{ab}	18.1 ^{ab}	1.19	0.035
[8] $NDF_{pe1.18}$	22.6	24.5	26.5	25.9	26.5	1.01	0.055
			Distribution of particles 24 h after feeding				
19 mm	20.8	28.5	27.7	24.5	26.7	2.88	0.356
8 mm	24.7 ^{b}	27.3 ^{ab}	25.9 ^{ab}	27.9 ^{ab}	29.6 ^{a}	1.07	0.036
1.8 mm	23.6	19.6	21.5	19.7	22.0	1.24	0.144
Bottom	30.7 ^{a}	24.3 ^{ab}	24.7 ^{ab}	27.8 ^{ab}	21.5 ^{b}	2.20	0.006
pef8	45.5	56.0	53.6	52.4	56.3	3.19	0.149
$pef_{1.18}$	69.2 ^{b}	75.5 ^{ab}	75.2 ^{ab}	72.1 ^{ab}	78.4 ^{a}	2.20	0.044
NDF_{fe8}	16.9	20.8	19.9	19.5	20.9	1.18	0.149
$NDF_{fe1.18}$	25.7 ^{b}	28.1 ^{ab}	27.9 ^{ab}	26.8 ^{ab}	29.1 ^{a}	0.82	0.045

[1] Cashew nut shell liquid. [2] SEM = standard error of the mean. [3] p-value significant at $p < 0.05$. Means followed by the same letter in the row are not statistically different according to Tukey's test at 5% probability. [4] Particle size distribution of the diet was measured using the Penn State Particle Separator (Kononoff et al. [37]). [5] pef8 and [6] pef1.18 = physical efficiency factor determined as the proportion of particles retained on 2 sieves (8 and 19 mm) (Lammers et al. [36]) and on 3 sieves (1.8, 8 and 19 mm) (Kononoff et al. [37]), respectively. [7] NDFpe8 and [8] NDFpe1.18 = effective neutral detergent fiber (NDF) of the sample multiplied by pef8 and pef1.18, respectively.

Table 6. Physiological and thermographic parameters according to the time of evaluation of sheep fed diets containing mixtures of CNSL (0.5% of total DM) and different vegetable oils rich in unsaturated fatty acids (1.5% of total DM).

Hour	Oils Mixed with CNSL [1]					SEM [2]	p-Value		
	MUFA	PUFA							
	Canola	Cottonseed	Sunflower	Corn	Soybean		Oil [3]	Time [4]	Oil × Time [5]
					Respiratory rate				
0	52.0 ^{Aa}	44.5 ^{Aa}	58.2 ^{Aa}	55.2 ^{Aa}	58.7 ^{Aa}	7.36	0.5363	0.0099	0.5043
3	76.0 ^{Ab}	63.7 ^{Ab}	60.2 ^{Aa}	59.0 ^{Aa}	71.5 ^{Ab}				
					Rectal temperature				
0	38.8 ^{Aa}	38.6 ^{Aa}	38.7 ^{Aa}	38.9 ^{Aa}	39.0 ^{Aa}	0.17	0.0737	0.0748	0.6004
3	39.1 ^{Aa}	39.0 ^{Aa}	38.6 ^{Aa}	39.2 ^{Aa}	39.1 ^{Aa}				
					Left flank surface temperature				
0	38.5 ^{Aa}	38.4 ^{Aa}	38.3 ^{Aa}	38.4 ^{Aa}	37.6 ^{Aa}	0.25	0.1665	<0.0001	0.4974
3	40.0 ^{Ab}	39.8 ^{Ab}	39.9 ^{Ab}	39.5 ^{Ab}	39.7 ^{Ab}				
					Right flank surface temperature				
0	38.2 ^{Aa}	38.4 ^{Aa}	38.3 ^{Aa}	38.2 ^{Aa}	37.6 ^{Aa}	0.25	0.2114	<0.0001	0.2839
3	39.9 ^{Ab}	39.8 ^{Ab}	39.8 ^{Ab}	39.2 ^{Ab}	39.7 ^{Ab}				

[1] Cashew nut shell liquid. [2] SEM = standard error of the mean. [3] Means followed by different letters in the row are statistically different among treatments according to Tukey's test at 5% probability. [4] Means followed by different lowercase letters in the columns are statistically different for the time factor according to Tukey's test $p < 0.05$. [5] Interaction (Oil × Time) effect by Tukey's test when $p < 0.05$.

Table 7. Blood metabolites in sheep fed diets containing mixtures of CNSL (0.5% of total DM) and different vegetable oils rich in unsaturated fatty acids (1.5% of total DM).

Serum Metabolites	Oils Mixed with CNSL [1]					SEM [2]	*p*-Value [3]
	MUFA	PUFA					
	Canola	Cottonseed	Sunflower	Corn	Soybean		
Albumin (g/dL)	2.59	2.59	2.61	2.68	2.62	0.07	0.892
Total proteins (g/dL)	5.64	5.42	5.38	5.63	5.52	0.14	0.621
Urea (mg/dL)	25.7	24.8	20.4	24.3	24.8	1.59	0.169
Creatinine (mg/dL)	0.83 ^{a}	0.80 ^{ab}	0.70 ^{b}	0.87 ^{a}	0.78 ^{ab}	0.03	0.007
Cholesterol (mg/dL)	53.5 ^{a}	46.5 ^{ab}	43.5 ^{b}	49.1 ^{ab}	45.9 ^{ab}	2.35	0.050
Triglycerides (mg/dL)	13.5	11.7	13.1	14.2	14.5	1.03	0.345
AST [4] (mg/dL)	91.4	97.6	87.7	95.5	86.5	6.51	0.704
GGT [5] (mg/dL)	34.3	33.6	38.2	39.1	34.1	1.92	0.155
Calcium (mg/dL)	8.43 ^{b}	8.84 ^{ab}	11.1 ^{a}	11.1 ^{a}	8.53 ^{b}	0.59	0.001
Phosphorus (mg/dL)	8.56	9.15	9.01	8.84	9.17	0.47	0.891
Magnesium (mmol/L)	1.86	1.71	2.08	1.59	1.57	0.31	0.761

[1] Cashew nut shell liquid. [2] SEM = standard error of the mean. [3] Means followed by different letters on the same line are statistically different according to the Tukey's test at 5% probability. [4] Aspartate aminotransferase. [5] Gamma glutamyl transferase.

4. Discussion

Cottonseed oil had the highest concentration of SFA (24.3%), followed by sunflower oil (17.1%) and corn oil (16.2%), while canola oil had the highest concentration of MUFA (63.7%), which was much higher than the other oils. Corn oil (59.5%) and sunflower oil (54.15%) had the highest concentrations of PUFA. Consequently, despite the intense similarity in composition, the diets allowed differences in the consumption of different fatty acids (FA). Sheep fed canola oil, which was more concentrated in MUFA than other oils, showed lower consumption of SFA and PUFA and higher consumption of UFA and total MUFA, which promoted differences in rumen fermentation.

The microbial ecosystem in the rumen responds to variations in diet type and composition, which are main factors influencing microbial activity and function in the rumen [14–17]. Therefore, even though the diets were very similar in overall composition (Table 1), altering only the oils' fatty acid (FA) composition demonstrated that the FA saturation level was sufficient to modify the animals' feeding behavior. This highlights the significance of comprehending the intricate populations of ruminal microbes, their interplay, and their responses to diverse dietary compositions [40]. Additionally, it is crucial to elucidate responses to various dietary FA compositions based on proposed differences in lipid metabolism [41].

The similarity between diets in terms of fiber, energy, and protein content may explain the intake of aNDF and DM and, consequently, the times of the activities related to ingestive behavior since dietary fiber is one of the main factors that interferes with feeding and rumination activities. In addition, vegetable oils rich in unsaturated fatty acids were mixed with CNSL in the same proportion and in quantities that did not cause negative effects on the digestive process. Several studies using vegetable oils in sheep diets [42–44] have shown that DM and aNDF intake can be affected by the addition of oil sources to the diet. However, the extent of this interference depends on both the oil source and the level at which it is added to the feed. Most studies confirm that the use of lipids has little or no effect on the activities of the microbial flora and the other characteristics of the rumen environment as long as it does not exceed 7% of the total DM of the diet.

However, adding CNSL to the oils reduced its astringent effect since DM intakes were above 1300 g/d. This can be explained by the fact that the diets contained 0.5% CNSL in the total DM, which is a level below the recommendation of Ramos et al. [16] and Araújo et al. [17], who indicated a maximum of 0.75% CNSL in sheep diets, given the risk of toxicity and astringency, which negatively affects intake and digestibility of nutrients. The combination of oil and CNSL could be an interesting alternative to increase the energy content of the diet without affecting intake.

Despite not affecting rumination time and intake, the inclusion of the CNSL mixed with soybean and sunflower oils increased the number of boluses ruminated (No./day), which may be associated with the distribution of particles, especially in the 1.8 mm sieves (Table 3), indicating a greater amount of concentrate in the leftovers, and consequently allowing more selection of the fiber fractions [45].

Particle size analysis tries to determine the actual frequency distribution of the particles according to the size and selection of the animal's diet. The particle size distribution of the diets offered to the sheep 12 h after feeding showed that the animals fed the mixture of CNSL and canola oil had higher intake of larger particles (pef8), corresponding to a higher intake of NDFpe8. In the distribution of particles twenty-four hours after feeding, the animals fed the CNSL and canola oil blend continued to consume more of the roughage and rejected the concentrate, with a greater volume of particles at the base, confirmed by a higher intake of $NDF_{pe1.18}$. The selection of the diet by the animals that consumed CNSL mixed with canola oil can be attributed to the direct action on the rumen caused by the higher composition of unsaturated fatty acids in canola oil, causing a reduction in intake due to the loss of palatability of the diet and the change in rumen fermentation [46,47].

The combination of CNSL with various oils did not induce changes in the physiological variables, suggesting that the oil type in the diet does not necessarily correlate with its effectiveness in enhancing animal thermal comfort. In regions like semi-arid areas, temperatures often surpass the thermal comfort zone for small ruminants during the hottest hours of the day, prompting them to seek shade, reduce dry matter intake (DMI), and increase water consumption. In such conditions, diets with higher energy density can aid animals by enabling quicker eating and reducing fermentation rates.

Studies by Gomes et al. [48], Neiva et al. [49], and Nobre et al. [7] have reported similar environmental conditions, with temperatures exceeding the thermal comfort zone, especially from midday onward. The average BGHI (Black Globe Humidity Index) values reflect a correlation between feeding time and environmental factors, with the highest averages occurring three hours post-feeding, indicating elevated environmental temperatures that expose animals, even those in confinement, to thermal discomfort.

The BGHI levels for sheep and goats categorize heat stress severity: values below 82 indicate no heat stress, while values above 82 and below 86 denote severe heat stress, and those above 86 signify extremely severe heat stress. The climate data and BGHI indicate that the animals experienced extremely severe heat stress. Despite these adverse environmental conditions, physiological variables displayed average rectal temperatures (RT), typically ranging from 38.3 °C to 39.9 °C in sheep.

Respiratory rate (RR) values can quantify heat stress severity, with ranges indicating low, medium–high, and high stress levels for ruminants. The observed high RR values suggest medium–high stress levels in the study animals, likely influenced by the elevated BGHI values, mainly three hours post-feeding. Sheep fed CNSL mixed with sunflower and corn oil exhibited lower RR values, possibly due to a lesser calorie increase from their diet. These oils contain higher amounts of linoleic acid, which can inhibit fat synthesis by lysogenic enzymes.

Thermographic measurements of the left flank surface temperature (ST) suggest higher fermentation heat in animals fed CNSL mixed with sunflower or corn oil, resulting in elevated RR values. Conversely, sheep fed CNSL mixed with canola oil displayed the lowest RR values, likely due to reduced biohydrogenation and less heat production during rumen fermentation. This indicates that fermentation heat correlates more strongly with unsaturation levels and double bond presence in the diet's fat content rather than the amount of vegetable fat consumed.

Flank ST is also significantly correlated with climatic parameters, particularly environmental temperature. Previous studies have highlighted positive correlations between physiological parameters and thermographic measurements, indicating thermography's efficacy as a thermal comfort indicator. These correlations are attributed to the efficiency of thermoregulation mechanisms, which rely on the thermal gradient between the animal's body and its surroundings. A more significant thermal gradient facilitates heat dissipation through conduction, convection, and radiation, enhancing thermal comfort.

Plasma creatinine concentrations show the efficiency of glomerular filtration, with total excretion by the kidneys. High levels of this metabolite suggest impaired kidney function [50]. Among the treatments, creatinine concentrations in the sheep blood were below the reference values (1.2–1.9 mg/dL) for the sheep [51], indicating low physical activity and low proteolysis by muscle tissues, which reflects the confinement. Creatinine concentrations in the body are dependent on muscle mass and endogenous nitrogen utilization [52].

For cholesterol levels, the range considered normal for sheep is 50–76 mg/dL [50,51] and can be influenced by the diet, intake, age, and physiological state of the animals. The reduction in the concentration of cholesterol in the blood of the sheep in treatments that used CNSL associated with soybean, cottonseed, sunflower, and corn oil in comparison to canola oil may be related to the composition of polyunsaturated fatty acids (Table 2). According to Brzozowska and Oprządek [53], PUFA and non-esterified PUFA appear to be more potent inhibitors of intake than monounsaturated and esterified PUFA, respectively. These data indicate that the use of lipid sources known to be rich in PUFA and non-esterified can influence dry matter intake. Despite the similar DMI, animals that consumed canola oil continued to consume more of the roughage and rejected the concentrate, with a greater volume of particles at the bottom, as observed in the sieves (Table 4). This may have been due to the FA composition of the oils since canola oil had a lower amount of PUFA (32%) while the other oils had more than 50% PUFA. Therefore, the more selective roughage diet may have contributed to the increase in blood cholesterol.

5. Conclusions

Including CNSL combined with different vegetable oils rich in unsaturated fatty acids in the diet of confined sheep did not affect intake or physiological and thermographic variables, which were more susceptible to weather conditions and the time of day when the data were recorded. The mixture of CNSL and canola oil caused the sheep to select the roughage over the concentrate and increased the animals' serum cholesterol compared to the other oils. Combining canola oil with CNSL reduced PUFA intake and minimized the effects of phenolic compounds (CNSL) on the animal's health but showed greater rejection to concentrate with this supplementation.

Author Contributions: É.L.G.A.: conceptualization, data curation, formal analyses, investigation, writing—original draft; K.H.d.O.S.d.L. and Y.C.S.B.: data curation, formal analyses; J.M.P.F. and B.B.d.S.: methodology, supervision; J.P.F.d.O. and M.A.F.: formal analyses, software, validation; A.F.d.M.V.: data curation, formal analyses, investigation; R.L.O.: investigation, funding acquisition, supervision; L.R.B.: conceptualization, project administration, supervision, funding acquisition, writing—review and editing. All authors have read and agreed to the published version of the manuscript.

Funding: This study was partially supported by the National Council for Scientific and Technological Development (Brazil) agency from the PQ-2023 (Grant #306920/2023-9) and MCTIC/CNPq (Grant #406734/2022-4).

Institutional Review Board Statement: The animal study protocol was approved by the Ethics Committee of Ethics Committee for Animal Experimentation Use (protocol code 55/2022, 1 November 2022).

Informed Consent Statement: Not applicable.

Data Availability Statement: The datasets that support the findings of this study are available from https://doi.org/10.17632/tj6yzp9rzy.1, accessed on 6 June 2024.

Acknowledgments: This research was supported by the Research Support Foundation of the State of Paraiba and the Federal University of Campina Grande through facilities support.

Conflicts of Interest: The authors declare no conflicts of interest.

References

1. National Research Council-NRC. *Nutrient Requirements of Small Ruminants: Sheep, Goats, Cervids, and New World Camelids*; The National Academies Press: Washington, DC, USA, 2007. [CrossRef]
2. Oliveira, F.d.S.; Fernandes Neto, V.d.P.; Nascimento e Silva, M.N.d.; Cardoso, F.S.; Costa, A.P.R. Effect of heat stress on physiological and biochemical parameters of sheep raised in tropical climate. *Pubvet* **2012**, *6*, 1359. [CrossRef]
3. Maia, A.S.C.; Nascimento, S.T.; Nascimento, C.C.N.; Gebremedhin, K.G. Thermal equilibrium of goats. *J. Therm. Biol.* **2016**, *58*, 43–52. [CrossRef] [PubMed]
4. Sejian, V.; Bagath, M.; Krishnan, G.; Rashamol, V.P.; Pragna, P.; Devaraj, C.; Bhatta, R. Genes for resilience to heat stress in small ruminants: A review. *Small Rum. Res.* **2019**, *173*, 42–53. [CrossRef]
5. Santos, M.L.P.; Dada, J.M.V.; Muniz, P.C.; Nunes-Zotti, M.L.A.; Barros, F.R.O.d.; Vieira, F.M.C. Physiological responses of Santa Inês x Dorper ewes and lambs to thermal environment of silvopasture and open pasture systems. *Small Rum. Res.* **2021**, *205*, 106–565. [CrossRef]
6. Mascarenhas, N.M.H.; Furtado, D.A.; Fonsêca, V.F.C.; Souza, B.B.d.; Oliveira, A.G.; Morais, F.T.L.; Silva, R.d.S.; Silva, M.R.d.; Batista, L.F.; Dornelas, K.C.; et al. Thermal stress index for native sheep. *J. Therm. Biol.* **2023**, *115*, 103607. [CrossRef] [PubMed]
7. Nobre, I.S.; Souza, B.B.; Marques BA, A.; Azevedo, A.M.; Araújo, R.P.; Gomes TL, S.; Batista, L.F.; Silva, G.A. Evaluation of Levels of fat protected and concentrate on productive performance and sheep thermoregulation. *Rev. Bras. de Saúde e Produção Anim.* **2016**, *17*, 116–126. [CrossRef]
8. Baldwin, R.L.; Smith, N.E.; Taylor, J.; Sharp, M. Manipulating metabolic parameters to improve growth rate and milk secretion. *J. Anim. Sci.* **1980**, *51*, 1416–1428. [CrossRef]
9. Jin, Y.; Jiang, B.; Wang, H. Growth performance, meat quality and lipid metabolism in finishing lambs fed diets containing rumen-unprotected and rumen-protected betaine. Italian. *J. Anim. Sci.* **2021**, *20*, 2041–2050. [CrossRef]
10. Tedeschi, L.O.; Cannas, A.; Fox, D.G. A nutrition mathematical model to account for dietary supply and requirements of energy and other nutrients for domesticated small ruminants: The development and evaluation of the Small Ruminant Nutrition System. *Small Rum. Res.* **2010**, *89*, 174–184. [CrossRef]
11. Palmquist, D.L. The role of dietary fats in efficiency of ruminants. *J. Nutr.* **1994**, *124* (Suppl. S8), 1377S–1382S. [CrossRef]
12. Behan, A.A.; Loh, T.C.; Fakurazi, S.; Kaka, U.; Kaka, A.; Samsudin, A.A. Effects of Supplementation of Rumen Protected Fats on Rumen Ecology and Digestibility of Nutrients in Sheep. *Animals* **2019**, *9*, 400. [CrossRef]
13. Wang, Y.; Wang, Q.; Dai, C.; Li, J.; Huang, P.; Li, Y.; Ding, X.; Huang, J.; Hussain, T.; Yang, H. Effects of dietary energy on growth performance, carcass characteristics, serum biochemical index, and meat quality of female Hu lambs. *Anim. Nutr.* **2020**, *6*, 499–506. [CrossRef] [PubMed]
14. Himejima, M.; Kubo, I. Antibacterial agents from the cashew Anacardium occidentale (*Anacardiaceae*) nut shell oil. *J. Agric. Food Chem.* **1991**, *39*, 418–421. [CrossRef]
15. Osmari, M.P.; Matos, L.F.d.; Salab, B.L.; Diaz, T.G.; Giotto, F.M. Cashew nut shell liquid: Characteristics and applicability in animal production. *Pubvet* **2015**, *9*, 143–149.
16. Ramos, L.M.G.; Bezerra, L.R.; Oliveira, J.P.F.d.; Souza, M.P.d.; Silva, A.L.d.; Sales, E.P.; Mazzetto, S.E.; Pereira Filho, J.M.; Oliveira, R.L. Effects of feeding growing-finishing lambs with cashew nut shell liquid on the growth performance, physicochemical attributes, lipid peroxidation and sensorial parameters of burger. *Small Ruminant Res.* **2021**, *202*, 106–468. [CrossRef]
17. Araújo, D.; Araújo, M.; Silva, S.; Pereira Filho, J.; Parente, M.; Oliveira, R.; Mazzetto, S.; Oliveira, J.; Edvan, R.; Bezerra, L. Effect of technical cashew nut shell liquid on growth, physicochemical and fatty acid composition of lamb meat. *Small Rum. Res.* **2023**, *227*, 107070. [CrossRef]
18. Watanabe, Y.; Suzuki, R.; Koike, S.; Nagashima, K.; Mochizuki, M.; Forester, R.J.; Kobayashi, Y. In vitro evaluation of cashew nut shell liquid as a methane-enhancing agent for ruminants. *J. Dairy Sci.* **2010**, *93*, 5258–5267. [CrossRef]
19. Shinkai, T.; Enishi, O.; Mitsumori, M.; Higuchi, K.; Kobayashi, Y.; Takenaka, A.; Nagashima, K.; Mochizuki, M.; Kobayashi, Y. Mitigation of methane production from cattle by feeding cashew nut shell liquid. *J. Dairy Sci.* **2012**, *95*, 5308–5316. [CrossRef]
20. Compton, C.; Peña, O.M.; Hikita, C.; Watanabe, T.; Jenkins, T.C.; Lascano, G.J.; Aguerre, M.J. Effects of cashew nut shell extract on ruminal fermentation and nutrient digestibility under continuous culture. *Ruminants* **2023**, *3*, 92–99. [CrossRef]
21. Lup, F.; Pop, I.M.; Simeanu, D.; Vicas, S.; Simeanu, C.; Mierlita, D. Research regarding fatty acid profile and health lipid indices in the lambs meat of employing feed supplemented with different vegetable oils. *Rev. Chim.* **2018**, *69*, 222–227. [CrossRef]
22. Instituto Nacional de Meteorologia. *Banco de Dados Meteorológicos para Ensino e Pesquisa*; INMET/BDMEP: Brasília, DF, Brazil, 2023.
23. Lubi, M.C.; Thachil, E.T. Cashew nut shell liquid (CNSL)-a versatile monomer for polymer synthesis. *Des. Monomers Polym.* **2000**, *3*, 123–153. [CrossRef]
24. Mazzetto, S.E.; Lomonaco, D.; Mele, G. Cashew nut oil: Opportunities and challenges in the context of sustainable industrial development. *Quím. Nova* **2009**, *32*, 732–741. [CrossRef]
25. Association of Official Analytical Chemists—AOAC. *Official Methods of Analysis*, 19th ed.; AOAC International: Gaithersburg, MD, USA, 2012.
26. Van Soest, P.J.; Robertson, J.B.; Lewis, B.A. Methods for dietary fiber, neutral detergent fiber, and nonstarch polysaccharides in relation to animal nutrition. *J. Dairy Sci.* **1991**, *74*, 3583–3597. [CrossRef] [PubMed]

27. Senger, C.C.D.; Kozloski, G.V.; Bonnecarrère Sanchez, L.M.; Mesquita, F.R.; Alves, T.P.; Castagnino, D.S. Evaluation of autoclave procedures for fibre analysis in forage and concentrate feedstuffs. *Anim. Feed Sci. Technol.* **2008**, *146*, 169–174. [CrossRef]
28. Licitra, G.; Hernandez, T.M.; Van Soest, P.J. Standardization of procedures for nitrogen fractionation of ruminant feeds. *Anim. Feed Sci. Technol.* **1996**, *57*, 347–358. [CrossRef]
29. Mertens, D.R. Creating a system for meeting the fiber requirements of dairy cows. *J. Dairy Sci.* **1997**, *80*, 1463–1481. [CrossRef]
30. Weiss, W.P. Energy Prediction Equations for Ruminant Feeds. In *Cornell Nutrition Conference Feed Manufactures*; Proceedings; Cornell University: Ithaca, NY, USA, 1999; pp. 176–185. [CrossRef]
31. Bligh, E.G.; Dyer, W.J. A rapid method of total lipid extraction and purification. *Can. J. Biochem.* **1959**, *37*, 911–917. [CrossRef]
32. Hartman, L.; Lago, R.C. Rapid preparation of fatty acid methyl esters from lipids. *Lab. Pract.* **1973**, *22*, 475–481.
33. Martin, P.; Bateson, P. *Measuring Behaviour*, 2nd ed.; Cambridge University Press: Cambridge, UK, 1993; Volume 222.
34. Bürger, P.J.; Pereira, J.C.; Queiroz, A.C.d.; Silva, J.F.C.d.; Valadares Filho, S.d.C.; Cecon, P.R.; Casali, A.D.P. Ingestive behavior in Holstein calves fed diets with different concentrate levels. *Rev. Bras. Zootec.* **2000**, *29*, 236–242. [CrossRef]
35. Polli, V.A.; Restle, J.; Senna, D.B.; Almeida, S.R.S.d. Aspectos relativos à ruminação de bovinos e bubalinos em regime de confinamento. *Rev. Bras. Zootec.* **1996**, *25*, 987–993.
36. Lammers, B.P.; Buckmaster, D.R.; Heinrichs, A.J. A simple method for the analysis of particle sizes of forage and total mixed rations. *J. Dairy Sci.* **1996**, *79*, 922–928. [CrossRef] [PubMed]
37. Kononoff, P.J.; Heinrichs, A.J.; Buckmaster, D.R. Modification of Penn State forage and total mixed ration particle separator and the effects of moisture content on its measurements. *J. Dairy Sci.* **2003**, *86*, 1858–1863. [CrossRef]
38. Buffington, D.E.; Collazo-Arocho, A.; Canton, G.H.; Pitt, D. Black Globe-humidity index (BGHI) as Comfort Equation for Dziry Cows. *Trans. Asae* **1981**, *24*, 0711–0714. [CrossRef]
39. Statistical Analysis System, S.A.S. *User's Guide: Statistics*; SAS Institute Inc.: Cary, NC, USA, 2002.
40. Maia MR, G.; Chaudhary, L.C.; Figueres, L.; Wallace, R.J. Metabolism of polyunsaturated fatty acids and their toxicity to the microflora of the rumen. *Antonie van Leeuwenhoek* **2007**, *91*, 303–314. [CrossRef]
41. Shingfield, K.J.; Bernard, L.; Leroux, C.; Chilliard, Y. Role of trans fatty acids in the nutritional regulation of mammary lipogenesis in ruminants. *Animal* **2010**, *4*, 1140–1166. [CrossRef] [PubMed]
42. Francisco, A.; Dentinho, M.T.; Alves, S.P.; Portugal, P.V.; Fernandes, F.; Sengo, S.; Jerónimo, E.; Oliveira, M.A.; Costa, P.; Sequeira, A.; et al. Growth performance, carcass and meat quality of lambs supplemented with increasing levels of a tanniferous bush (*Cistus ladanifer* L.) and vegetable oils. *Meat Sci.* **2015**, *100*, 275–282. [CrossRef] [PubMed]
43. Diógenes, L.; Bezerra, L.; Pereira Filho, J.; Silva Junior, J.; Oliveira, J.; Moura, J.; Barbosa, A.; Souza, M.; Sousa, S.; Pereira, E.; et al. Effects of the dietary inclusion of buriti oil on lamb performance, carcass traits, digestibility, nitrogen balance, ingestive behavior and blood metabolites. *Animals* **2020**, *10*, 1973. [CrossRef]
44. de Oliveira Maia Parente, M.; Rocha, K.S.; Bessa, R.J.B.; Parente, H.N.; de Moura Zanine, A.; Machado, N.A.F.; de Brito Lourenço Júnior, J.; Bezerra, L.R.; Landim, A.V.; Alves, S.P. Effects of the dietary inclusion of babassu oil or buriti oil on lamb performance, meat quality and fatty acid composition. *Meat Sci.* **2020**, *160*, 107971. [CrossRef]
45. Ackermans, N.L.; Martin, L.F.; Hummel, J.; Müller, D.W.H.; Clauss, M.; Hatt, J.-M. Feeding selectivity for diet abrasiveness in sheep and goats. *Small Rum. Res.* **2019**, *175*, 160–164. [CrossRef]
46. Nguyen, D.V.; Malau-Aduli, B.S.; Nichols, P.D.; Malau-Aduli, A.E. Growth performance and carcass characteristics of Australian prime lambs supplemented with pellets containing canola oil or flaxseed oil. *Anim. Prod. Sci.* **2017**, *58*, 2100–2108. [CrossRef]
47. Lima, P.R.; Apdini, T.; Freire, A.S.; Santana, A.S.; Moura LM, L.; Nascimento JC, S.; Rodrigues RT, S.; Dijkstra, J.; Garcez Neto, A.F.; Queiroz, M.A.Á.; et al. Dietary supplementation with tannin and soybean oil on intake, digestibility, feeding behavior, ruminal protozoa and methane emission in sheep. *Anim. Feed Sci. Technol.* **2019**, *249*, 10–17. [CrossRef]
48. Gomes, C.A.V.; Furtado, D.A.; Medeiros, A.N.; Silva, D.S.; Pimenta Filho, E.C.; Lima Junior, V. Effect of thermal ambient and feed supplementation levels on physiologic parameters of Moxotó goats. *Rev. Bras. de Eng. Agrícola e Ambient.* **2008**, *12*, 213–219. [CrossRef]
49. Neiva, J.N.; Teixeira, M.; Turco, H.N.; Oliveira, S.M.; Moura, A.D. Effects of environmental stress on physiological parameters of feedlot sheep in the Northeast of Brazil. *Rev. Bras. Zootec.* **2004**, *33*, 668–678. [CrossRef]
50. Kaneko, J.J.; Harve, J.W.; Bruss, M.L. *Clinical Biochemistry of Domestic Animals*, 6th ed.; Academic Press: San Diego, CA, USA, 2008. [CrossRef]
51. Meyer, D.J.; Harvey, J.W. *Veterinary Laboratory Medicine: Interpretation and Diagnosis*; WB Saunders: St. Louis, MO, USA, 2004.
52. Caldeira, R.M.; Belo, A.T.; Santos, C.C.; Vasquez, M.I.; Portugal, A.V. The effect of long-term feed restriction and over-nutrition on body condition score, blood metabolites and hormonal profiles in ewes. *Small Rum. Res.* **2007**, *68*, 233–241. [CrossRef]
53. Brzozowska, A.M.; Oprządek, J. Metabolism of fatty acids in tissues and organs of the ruminants-a review. *Anim. Sci. Pap. Rep.* **2016**, *34*, 211–219.

ruminants

MDPI

Review

Impact of Mixed Rations on Rumen Fermentation, Microbial Activity and Animal Performance: Enhancing Livestock Health and Productivity—Invited Review

Methun C. Dey [1], Gauri Jairath [2], Ishaya U. Gadzama [3,*], Susana P. Alves [4] and Eric N. Ponnampalam [5,*]

1 Institute of Livestock Science and Technology, Netrokona 2400, Bangladesh
2 Livestock Products Technology, ICAR-Indian Veterinary Research Institute, Regional Station, Palampur 176061, India
3 School of Animal and Veterinary Sciences, University of Adelaide, Roseworthy, SA 5371, Australia
4 CIISA—Centre for Interdisciplinary Research in Animal Health, Faculty of Veterinary Medicine, University of Lisbon, 1649-004 Lisboa, Portugal
5 School of Agriculture, Food and Ecosystem Sciences, Faculty of Science, The University of Melbourne, Parkville, VIC 3010, Australia
* Correspondence: ishaya.gadzama@adelaide.edu.au (I.U.G.); eric.ponnampalam@unimelb.edu.au (E.N.P.)

Simple Summary: Ruminant animals reared under extensive systems spend a substantial amount of time grazing and selecting vegetative parts, which may be considered less effective in terms of feed efficiency, nutrient partition, milk and meat production. Within this context, total mixed ration (TMR) is a feeding method where all ingredients, such as hay, grains, and supplements, are blended into a single, balanced meal for livestock. This approach provides animals with consistent nutrition, enhances digestion, and boosts productivity, resulting in higher milk yield and improved growth, when compared with rearing animals under extensive systems. Studies show TMR increases feed efficiency, supports healthier rumen function, and reduces food waste by preventing animals from picking only their favorite parts. However, challenges include the risk of digestive issues if the mix is not balanced and potential contamination from mold toxins in poor-quality feed associated with feed formulation and storage. Innovations like fermented TMR can further enhance nutrient absorption and reduce methane emissions, benefiting both farms and the environment. Overall, TMR offers a sustainable approach to enhancing livestock health and animal productivity, but careful management is crucial to avoiding potential pitfalls. Future research should focus on optimizing recipes and reducing environmental impacts.

Abstract: Feeding a balanced diet such as total mixed ration (TMR) is a widely adopted feeding strategy providing a uniformly blended diet of roughages, concentrates, and supplements that enhances ruminant productivity by optimizing nutrient utilization, stabilizing rumen fermentation, and improving microbial activity. Scientific studies have confirmed that TMR increases dry matter intake (DMI), milk yield, and growth performance in dairy and beef cattle, as well as in sheep and goats. TMR's advantages include consistent feed quality, reduced selective feeding, and improved feed efficiency. A key benefit of TMR is its ability to promote the production of volatile fatty acids (VFAs), which are the primary energy source for ruminants, particularly propionate. This enhances energy metabolism, resulting in higher carcass yields, increased milk production, and economic benefits compared to conventional or supplementary feeding systems. However, TMR feeding is also susceptible to mycotoxin contamination (e.g., aflatoxins, zearalenone), potential effects on methane emissions, and the need for precise formulation to maintain consistency and optimise profitability. Prevention and good practices, including routine inspection of feed for pathogens and vulnerable ingredients, as well as careful management of particle size and forage-to-concentrate ratios, are crucial in preventing subacute ruminal acidosis (SARA) and the development of other subclinical diseases. Mycotoxin binders, such as hydrated sodium calcium aluminosilicate, can also reduce mycotoxin absorption. Another advantage of practicing TMR is that it can support sustainable farming by integrating agro-industrial byproducts, which minimises environmental impact. In conclusion, TMR is a widely adopted feeding strategy that significantly enhances ruminant productivity by optimizing nutrient utilization, stabilizing rumen fermentation, and improving microbial activity, leading to increased dry matter intake, milk yield, and growth performance. It offers key benefits such as consistent feed quality, reduced selective feeding, improved feed efficiency, and enhanced energy metabolism, providing economic advantages and supporting sustainable farming through agro-industrial byproduct integration. However, its implementation requires careful management to mitigate risks, including mycotoxin contamination, potential impacts on methane emissions, and digestive issues like SARA if formulation is not precise. Therefore, for sustainable production, future research should focus on optimizing TMR formulations with alternative ingredients (e.g., agro-industrial byproducts) and precision feeding strategies to enhance livestock health and animal productivity while minimizing environmental impacts.

Keywords: agro-industrial byproducts; feed additives; forage-concentrate; nutrient digestibility; growth performance; milk production; carcass yield; meat quality; mycotoxin; methane emissions; sustainability

1. Introduction

The rapidly expanding global population necessitates critical advancements in sustainable livestock production to meet the rising demand for animal proteins [1,2]. Optimal nutrition is paramount for achieving high performance, improved animal health, and economic viability in modern ruminant farming for species such as dairy cows, beef cattle, buffaloes, goats, and sheep [3,4]. In this context, the formulation of balanced diets, specifically the Total Mixed Ration (TMR), has become a cornerstone of modern ruminant nutrition. TMR is a scientifically formulated feeding strategy that combines all dietary components—including forages, concentrates, vitamins, and minerals—into a single, uniform mix [3,5,6]. This approach is designed to ensure the consumed diet is nutritionally complete, minimize selective feeding, and provide numerous benefits for herd health and productivity [7–10]. The widespread adoption of TMR is driven by its significant benefits. It provides a consistent nutrient supply that stabilizes rumen pH and promotes efficient microbial activity, which is crucial for enhanced nutrient digestibility and feed efficiency [8,11,12]. Studies consistently demonstrate that TMR leads to maximized production outcomes, including higher milk yield and improved milk components (e.g., fat, protein) in dairy cows [13–16], as well as improved average daily gain (ADG), body weight, and marbling scores in beef cattle [17–21] (Figure 1). Furthermore, TMR systems can reduce feed costs and support sustainable practices by facilitating the efficient use of locally available agricultural by-products while minimizing feed wastage [22]. However, the implementation of TMR is accompanied by significant complexities and challenges that necessitate continuous research and refinement. High-concentration formulations, particularly in pelleted forms, can disrupt rumen pH and increase the risk of subacute ruminal acidosis (SARA) [17]. Animals may still sort feed based on particle size or palatability, undermining nutritional consistency and leading to digestive issues and reduced performance [23–25]. Susceptibility to aerobic deterioration and mycotoxin contamination from ingredients like silage also poses substantial animal health risks [25]. Additionally, the environmental footprint of TMR, particularly concerning methane emissions, varies considerably with diet composition and digestibility [24–26]. These challenges have spurred innovations aimed at optimizing TMR systems. Variations like Ensiled TMR (ETMR) enhance nutrient preservation [23,24], while Fermented TMR (FTMR) has shown promise in improving nutrient digestibility and rumen fermentation, especially when incorporating agricultural by-products [27]. The use of additives, such as yeast cultures or lactic acid bacteria with high antioxidant activity, is also being explored to enhance fibre breakdown and rumen stability, and mitigate nutrient degradation [28]. Given the complex interplay of formulation, animal physiology, economic viability, and environmental impact, a comprehensive synthesis of existing research is essential. The extensive, and sometimes conflicting, findings across diverse ruminant species and production systems require critical evaluation. Therefore, this review aims to consolidate the current understanding of TMR, focusing on its formulation, benefits, inherent challenges, and recent innovations. By identifying optimal strategies and research gaps, this work provides practical insights for farmers and researchers, guiding future efforts toward more efficient, sustainable, and profitable ruminant production systems. This consolidated knowledge is fundamental for translating scientific insights into practical applications that enhance on-farm success and contribute to global food security.

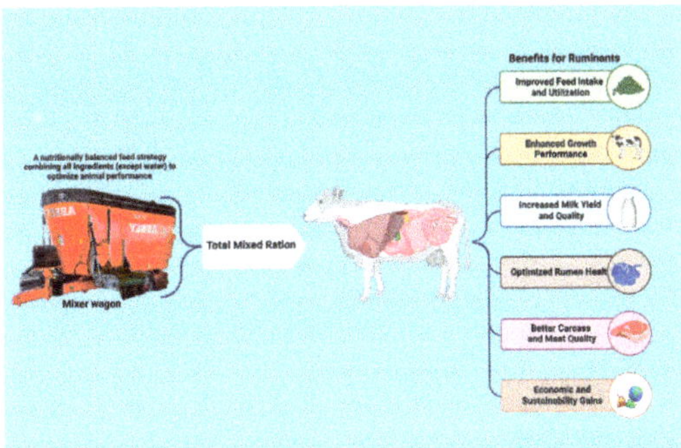

Figure 1. A total mixed ration represents a nutritionally balanced feeding strategy blended using a mixer wagon, offering numerous benefits, including increased dry matter intake, higher milk yield, and improved growth performance. It also stabilizes the rumen environment, reducing the risk of acidosis, and enhances nutrient digestibility, leading to better animal health.

2. Methodology

This review consolidates current knowledge on the formulation, benefits, and challenges of Total Mixed Ration in ruminant nutrition, with a focus on its impact on productivity, rumen health, and environmental sustainability. To ensure a comprehensive assessment, the literature search was conducted across multiple academic databases, including Web of Science, PubMed, Google Scholar, and Scopus, using targeted search terms such as "Total Mixed Ration AND ruminant performance," "TMR AND rumen fermentation," "fermented TMR AND meat quality," and "TMR formulation AND feed efficiency." The inclusion criteria prioritized peer-reviewed studies, in vivo studies, and selected in vitro studies, while excluding non-ruminant research and non-scientific sources.

The extracted data encompassed TMR composition (e.g., forage-to-concentrate ratios, inclusion of wet brewers' grains, yeast cultures, and fermented additives), processing methods (ensiling techniques), and their effects on feed intake, growth performance, milk yield, rumen fermentation parameters (volatile fatty acid profiles, pH stability, microbial diversity), and methane mitigation, as well as economic and environmental considerations. Given the narrative structure of this review, no formal statistical analysis was conducted. Instead, findings were synthesized to provide a balanced interpretation of TMR's role in modern ruminant production systems, emphasizing practical implications for farmers and researchers. The discussion integrates mechanistic insights, such as how fiber digestibility is influenced by particle size or how microbial additives improve aerobic stability, to clarify the functional basis of TMR's benefits and limitations. References were selected based on relevance, scientific rigor, and applicability across dairy and beef cattle, sheep, and goats, ensuring a representative overview of global TMR practices.

3. Total Mixed Ration: Composition, Advantages, and Fermentation Technologies

The Total Mixed Ration is a cornerstone of modern ruminant production, designed to deliver a balanced intake of nutrients by homogeneously blending forages, concentrates, and supplements [18,25,29]. This method prevents selective feeding, stabilizes rumen fermentation, and enhances overall feed efficiency, making it particularly beneficial for high-producing dairy cattle and intensively reared beef and sheep [18,30].

3.1. Typical Composition and Formulation

The formulation of TMR is tailored to the animal's species, production stage, and available feed resources. A typical TMR consists of forages (50–70%) and concentrates (30–50%) on a dry matter basis, with minor inclusions (1–2%) of additives such as yeast or molasses to enhance palatability and digestion [14,31,32] (Figure 2). Dairy cows often receive a forage-to-concentrate ratio between 50:50 and 60:40, while beef cattle and growing lambs may be fed higher forage proportions [33].

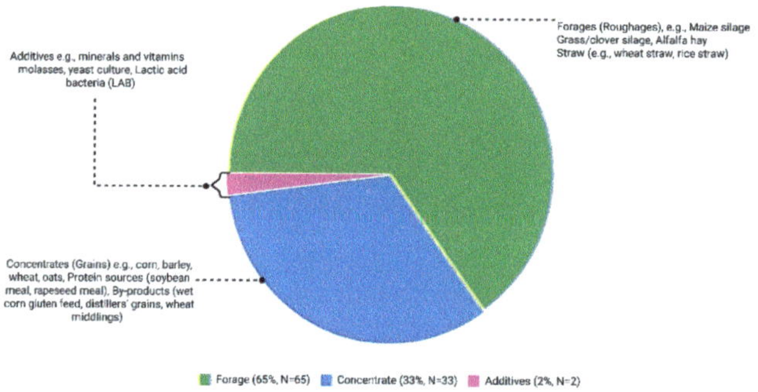

Figure 2. The typical composition of TMR for ruminants includes a balanced mix of forage, concentrates, and additives, tailored to meet the specific nutritional needs of the animals. Note that the ratios may vary slightly from species to species, depending on the selection and availability of forages and concentrates, which affect their economic viability.

3.2. Forage Components

Common forages include corn silage, alfalfa hay, and grass/clover silage, which provide essential structural fiber to promote rumination and saliva production for rumen pH buffering [34,35]. In tropical regions, grasses like Napier (*Pennisetum purpureum*) and *Megathyrsus maximus* are frequently used [35,36].

3.3. Concentrate Components

Concentrates typically include energy sources, such as corn meals, and protein supplements, like soybean meals, often supplemented with minerals and vitamins [37]. To improve economic viability, by-products like wet corn gluten feed (WCGF) and distillers' grains are increasingly incorporated into formulations [21]. Processing methods are critical to TMR effectiveness. Pelleting can improve starch digestibility, while fermentation technologies (e.g., Fermented TMR) enhance preservation and aerobic stability [26,38].

3.4. Key Advantages and Innovations

The primary advantages of TMR include enhanced rumen function, improved nutrient utilization, optimized production performance, and cost-effective sustainability (Figure 3). By preventing selective consumption, TMR ensures a consistent nutrient supply, which stabilizes rumen pH and promotes efficient microbial activity. This leads to higher average daily gain (ADG), improved milk yield (with reported increases of 3–5%), and enhanced feed conversion efficiency [18,30]. Furthermore, TMR minimizes feed wastage and allows for the precision formulation of nutrients using locally available by-products, supporting sustainable farming practices [21]. Innovations continue to optimize the TMR system. Ensiled TMR (ETMR) and Fermented TMR (FTMR) leverage preservation technologies to extend shelf-life, improve nutrient digestibility, and enhance aerobic stability [39–42]. These approaches create an anaerobic environment that promotes lactic acid fermentation, thereby preserving nutrients and inhibiting the growth of spoilage microorganisms [30,42]. The fermentation process can break down complex fibers and proteins, improving volatile fatty acid (VFA) production and increasing energy availability for the animal [42–44]. This not only supports higher dry matter intake and better growth performance but may also reduce methane emissions by shifting rumen fermentation toward propionate production [30,44]. Compact TMR (CTMR) is another innovation aimed at reducing labor costs and improving feed utilization [26].

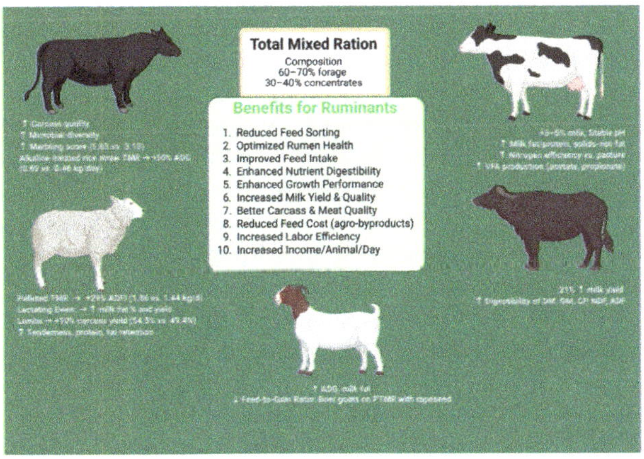

Figure 3. Key Advantages of TMR in Ruminants: Promoting higher feed intake, milk yield, and growth performance, while stabilizing rumen health and improving nutrient digestibility. (ADG: Average Daily Gain; ADF: Acid Detergent Fiber; ADFI: Average Daily Feed Intake; CP: Crude Protein; DM: Dry Matter; NDF: Neutral Detergent Fiber; OM: Organic Matter; VFA: Volatile Fatty Acids; ↑: increase; ↓: decrease).

3.5. *Fermentation Quality and Implications*

The fermentation quality of silage components is a critical determinant of TMR's nutritional value and stability. Successful ensiling, driven by lactic acid bacteria (LAB), lowers the pH below 4.2 and results in a high lactic acid content with minimal undesirable byproducts, such as butyric acid or ammonia-N [28,42–51]. High-quality fermentation enhances palatability, supports a stable rumen environment by maintaining optimal pH, and promotes the growth of cellulolytic bacteria essential for fiber degradation [24,45]. The strategic use of microbial inoculants (e.g., *Lactobacillus acidophilus*, *Saccharomyces cerevisiae*) can significantly improve fermentation efficiency by accelerating acid production and suppressing undesirable microbes [43,52]. Additives like molasses provide substrates for LAB, while organic acids (e.g., propionic acid) improve aerobic stability, which is a particular challenge in hot and humid climates [42,53]. Poor fermentation quality has direct negative implications, potentially leading to reduced feed intake, impaired rumen function, and metabolic disorders such as subacute ruminal acidosis (SARA) [17,42]. Conversely, well-fermented TMR silage promotes efficient nutrient utilization, better growth rates, and improved overall health and productivity [45]. Therefore, careful management of factors like compaction, sealing, storage conditions, and particle size is essential to maximize the benefits of TMR systems [26,39–41].

4. Effects of Total Mixed Ration on Ruminant Diets

4.1. *Effects of Total Mixed Ration on Rumen Fermentation*

The composition of TMR plays a crucial role in determining rumen fermentation dynamics, which affect pH stability, VFA production, and microbial diversity (Table 1). One of the most critical aspects of TMR is its influence on rumen pH stability. A well-balanced TMR improves nutrient utilization, enhances animal performance, and reduces metabolic disorders, whereas a poorly formulated TMR can lead to acidosis and reduces feed efficiency. Research indicates that TMR formulations with balanced roughage-to-concentrate ratios help maintain optimal rumen pH (6.0–6.5), reducing the risk of SARA [26]. However, high-concentrate TMR diets (exceeding 60% concentrate) can lead to excessive production of lactic acid, lowering pH and impairing fiber-digesting microbes [17]. This is particularly evident in pelleted TMR, where smaller particle sizes accelerate fermentation, further exacerbating pH fluctuations [54]. Conversely, TMR containing adequate roughage, such as maize silage or alfalfa hay, promotes saliva production during chewing, which buffers rumen acidity and stabilizes pH [45].

The production of VFAs is another key indicator of rumen fermentation efficiency. TMR formulations rich in fermentable carbohydrates, such as grains and molasses, enhance propionate and butyrate production, which are crucial for energy metabolism in ruminants [55]. However, excessive concentrate inclusion can shift VFA profiles toward higher propionate at the expense of acetate, reducing fiber digestibility [26]. FTMR has been shown to improve VFA production by promoting microbial activity through the inclusion of probiotics like *Lactobacillus acidophilus* and *Saccharomyces cerevisiae* [42]. Moreover, yeast-supplemented TMR enhances fiber degradation, leading to a more balanced acetate-to-propionate ratio, which supports both energy supply and rumen health [24].

Table 1. Effects of Total Mixed Ration on Rumen Fermentation (pH, VFA, Microbial Diversity).

Species/Breed	TMR Type/Modification	Summary of Results	Reference
Holstein calves	WPCS-based TMR (CTMR)	Higher rumen pH, total VFA, and propionate vs. CSCS (15% WPCS)	[56]

Table 1. *Cont.*

Species/Breed	TMR Type/Modification	Summary of Results	Reference
Crossbred lambs	Pelleted TMR (PTMR)	Higher acetate (49.8 vs. 45.7 mmol/L), propionate (24.8 vs. 21.4 mmol/L)	[48]
Hu sheep lambs	High-grain pelleted TMR	Lower pH, Increased lactate, Reduced Fibrobacteres	[26]
Mixed ruminal microbes	TMR + ryegrass pasture	↑ Butyrate/valerate, ↓ Methane, ↑ Microbial biomass N	[57]
Holstein dairy cows	TMR + FF	Stable pH/VFA; ↓ N-NH_3 in 50% FF	[58]
Red Chittagong Cows	Maize stover-based TMR	↑ TVFA/NH_3-N; stable rumen pH	[16]
Dairy cows (in vitro)	Varied TMR compositions	CP fermentation: 25–60%; CPM synthesis: 677–1778 mg/day	[59]
Red Chittagong cows	Maize stover-based TMR	↑ TVFA and NH3-N	[16]
Buffalo (in vitro)	HFA-supplemented TMR	↑ VFAs, ↓ methane (Shatavari @ 3% most effective)	[60]
Dairy cows (in vitro)	100% TMR (69:31 forage/concentrate)	↑ Total VFA (+16.6 mmol/L), ↑ Acetate/propionate ratio	[33]
German Holstein cows	Pasture transition (from TMR to pasture)	Lower pH in the pasture group (SARA risk in wk 9–10). No adverse LPS effects.	[11]
Holstein-Zebu steers	Fermented TMR (FTMR; pH 3.5)	Stable ruminal pH despite low TMR pH. No acidosis observed.	[61]
Angus beef cattle	High-concentrate TMR	Significantly decreased rumen pH	[18]
Wethers (Sheep)	Pelleted TMR + 1% yeast culture	Higher mean pH, reduced time below pH 5.8.	[24]
Simmental bulls	Multi-silage TMR (MS)	Improved rumen pH stability linked to higher VFA production.	[5]
German Holstein cows	Pasture transition	↓ Acetate (C2%), ↑ butyrate (C4%), ↓ C2/C3 ratio	[11]
Cattle (Bulls)	60:40 Roughage/Concentrate	Highest TVFA production (optimal fermentation)	[62]
Holstein-Zebu steers	FTMR (pH 3.5)	↑ Acetic/butyric acid; ↓ propionate.	[61]
Crossbred lambs	Pelleted TMR (PTMR)	↑ Acetate (49.8 vs. 45.7 mmol/L) and propionate (24.8 vs. 21.4 mmol/L).	[48]
Wethers (Sheep)	Pelleted TMR + 1% yeast	↑ Total VFA, propionate, and n-butyrate.	[24]
Simmental bulls	Multi-silage TMR (MS)	↑ TVFA (62.49 vs. 56.09 mmol/L) and acetate.	[63]
German Holstein cows	Pasture transition	↓ Rumen papillae surface area (recovered by wk 10).	[11]
Angus beef cattle	High-concentrate TMR	↑ Starch-degraders (Bacteroidota), ↓ fiber-degraders (Ruminococcus).	[18]
Crossbred lambs	Pelleted TMR (PTMR)	↑ Prevotellaceae (rumen), ↓ Ruminococcaceae.	[48]
Holstein cows	Fermented TMR (FTMR)	↑ Unclassified_Bacteroidales, ↓ Candida (fungi).	[64]
Wethers (Sheep)	Pelleted TMR + 1% yeast	↑ Fibrolytic bacteria (NK4A214, FD2005).	[24]
Yellow cattle (in vitro)	Fermented TMR (FTMR)	↓ Methane production, ↓ Methanobrevibacter abundance.	[46]
Holstein cows	Fermented TMR (FTMR)	↑ Methanobrevibacter (due to higher H_2 availability).	[64]

Table 1. *Cont.*

Species/Breed	TMR Type/Modification	Summary of Results	Reference
Holstein (Dairy)	Pelleted TMR	Lower rumen pH (6.10 vs. 6.48), higher propionate	[65]
Cattle (Bulls)	60:40 Roughage/Concentrate	Optimal TVFA, stable pH	[62]
Hanwoo Heifers	Italian Ryegrass TMR	Increased propionate, enriched Ruminococcus bromii	[66]
Holstein dairy cows	Grass silage + concentrate + hay	Higher early GP (2–4 h) with particle-associated inocula (PAL). Declined later.	[67]
Brown Swiss cows	TMR + Saccharomyces cerevisiae (CE/LC)	CE improved asymptotic GP more than LC; low/intermediate doses are most effective.	[68]
Suffolk sheep	Fermented TMR (FTMR)	Higher GP due to enhanced microbial activity from lactic acid fermentation.	[44]
Hanwoo steers	TMR with fermented feed (TMRF)	Improved GP linked to higher acetate/propionate production.	[13]
Holstein cows	TMR with varying particle sizes (5.5–25 mm)	Smaller particles ↑ SCFA in dorsal rumen; ↓ acetate/propionate ratio.	[69]
Suffolk sheep	Fermented TMR (FTMR)	↑ Propionate (392.4 mmol/mol), ↓ butyrate (86.6 mmol/mol).	[30]
Nellore bulls	TMR with pefNDF	Optimal SCFA at 20.5 g pefNDF/kg DM; higher fiber ↑ butyrate.	[70]
Holstein steers	TMR vs. separate feeding	Higher total VFA and propionate at 1.5 h post-feeding.	[9]
Holstein cows	TMR (5.5–25 mm particle size)	5.5 mm: ↓ rumen pH; 11 mm: maintained pH and ↑ protozoa.	[69]
Montbéliarde cattle	TMR (high concentrate)	Lower rumen pH (5.58 vs. 5.87 in control), higher acidosis risk.	[17]
Suffolk sheep	FTMR in varying pH media	Lower pH (5.62–5.66) ↓ CH_4 and ↑ propionate.	[44]
Nellore steers	High-concentrate TMR	Faster microbial adaptation in preconditioned cattle.	[71]

TMR = total mixed ration, WPCS = whole plant corn silage, CTMR = corn silage-based TMR, CSCS = corn stover and concentrate supplement, PTMR = pelleted TMR, FF = fresh forage, CP = crude protein, CPM = microbial protein, HFA = herbal feed additive, FTMR = fermented TMR, MS = multi-silage, GP = gas production, SCFA = short-chain fatty acids, pefNDF = physically effective neutral detergent fiber, DM = dry matter, PAL = particle-associated inocula, CE = commercial extract, LC = live culture, TMRF = TMR with fermented feed. ↑ = increase, ↓ = decrease, pH = measure of acidity, VFA = volatile fatty acids, mmol/L = millimoles per liter, N-NH_3 = ammonia nitrogen, NH_3-N = ammonia nitrogen, TVFA = total volatile fatty acids, LPS = lipopolysaccharides, SARA = subacute ruminal acidosis, C2 = acetate, C3 = propionate, C4 = butyrate, H_2 = hydrogen, CH_4 = methane, g/kg = grams per kilogram, h = hour, wk = week.

Microbial diversity in the rumen is significantly influenced by TMR composition. A well-formulated TMR supports a diverse microbial population, including cellulolytic bacteria (*Fibrobacter succinogenes*, *Ruminococcus flavefaciens*) and amylolytic species (*Streptococcus bovis*) [43]. However, high-grain TMR can reduce microbial diversity by favoring lactate-producing bacteria (*Megasphaera elsdenii*) over fiber degraders, leading to dysbiosis [17]. Fermented TMR, particularly that incorporating lactic acid bacteria, has been found to enhance microbial stability by suppressing pathogenic bacteria and promoting beneficial microbes [52]. Additionally, the inclusion of wet brewers' grain (WBG) as a roughage substitute in TMR has been reported to maintain microbial balance while improving nutrient utilization [64].

The implications of these findings are significant for both animal performance and feed efficiency. Stable rumen pH ensures optimal digestion and minimizes metabolic disorders, while balanced VFA production maximizes energy availability for growth and milk production (Figure 4). Furthermore, maintaining microbial diversity enhances feed efficiency and reduces methane emissions, contributing to more sustainable livestock production [55]. However, improper TMR formulation, such as excessive concentrate or insufficient fiber, can lead to digestive disturbances, reduced feed intake, and lower productivity [26].

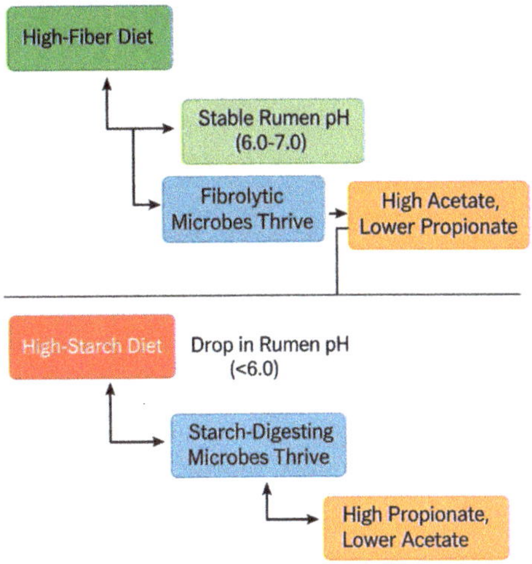

Figure 4. Summary of the complex relationship between diet, ruminal pH, microbes and VFA production.

4.2. Effects of Total Mixed Ration on Nutrient Digestibility

The adoption of TMR in ruminant feeding systems significantly improves the digestibility of nutrients—such as dry matter, organic matter, crude protein, neutral detergent fiber, and acid detergent fiber—by ensuring a balanced and consistent diet (Table 2). Proper formulation, including optimal roughage-to-concentrate ratios and strategic use of additives, enhances rumen function and overall animal performance. This approach not only boosts productivity but also supports sustainable livestock farming by improving feed efficiency and reducing environmental impacts. Several studies have demonstrated that TMR enhances the digestibility of all nutrients compared to conventional feeding systems. For instance, Arbaoui et al. [17] found that TMR formulations with optimized concentrate-to-roughage ratios significantly improved DM and OM digestibility by ensuring a consistent nutrient supply, which stabilizes rumen fermentation. This is primarily because a finely balanced TMR prevents selective feeding, ensuring that animals consume all dietary components in the intended proportions [54]. Moreover, the inclusion of high-quality roughages such as alfalfa hay and maize silage in TMR has been shown to enhance fiber digestibility (NDF and ADF) by promoting rumen microbial activity [45]. Hamidan et al. [53] reported that diets with adequate NDF content support rumen health by maintaining proper rumen motility and buffering capacity, which in turn improves fiber degradation. However, excessive inclusion of roughage may reduce digestibility due to the increased lignin content, particularly in low-quality forages [70]. Bo Trabi et al. [26] observed that smaller forage particle sizes in TMR increase passage rates but may reduce fiber digestibility due to insufficient rumen retention time for microbial degradation.

Table 2. Effects of Total Mixed Ration on Nutrient Digestibility (DM, OM, CP, NDF, ADF).

Species/Breed	TMR Type/Modification	Summary of Results	Reference
Holstein calves	WPCS-based TMR (CTMR)	Lower in vitro DM/CP/NDF digestibility vs. starter (CONS)	[56]
Karakul sheep	40% SS-AF silage TMR	Higher DM, CP, and NDF digestibility	[5]
Fattening lambs	Pelleted TMR + LY	Increased DM (38 g/kg), OM (41 g/kg), and NDF (193 g/kg) digestibility	[72]
Comisana lambs	WM-based TMR	Higher aNDF/ADF digestibility No difference in DM/OM/CP	[73]
Red Chittagong cows	Maize stover-based TMR	↑ DM/CP/NDF digestibility	[16]
Dorper lambs	LB-inoculated PH-TMR	Higher DM intake and nutrient digestibility vs. untreated PH-TMR	[74]
Dairy ewes	Wheat middling-based TMR	↑ NDF digestibility	[73]
Dorper lambs	Cactus pear + cottonseed cake	↑ DMD, OMD, EED in 20–30% cottonseed TMR ↓ Rumination time	[7]
Crossbred cows	FTMR with 25% peNDF	Improved nutrient digestibility (CP, NDF, ADF)	[75]
Red Chittagong Cows	Maize stover-based TMR	↑ Digestibility of DM, CP, OM	[16]
Korean native goats	TMR with varying peNDF	No difference in DM, CF, or other nutrient digestibility	[76]
Dairy cows (in vitro)	16 TMR formulations	OM fermentation: 35–47%; NDF fermentation: 3–28%	[59]
Buffaloes	TMR vs. conventional	↑ DM, OM, NDF digestibility	[77]
Holstein dairy cows	TMR + FF	↑ Nitrogen efficiency in 50% FF ↓ urinary N excretion	[58]
Korean native goats	TMR with varying peNDF	No difference in Nitrogen balance	[76]
Holstein-Zebu steers	FTMR (pH 3.5)	↑ Crude protein digestibility ↓ fat digestibility in silage-TMR.	[61]
Angus beef cattle	High-concentrate TMR	↓ DM, CP, and NDF digestibility	[18]
Yellow cattle (in vitro)	Fermented TMR (FTMR)	↓ NDF/ADF; ↑ lactic acid and soluble carbohydrates.	[46]
Simmental bulls	Multi-silage TMR (MS)	Improved fiber degradation linked to ↑ Prevotella-1.	[63]
Murrah buffaloes	TMR (maize silage/concentrate ratios)	Highest DMI (14.35 kg/d) at 50:50 ratio; ME content ↓ with ↑ silage.	[78]
Hanwoo steers	TMR with fermented feed (TMRF)	Higher DM disappearance (3–12 h) and weight gain (308 kg vs. 284 kg control).	[13]
Suffolk sheep	Fermented TMR (FTMR)	↑ CP, EE, ADF, and GE digestibility vs. non-fermented TMR.	[30]
Montbéliarde cattle	TMR (90% concentrate)	No difference in DMI or digestibility vs. separate feeding.	[17]
Nellore bulls	TMR with pefNDF	Optimal fiber digestion at 20.5 g pefNDF/kg DM.	[70]

TMR = total mixed ration, DM = dry matter, OM = organic matter, CP = crude protein, NDF = neutral detergent fiber, ADF = acid detergent fiber, ME= metabolizable energy, WPCS = whole-plant corn silage, CTMR = corn silage-based TMR, CONS = conventional starter, SS-AF = sweet sorghum-alfalfa, LY = live yeast, WM = wheat middlings, aNDF = amylase-treated NDF, LB = lactic acid bacteria, PH-TMR = pineapple husk TMR, DMD = dry matter digestibility, OMD = organic matter digestibility, EED = energy efficiency of digestion; g/kg = grams per kilogram, ↑ = increased, ↓ = decreased.

Crude protein digestibility is also positively influenced by TMR, particularly when protein sources are well-balanced between degradable and undegradable fractions. Lakhani & Tyagi [78] reported that TMR containing 11–14% CP optimizes microbial protein synthesis while minimizing nitrogen wastage. Additionally, FTMR containing microbial additives, such as *Lactobacillus acidophilus* and *Saccharomyces cerevisiae* further enhances CP digestibility by promoting proteolytic activity in the rumen [42]. Wang et al. [24] also demonstrated that yeast supplementation in TMR improves fiber digestibility by stimulating cellulolytic bacteria, leading to better NDF and ADF breakdown. The implications of these findings are significant for both animal productivity and feed efficiency. Improved DM and OM digestibility indicate better energy utilization, which translates to higher milk yield in dairy cows and improved weight gain in beef cattle [79]. Enhanced fiber digestibility ensures optimal rumen function, reducing the risk of metabolic disorders such as SARA. Furthermore, the use of FTMR and additives like wet brewers' grains [38] or yeast culture [24] offers a sustainable approach to enhancing nutrient utilization while reducing feed costs. Farmers and nutritionists should focus on ingredient quality, forage particle size, and fermentation techniques to maximize digestibility while minimizing metabolic risks.

4.3. Effects of Total Mixed Ration on Growth Performance

Total Mixed Ration represents a significant advancement in ruminant nutrition, integrating forage, concentrates, and additives into a homogeneous mixture that enhances growth performance metrics, including BW, ADG, DMI, and feed efficiency (Table 3). Early research by Lailer et al. [80] established that TMR ensures uniform nutrient delivery, minimizing selective feeding and stabilizing rumen fermentation. This consistency prevents fluctuations in ruminal pH and VFAs production, both of which are critical for efficient nutrient utilization. For instance, Liu et al. [45] reported that lambs fed TMR with optimal oat-to-alfalfa ratios exhibited higher DMI and ADG due to improved palatability and balanced energy-to-protein ratios. Similarly, Nguyen et al. [14] observed increased BW in cattle fed alkaline-treated rice straw-based TMR, attributing it to enhanced digestibility from reduced lignin content, which facilitated microbial access to cellulose.

The scientific basis for improved ADG lies in TMR's ability to synchronize carbohydrate and protein degradation in the rumen. Wang et al. [5] found that lambs fed TMR with 40% sweet sorghum and 60% alfalfa silage achieved the highest ADG, as the blend provided readily fermentable carbohydrates alongside bypass protein, promoting microbial protein synthesis. Furthermore, Sun et al. [72] demonstrated that pelleting TMR with live yeast increased ADG in lambs by 11%, as yeast metabolites enhanced fibrolytic bacteria activity, boosting neutral detergent fiber digestibility. This aligns with Bo Trabi et al. [26], who noted that smaller forage particles in high-grain TMR accelerated ruminal turnover but cautioned that excessive starch could reduce fiber digestibility and increase the risk of acidosis.

Improvements in dry matter intake under TMR systems are often linked to the physical and chemical properties of the feed. Koch et al. [81] observed that dry TMR formulations incorporating by-products like straw maintained DMI comparable to conventional silage-based rations, as the mix's density and moisture content reduced sorting behavior. Ferrari et al. [21] further noted that replacing modified distillers' grains with dry-rolled corn in beef steer TMR linearly decreased DMI, likely due to lower palatability and energy density. However, wet brewers' grains in TMR countered this by providing soluble fibers that stimulated rumination and saliva production, buffering rumen pH and supporting consistent intake [38].

Feed efficiency improvements stem from TMR's optimization of nutrient absorption. Tufarelli et al. [73] reported a 15% better FCR in lambs fed wheat middling-based TMR, as the diet's balanced degradable protein and energy reduced nitrogen wastage and directed nutrients toward muscle deposition. Additionally, fermented TMR with lactic acid bacteria, as studied by Chen et al. [28], increased propionate production through enhanced carbohydrate fermentation, thereby supplying more glucogenic precursors for growth. This was corroborated by Zhang et al. [56] in dairy calves, where TMR enriched cellulolytic bacteria like Rikenellaceae, improving fiber degradation and metabolic energy availability. Challenges persist, however, high-grain pelleted TMR can lower rumen pH and increase lactate production, predisposing animals to subacute acidosis [26]. Martins et al. [25] also highlighted the risks of mycotoxin contamination in TMR, particularly from maize silage, which can suppress feed intake and impair hepatic function. Nonetheless, innovations like yeast supplementation [24] and moisture-stabilizing additives [28] mitigate these issues by enhancing aerobic stability and detoxification. Total Mixed Ration significantly enhances ruminant growth performance by ensuring balanced nutrient intake, stabilizing rumen function, and improving feed efficiency. Its adaptability to incorporate agricultural by-products reduces costs while maintaining productivity, making it indispensable for sustainable livestock farming. Farmers should prioritize consistent mixing, moisture control and mycotoxin management to maximize benefits, as these factors directly influence animal health and economic returns.

Table 3. Effects of Total Mixed Ration on Growth Performance (BW, ADG, DMI, Feed Efficiency).

Species/Breed	TMR Type/Modification	Summary of Results	Reference
Holstein dairy calves	WPCS-based TMR (CTMR)	No difference in BW, ADG, or feed efficiency vs. starter (CONS)	[56]
Karakul sheep	40% SS-AF silage TMR	Highest BW, ADG, and DMI	[5]
Fattening lambs	Pelleted TMR + LY (0.8 g/kg)	11% higher ADG (+36 g/d)	[72]
Jersey cows	60% grass hay + 40% concentrate	Higher DMI (12.82 vs. 10.55 kg/day; ↓ FCR (1.36 vs. 1.72)	[82]
Red Chittagong cows	Maize stover-based TMR (50:50)	Higher DMI in block form (T1) vs. mash (T2)	[16]
Holstein cows	MS or IRS TMR + grazing	Night grazing ↑ grass intake (8.53 vs. 5.65 kg DM/d)	[83]
Finnish Ayrshire cows	Grass silage + concentrate (FF1 vs. FF5)	FF1 ↑ DMI (20.9 vs. 19.9 kg/d) in multiparous cows	[84]
Holstein-Friesian cows	High-starch (27.7% DM) TMR	Starch content ↑ DMI and milk yield	[85]
Holstein dairy cows	100% TMR vs. TMR + fresh forage (FF)	No DMI reduction with ≤29% FF; 8% decrease at 47% FF	[58]
Red Chittagong Cows	Maize stover-based TMR (50:50)	Higher DMI vs. conventional feeding	[16]
Korean native goats	TMR with varying peNDF (grinding speeds)	No DMI differences despite reduced peNDF	[76]
Crossbred lambs	Pelleted TMR (PTMR)	Higher ADG (341 vs. 265 g/d) and ADFI (1.86 vs. 1.44 kg/d)	[48]
Comisana lambs	Wheat middlings (WM)-based TMR	Higher final BW (23.5 vs. 21.9 kg) and daily gain (199 vs. 174 g/d)	[73]
Hu sheep lambs	High-grain pelleted TMR (70% concentrate)	No ADG difference vs. high-grain non-pelleted; lower rumen pH	[26]
Aberdeen Angus cattle	60% grass silage + barley (MC TMR)	Highest carcass gain (967 g/d); Best feed conversion (11.1 kg DM/kg gain)	[34]
Crossbred lambs	FTMR with varying oat/alfalfa ratios	AH-300: Higher DMI, ADG, and total weight gain vs. CK and AH-400	[45]
Dorper lambs	Cactus pear + cottonseed cake (20–30%)	No effect on WG/ADG; all treatments met target ADG (200 g/day)	[7]
Beef steers	Hedge lucerne/leucaena TMR	Higher ADG and FCR vs. control (fresh grass + concentrate)	[86]

Table 3. *Cont.*

Species/Breed	TMR Type/Modification	Summary of Results	Reference
Dairy cattle (fattening)	Whole crop rice TMR	Higher BW and ADG in mid/late fattening stages	[87]
Goats (barn-fed)	TMR vs. mountainous pasture	ADG doubled in the TMR group	[88]
Sheep	Pelleted vs. unpelleted TMR	Higher feed intake and ADG with pelleted TMR	[46]
Crossbred lambs	Pelleted TMR (PTMR)	↑ ADG (341 vs. 265 g/d) and carcass yield (54.5% vs. 49.4%).	[48]
Holstein-Zebu steers	Grass silage-TMR (STMR)	↑ Early-phase ADG; better FCR in FTMR later.	[61]
Simmental bulls	Multi-silage TMR (MS)	↑ ADG (1.56 vs. 1.30 kg/day); ↓ FCR (10.96 vs. 12.36).	[63]
Angus beef cattle	High-concentrate TMR	↑ DMI but no improvement in ADG or feed efficiency.	[18]
Hanwoo Steers	T70 (70:30 forage/concentrate)	Compensatory growth in late fattening stage; slower initial growth	[66]
Hanwoo Steers	Fermented TMR	Higher DMI (7.17 kg) Increased BW (615.20 kg), improved ADG (0.56 kg)	[89]
Yak	High-energy TMR	Higher ADG (0.87 kg/day vs. −0.17 kg/day in grazing)	[6]
Karakul Sheep	40%SS–60%AF silage TMR	Higher BW, ADG, and DMI ($p < 0.05$)	[90]
Naemi Lambs	TMR + alfalfa hay (300 g/3 days)	Increased BW and feed conversion ratio	[10]
Boer Goats	TMR + 7.5% intact rapeseed	Reduced feed-to-gain ratio (improved efficiency)	[42]
Sindhi Crossbred	Alkaline-treated straw TMR	Higher ADG (0.69 vs. 0.46 kg/day) & BW (278.8 vs. 258.2 kg)	[14]
Dorper Lambs	Creep feed (18% CP TMR)	Higher weight gain (22.17 vs. 17.83 kg pre-weaning)	[22]
Hanwoo Steers	TMR + Medicinal Plants (30 g/kg)	Improved ADG & feed efficiency	[27]
Simmental Heifers	TMR vs. Free choice	No difference in ADG	[19]
Holstein (Dairy)	pTMR	Balanced nutrient intake, reduced purchased feeds	[91]
Hanwoo Heifers	Italian Ryegrass TMR	No difference in DMI/FCR; improved nitrogen efficiency	[66]
Dairy cows	Apple pomace TMR	Increased LWG	[15]
Sheep	TMR blocks vs. mash	Higher B:C ratio in TMR blocks	[92]
Hanwoo steers	TMRF (fermented feed)	↑ Weight gain (308 kg vs. 284 kg) and feed efficiency (0.16 vs. 0.12).	[13]
Holstein-Friesian	TMR (maize silage + concentrate)	↑ BW gain (0.54 kg/d vs. loss in control).	[93]

BW = body weight; ADG = average daily gain; DMI = dry matter intake; FCR = feed conversion ratio; TMR = total mixed ration; WPCS = whole plant corn silage; CONS = conventional starter; SS-AF = sunflower silage and alfalfa; LY = live yeast; MS = maize silage; IRS = Italian ryegrass silage; FF = fresh forage; peNDF = physically effective neutral detergent fiber; ADFI = average daily feed intake; PTMR = pelleted total mixed ration; FTMR = fermented total mixed ration; WG = weight gain; STMR = silage-based total mixed ration; CP = crude protein; LWG = live weight gain; B:C ratio = benefit-to-cost ratio. ↑ = increase; ↓ = decrease; vs. = versus; % = percent; g/kg = grams per kilogram; kg/day = kilograms per day; g/d = grams per day; DM = dry matter; $p < 0.05$ = statistically significant difference.

4.4. *Effects of Total Mixed Ration on Milk Yield and Composition*

Total Mixed Ration has been widely adopted in ruminant nutrition due to its ability to provide a balanced diet, ensuring consistent nutrient intake and improving milk production (Table 4). Several studies have demonstrated that TMR enhances milk yield and modifies milk composition, primarily due to its well-balanced formulation of concentrates, roughages, and additives. Sunarso et al. [79] reported that dairy cows fed TMR exhibited higher milk yields compared to those on conventional feeding systems, likely due to the optimized energy and protein balance in TMR. Similarly, Lailer et al. [80] found that TMR improves feed efficiency, leading to increased milk production by maintaining stable rumen fermentation conditions. The nutritional quality of TMR also influences the composition of milk. A higher concentrate proportion in TMR (30–80%) elevates energy density, which directly impacts milk fat and protein content [26]. However, excessive concentrate inclusion may reduce fiber digestibility, negatively affecting milk fat synthesis due to altered rumen pH and volatile fatty acid production [53]. The fatty acid composition of milk can also be influenced by the high proportion of concentrate in TMR, particularly by reducing the levels of health-promoting fatty acids like vaccenic, rumenic, and alpha-linolenic acids, and increasing the formation of *trans*-10 fatty acids associated with milk fat depression in dairy ruminants [94].

Conversely, adequate roughage inclusion, such as maize silage or alfalfa hay, ensures sufficient fiber intake for rumen health, promoting optimal milk composition [45]. The inclusion of fermented components, such as molasses and microbial additives like *Saccharomyces cerevisiae*, further enhances nutrient availability, leading to improved milk yield and quality [24,42].

Moreover, the physical characteristics of TMR, such as particle size, play a crucial role in digestion and milk production. Smaller particle sizes increase feed intake and rumen turnover but may reduce fiber digestibility, potentially lowering milk fat content [54]. On the other hand, proper fermentation of TMR, as seen in FTMR, enhances nutrient preservation and digestibility, leading to better milk yield and composition [52]. The addition of yeast culture in TMR has also been shown to stabilize rumen pH and improve fiber degradation, further supporting milk production [24]. Despite these benefits, challenges such as feed sorting and SARA may arise, particularly with high-grain TMR formulations [17,84]. Proper management, including optimal roughage-to-concentrate ratios and the use of fermentation enhancers, is essential to mitigate these risks. Additionally, environmental factors, such as storage conditions, can affect TMR quality, emphasizing the need for proper ensiling techniques to maintain nutrient integrity [42,61].

The adoption of TMR in dairy farming significantly improves milk yield and composition by ensuring balanced nutrient intake, enhancing rumen function, and optimizing feed efficiency. However, careful formulation and management are necessary to prevent metabolic disorders and maintain consistent milk quality. Farmers should consider incorporating fermented additives and maintaining proper roughage levels to maximize the benefits of TMR while minimizing potential drawbacks. This approach not only enhances productivity but also supports sustainable dairy farming practices.

Table 4. Effects of Total Mixed Ration on Milk Yield and Composition.

Species/Breed	TMR Type/Intervention	Summary of Results	Reference
Lactating Holstein cows	Hay-based TMR (DM-adjusted)	Increased milk yield (26.99 → 27.29 kg/d)	[23]
Jersey cows	60% grass hay + 40% concentrate	↑ Milk yield (9.57 vs. 6.23 kg/day	[82]

Table 4. *Cont.*

Species/Breed	TMR Type/Intervention	Summary of Results	Reference
Red Chittagong cows	Maize stover-based TMR	↑ Milk yield (T1: 3.6 L/d; T2: 3.49 L/d vs. T0: 3.35 L/d)	[16]
Holstein cows	TMR vs. separate feeding	TMR ↑ milk yield (34.4 vs. 32.7 kg/d in R2X/R4X)	[95]
Finnish Ayrshire cows	Once vs. 5× daily feeding	No difference (32.8 kg/d ECM FF1 vs. 32.5 kg/d FF5)	[84]
Jersey cows	60% grass hay + 40% concentrate	No difference in fat %; No difference in protein %	[82]
Red Chittagong cows	Maize stover-based TMR	↑ Fat % in T1/T2	[16]
Holstein cows	TMR vs. separate feeding	Separate feeding ↓ fat % (2.14–2.31% vs. 3.31%)	[95]
Dairy ewes	Wheat middling-based TMR	↑ Fat % and yield	[73]
Holstein (Dairy)	Confinement TMR	Highest yield (10,000 kg/cow)	[91]
Holstein (Dairy)	Pasture + Concentrate (PC)	Lowest yield (7500 kg/cow)	[91]
Holstein (Dairy)	Full TMR (confinement)	38.1 kg/day > pTMR (32.0 kg/day) > PC (28.5 kg/day)	[12]
Danish Black/White	Multi-group TMR	Higher yield at high feed levels vs. single-group TMR	[96]
Dairy cows	Apple pomace TMR	Increased milk yield	[15]
Holstein (Dairy)	Pelleted TMR	Higher milk protein (3.38% vs. 3.16%), lower fat	[65]
Holstein-Friesian	TMR (maize silage + concentrate)	↑ Milk yield (29.5 kg/d vs. 21.1 kg/d)	[93]
Crossbred cows	FTMR with 25% peNDF	↑ Milk fat % due to ↑ acetate	[75]
Aosta Red Pied cows	TMR vs. separate feeding	No difference in protein %	[97]
Holstein cows	TMR + night grazing	↑ PUFA (CLA, VA, ALA) in milk	[98]
Holstein dairy cows	100% TMR vs. TMR + FF	8.5% higher yield in 100% TMR; ↑ UFA in 50% FF	[58]
Red Chittagong Cows	Maize stover-based TMR	Higher milk yield, fat, and SNF vs. control	[16]
Holstein cows	Cracked cottonseed in FTMR	↑ C18:2 (linoleic acid) in milk	[99]
Dairy cows	Pasture vs. TMR (maize silage + concentrates)	TMR: 33% higher milk yield; No difference in fat/SCC	[100]
Buffaloes	Brewers' grain + rice straw TMR	Higher milk yield at 1.2% supplement No difference in composition	[101]
Dairy cows	Apple pomace TMR	Increased milk yield and protein; reduced lactose	[15]

TMR = total mixed ration, DM = dry matter, d = day, L = liter, ECM = energy-corrected milk, FF = feeding frequency, pTMR = partial TMR, PC = pasture + concentrate, FTMR = fermented TMR, peNDF = physically effective neutral detergent fiber, PUFA = polyunsaturated fatty acids, CLA = conjugated linoleic acid, VA = vaccenic acid, ALA = α-linolenic acid, FF = fresh forage, UFA = unsaturated fatty acids, SNF = solids-not-fat, SCC = somatic cell count. R2X/R4X refers to different feeding frequencies or rationing schedules for the "separate feeding" group, ↑ = increased, ↓ = decreased, → = change from baseline, vs. = versus, % = percent, kg = kilogram, g = gram, > = greater than.

4.5. Carcass Traits and Meat Quality in Livestock Fed Total Mixed Ration

The evaluation of carcass traits and meat quality in livestock fed TMR reveals significant insights into the nutritional and physiological impacts of this feeding strategy. Table 5 summarizes data from various studies that demonstrate how TMR formulations influence carcass yield and meat characteristics, with variations observed across species, breeds, and TMR compositions. Khy et al. [86] reported that beef steers fed a hedge lucerne-based TMR showed no difference in dressing percentage but had higher chilled carcass weight compared to controls. This suggests that TMR enhances muscle deposition without altering fat distribution, likely due to balanced nutrient intake. Similarly, Huuskonen et al. [34] found that Aberdeen Angus cattle fed a TMR with 60% grass silage achieved superior carcass gain (967 g/d) and conformation, attributed to the optimal fiber-to-energy ratio promoting lean tissue growth.

Marbling, a critical determinant of meat quality [102], was significantly improved in beef cattle fed whole crop rice-based TMR [87]. The higher marbling score indicates enhanced intramuscular fat deposition, which is linked to improved juiciness and flavor [1,103]. However, the absence of significant changes in rib eye area suggests that TMR primarily influences fat metabolism rather than muscle hypertrophy. This is corroborated by Ku et al. [66], who observed that Hanwoo steers fed a 70:30 forage-to-concentrate TMR had the highest intramuscular fat content, alongside favorable shear strength and drip loss. Moreover, the inclusion of fermented components in TMR, such as molasses and *Saccharomyces cerevisiae*, has been shown to positively influence meat quality by improving tenderness and reducing shear force [42]. This is attributed to enhanced protein metabolism and reduced oxidative stress in muscle tissues, which preserves meat structure during post-mortem aging [4]. Additionally, the use of wet brewers' grains as a partial replacement for traditional roughages contributes to higher IMF content, which is associated with improved juiciness and flavor [1,55]. However, excessive concentrate inclusion in TMR may lead to rapid fat deposition, which, while increasing marbling, could also result in undesirable fatty acid profiles if not adequately balanced with fiber sources [53]. The lower n-6/n-3 fatty acid ratio in this group further highlights the role of forage-rich TMR in promoting healthier lipid profiles, as omega-3 fatty acids are associated with cardiovascular benefits in consumers [55,103].

In small ruminants, Wang et al. [90] demonstrated that Karakul sheep fed a 40% sweet sorghum-alfalfa silage TMR exhibited increased carcass weight and subcutaneous fat thickness, and improved meat quality parameters such as water-holding capacity (WHC) and CP content. The enhanced WHC is particularly important, as it results in better moisture retention during cooking, which contributes to tenderness [4,55]. Furthermore, Zhang et al. [48] observed that crossbred lambs fed pelleted TMR (PTMR) had higher carcass yield (54.5% vs. 49.4%) compared to unpelleted TMR, likely due to improved feed efficiency and nutrient utilization. The scientific basis for these findings lies in the ability of TMR to synchronize nutrient release in the rumen, optimizing microbial fermentation and energy availability. For instance, Liu et al. [6] reported that yaks fed a high-energy TMR exhibited a 106.43% increase in carcass weight and 57.52% dressing percentage, alongside improved tenderness and reduced cooking loss. These outcomes are driven by the TMR's balanced energy-to-protein ratio, which supports muscle hypertrophy and fat deposition while minimizing metabolic waste.

On the other hand, Cooke et al. [29] found that beef heifers fed high-concentrate TMR had higher marbling but also an elevated n-6/n-3 ratio, which is less desirable from a human health perspective [3,55]. This highlights the importance of carefully formulating TMR to strike a balance between meat quality and nutritional value. The study by Horcada et al. [20] on Retinta cattle further emphasized that TMR with maize silage increased PUFA content (18.8% vs. 14.3%) and n-3 PUFA (0.47% vs. 0.35%) compared to a high-concentrate diet, suggesting that forage inclusion can enhance the nutritional quality of meat [1,2]. TMR ensures consistent nutrient intake, reducing selective feeding and metabolic disorders. For example, Iraira et al. [19] found that Simmental heifers fed TMR had no differences in meat tenderness compared to free-choice diets. However, the TMR group exhibited longer rumination times, indicating better rumen health. This aligns with the findings of Alhidary et al. [10], where Naemi lambs fed TMR supplemented with alfalfa hay showed improved meat color and reduced shear force, reflecting enhanced muscle fiber structure.

The impact of TMR on meat quality extends to its oxidative stability and shelf life. FTMR has been reported to enhance the antioxidant capacity of meat due to the presence of beneficial microbial metabolites, such as short-chain fatty acids and bioactive peptides [89,90]. This is particularly evident when FTMR includes *Lactobacillus acidophilus* and *Bacillus subtilis*, which modulate rumen microbiota and reduce lipid oxidation in meat [43]. On the other hand, improper TMR processing, such as inadequate particle size reduction, may lead to inconsistent nutrient absorption, negatively affecting meat texture and color stability [54]. From a practical standpoint, the adoption of TMR in livestock farming ensures consistent growth performance and meat quality; however, it requires careful formulation to avoid metabolic disorders, such as SARA. The inclusion of NDF at optimal levels is essential for maintaining rumen health and preventing excessive fat deposition, which can compromise carcass yield [6,78,86]. For farmers, TMR offers a practical solution to optimize growth and carcass quality while reducing feed wastage. For scientists, the data highlight the importance of ingredient selection and processing (e.g., pelleting, fermentation) in modulating meat quality. For instance, Chen et al. [42] demonstrated that Boer goats fed TMR with 7.5% intact rapeseed had altered fatty acid profiles, including higher linolenic acid and lower palmitic acid, which are beneficial for human health. Additionally, Santos-Silva et al. [104] found that feeding finishing young bulls a TMR with high forage content supplemented with sunflower seeds (10% DM) improved the meat fatty acid profile by increasing the levels of alpha-linolenic, vaccenic, and rumenic acids, while reducing the proportion of *t*10–18:1, a fatty acid that can be detrimental to human health and is often present in meat from animals fed high-concentrate diets [105].

Table 5. Effects of Total Mixed Ration on Carcass Traits and Meat Quality.

Species/Breed	TMR Type/Modification	Summary of Results	Reference
Beef steers	Hedge lucerne TMR	No difference in dressing %; Higher chilled carcass weight	[86]
Aberdeen Angus cattle	MC TMR (60% grass silage)	Best carcass gain (967 g/d) and conformation	[34]
Beef cattle	Whole crop rice TMR	Higher marbling score; No difference in carcass weight/rib eye area	[87]
Karakul sheep	40% SS-AF silage TMR	↑ Carcass weight, subcutaneous fat; Improved WHC, CP, EE, and shear force	[90]
Crossbred lambs	Pelleted TMR (PTMR)	Higher carcass yield (54.5% vs. 49.4%)	[48]
Comisana lambs	WM-based TMR	Higher cold-carcass dressing (10.5 vs. 9.7 kg)	[73]
Crossbred lambs	FTMR (AH-300)	Higher backfat thickness, intramuscular fat; Lower shear force	[45]
Hanwoo Steers	T50 (50:50 forage/concentrate)	Comparable carcass weight to control; higher IMF	[66]
Hanwoo Steers	FTMR	Higher marbling score (5.63 vs. 3.13), fat thickness (13.25 mm)	[89]

Table 5. *Cont.*

Species/Breed	TMR Type/Modification	Summary of Results	Reference
Yak	High-energy TMR	Increased carcass weight (106.43%), dressing percentage (57.52%)	[6]
Karakul Sheep	40%SS–60%AF silage TMR	Higher carcass weight and subcutaneous fat thickness.	[90]
Beef Cattle	TMR (pre-mixed)	Higher carcass weight (279.5 kg vs. 268.6 kg in control)	[29]
Hanwoo Steers	FTMR	Improved tenderness, juiciness, crude fat (18.39%); no pH/cooking loss differences	[89]
Yak	High-energy TMR	Improved tenderness (↓ shear force), reduced cooking loss (↓ 7.28%)	[6]
Karakul Sheep	40%SS–60%AF silage TMR	Improved WHC, CP, EE; reduced shear force	[90]
Naemi Lambs	TMR + alfalfa hay	Improved meat color (L*, a*, b*); reduced shear force	[10]
Beef Cattle (Retinta)	Maize silage TMR	Higher PUFA (18.8% vs. 14.3%) and n-3 PUFA (0.47% vs. 0.35%)	[20]
Boer Goats	TMR + 7.5% intact rapeseed	↑ Linolenic acid, eicosenoic acid; ↓ palmitic acid	[42]
Beef Cattle	High-concentrate TMR	Higher 18:1, lower 18:3; increased n-6/n-3 ratio (3.83 vs. 2.72)	[29]
Hanwoo Steers	TMR + Medicinal Plants	Improved meat quality grade	[27]
Simmental Heifers	TMR	No difference in meat tenderness	[19]

TMR = total mixed ration, SS-AF = sweet sorghum-alfalfa, WHC = water-holding capacity, CP = crude protein, EE = ether extract, PTMR = pelleted TMR, WM = wheat middlings, FTMR = fermented total mixed ration, MC = mixed concentrate; ↑ = increased, ↓ = decreased % = percentage, kg = kilograms, g/d = grams per day; Carcass yield = (carcass weight/live weight) × 100; Cold-carcass dressing = carcass weight after chilling; Marbling score = intramuscular fat content grading; Shear force = tenderness measurement (lower values indicate more tender meat); Rib eye area = cross-sectional area of longissimus dorsi muscle; Dressing % = (hot carcass weight/live weight) × 100.

5. Methane Emission and Environmental Impact

5.1. Enteric Methane Emissions

Methane (CH_4) is a potent greenhouse gas, and agriculture is a significant contributor to global CH_4 emissions, largely due to enteric fermentation in ruminants. Globally, agriculture accounts for 52% of CH_4 emissions, with 80 million tonnes of this being a product of enteric fermentation in ruminants [93]. This represents a considerable energy loss for the animal, typically ranging from 2% to 12% of gross energy intake [93,106]. Therefore, strategies to mitigate enteric CH_4 are crucial for both environmental sustainability and animal productivity. Methane production is highly dependent on the quantity of feed consumed and the composition of the diet [2,93,106]. Furthermore, methane emissions from ruminants are primarily driven by rumen microbial fermentation, where methanogenic archaea utilize hydrogen (H_2) and carbon dioxide (CO_2) to produce CH_4 [3]. Studies indicate that TMR formulations influence methane production by altering the rumen environment, revealing complex interactions between diet composition and microbial activity. Metzler-Zebeli et al. [67] found that TMR with a high concentrate content (490 g/kg DM) resulted in rapid gas production during the early fermentation stage (2–4 h), followed by a decline in gas production. This suggests that associative effects between feed components (e.g., grass silage, cereal concentrate) may transiently stimulate microbial activity, increasing methane output before stabilizing. Particle size also plays a critical role. Tafaj et al. [69] demonstrated that reducing TMR particle size (5.5 mm vs. 25 mm) lowered rumen pH and shifted VFA profiles toward propionate, which competitively reduces H_2 availability for methanogenesis. Smaller particles increase the surface area for microbial degradation, accelerating fermentation and potentially lowering methane yield due to improved energy utilization. Recently Jairath et al. [51] found significantly lower methane production under in vitro conditions when maize grain in the concentrate portion of TMR was replaced with fermented agro-waste at a 32% level. In contrast, feeding high-forage, low-starch TMR with 10% DM of sunflower seeds to finishing crossbred bulls increased digestive CH_4 emissions compared to a concentrate finishing diet. However, the carbon footprint did not differ between diets, 6.63 vs. 6.51 $kgCO_2e/kg$ LWG [104]).

5.2. Role of Rumen Inoculum and Microbial Populations

The source of rumen inoculum has a significant impact on methane emissions. Metzler-Zebeli et al. [67] observed higher gas production with particle-associated liquid (PAL) inoculum compared to free rumen liquid (FRL), likely due to greater microbial density and fibrolytic activity. This implies that in vitro studies using only FRL may underestimate methane potential, as PAL microbes are more representative of the rumen mat, where fiber degradation is most active. FTMR has shown promise in mitigating methane emissions. Cao et al. [30] reported a 25% reduction in methane when sheep were fed FTMR, attributed to increased propionate production from lactate fermentation. This metabolic shift consumes H_2, diverting it away from methanogenesis (the production of methane). Similarly, Li et al. [46] found that FTMR inoculated with lactic acid bacteria reduced methane by 15–20% in vitro, correlating with higher propionate and butyrate levels.

5.3. Impact of Dietary Composition

High-concentrate TMR typically reduces methane per unit of feed intake but may increase total emissions due to higher dry matter intake. O'Neill et al. [93] reported that dairy cows fed TMR produced more methane (397 g/day) than pasture-fed cows (251 g/day), despite higher milk yields. This reflects the trade-off between productivity and environmental impact, where energy-dense TMR improves efficiency but may elevate absolute emissions. Conversely, high-fiber TMR can lower methane but risk impairing digestibility. Zhong et al. [65] noted that pelleted TMR improved feed efficiency but had no significant effect on methane, likely due to balanced fiber and starch ratios. Including additives like yeast culture [24] or tannins [107] in TMR further reduces methane by inhibiting methanogens or altering fermentation pathways. Methane mitigation is crucial for environmental sustainability, and TMR strategies must simultaneously maintain optimal rumen health [65]. High-concentrate TMR, particularly in pelleted forms, significantly increases the risk of ruminal acidosis, specifically SARA. This occurs because such formulations can lead to excessive lactic acid production, thereby lowering rumen pH (e.g., to 5.58 compared to 5.87 in separate feeding) and impairing the activity of fiber-digesting microbes [17]. Conversely, fermented Total Mixed Ration offers a promising solution by stabilizing rumen pH and simultaneously mitigating methane emissions. FTMR achieves pH stability by promoting the growth of lactate-utilizing bacteria, such as *Megasphaera elsdenii*, which convert lactic acid into propionate [30,44]. This metabolic shift not only enhances energy availability for the animal but also consumes hydrogen, thereby diverting it away from methanogenesis, leading to a reported 15–25% reduction in methane production. This dual benefit of FTMR, supporting rumen function while reducing environmental impact, makes it a viable strategy for sustainable livestock production [44]. Therefore, sustainable TMR strategies necessitate careful formulation and management to strike a balance between methane reduction, feed efficiency, and rumen stability to meet both environmental and production goals.

5.4. Feed Utilisation

The environmental implications of TMR are closely linked to its composition, processing methods, and efficiency in nutrient utilization. Scientific evidence suggests that TMR reduces feed wastage and improves rumen fermentation stability, which in turn enhances productivity while mitigating methane emissions [17]. This is particularly significant given that methane is a potent greenhouse gas and optimizing feed efficiency can contribute to more sustainable livestock production. One of the key environmental benefits of TMR lies in its ability to incorporate agricultural by-products such as wet brewers' grains (WBG) and rice straw, reducing reliance on conventional feedstuffs while minimizing waste [45,90]. Furthermore, FTMR has been shown to improve aerobic stability and nutrient retention, which reduces spoilage and further enhances [42]. The inclusion of microbial additives, such as *Saccharomyces cerevisiae* and *Lactobacillus acidophilus*, in FTMR not only improves fibre degradation but also stabilises rumen pH, resulting in more efficient digestion and lower methane output [24]. These findings suggest that TMR, particularly when fermented, can play a vital role in reducing the carbon footprint of ruminant production systems (Table 6).

Table 6. Environmental Impact of Feeding Total Mixed Ration to Ruminants.

Species/Breed	TMR Type/Modification	Summary of Results	Reference
Brown Swiss cows	TMR + Saccharomyces cerevisiae (CE)	CE (0.6 mg/g DM) increased CH_4; LC had no effect.	[68]
Suffolk sheep	Fermented TMR (FTMR)	25% lower CH_4 compared to the control due to a propionate shift.	[30,44]
Holstein steers	High-concentrate TMR	Higher CH_4 (138.5 L/day) vs. separate feeding (118.2 L/day).	[9]
Holstein-Friesian	TMR (maize silage + concentrate)	Higher CH_4 (397 g/d) vs. grass diet (251 g/d).	[93]
Holstein (Dairy)	pTMR	Lower N leaching (21 kg/ha) but higher volatilization (98 kg/ha)	[91]
Holstein (Dairy)	Automatic TMR (AFS)	67.5% lower CO_2e emissions	[108]
Holstein (Dairy)	Pasture + Concentrate	Highest N volatilization (116 kg/ha)	[91]
Aberdeen Angus	MC TMR (60% grass silage)	Lowest GWP (19.1 kg CO_2 eq/kg beef); best feed conversion	[34]

TMR = total mixed ration; CE = Saccharomyces cerevisiae; DM = dry matter; LC = low concentrate; FTMR = fermented total mixed ration; pTMR = precision total mixed ration; AFS = automatic feeding system; MC = medium concentrate; GWP = global warming potential. CH_4 = methane; CO_2 = carbon dioxide; CO_2e = carbon dioxide equivalent; N = nitrogen; L/day = liters per day; g/d = grams per day; kg/ha = kilograms per hectare; kg CO_2 eq/kg beef = kilograms of carbon dioxide equivalent per kilogram of beef.

6. Mycotoxin Contamination and Mitigation Costs

Mycotoxin contamination in TMR poses significant risks to ruminant health and productivity, with scientific evidence linking it to poor feed quality, storage conditions, and ingredient sourcing (Table 7). Mycotoxins are toxic secondary metabolites produced by fungi, primarily *Aspergillus*, *Fusarium*, and *Penicillium* species, which can proliferate in feed components under favorable conditions [43].

6.1. Mycotoxins in Total Mixed Ration

The presence of mycotoxins in TMR is particularly concerning because TMR combines multiple ingredients, increasing the likelihood of contamination if even one component is compromised. One of the primary reasons for mycotoxin contamination in TMR is the inclusion of mold-infected roughages such as maize silage, rice straw, and alfalfa hay [45]. These forages are highly susceptible to fungal growth, especially when harvested or stored at high moisture levels [25]. Moreover, the fermentation process in ETMR does not always eliminate mycotoxins, as some fungal strains remain viable and continue to produce toxins even under anaerobic conditions [42]. This is further exacerbated when wet by-products, such as brewers' grains, are incorporated, as improper drying or storage can introduce additional fungal contamination [37,38].

The primary source of mycotoxins in TMR is often maize silage, which is highly susceptible to fungal colonization. Cogan et al. [109] reported that 90% of maize silage-based TMR samples contained deoxynivalenol (DON) and zearalenone (ZON). These are toxic secondary metabolites produced by fungi, and their presence in TMR, particularly in maize silage-based TMR, is a significant concern. In contrast, grass silage-based TMR showed no detectable mycotoxins. This contrast highlights the role of feed composition in contamination risk. Additionally, González-Jartín et al. [110] observed that fumonisins (74%), DON (42.5%), and ZEN (39.1%) were the most prevalent mycotoxins in maize silage TMR, with co-occurrence increasing health risks due to potential synergistic effects. Rodríguez-Blanco et al. [111] further noted that 58% of TMR samples were contaminated with *Fusarium*-derived mycotoxins, emphasizing the need for rigorous feed quality control. Sultana et al. [112] found that 100% of TMR samples tested were contaminated with aflatoxin B1 (AFB1) and ochratoxin A (OTA), exceeding European Union regulatory limits. AFB1, a potent hepatotoxin, was detected at 30 ng/g, while OTA averaged 48.5 ng/g, raising concerns about chronic exposure in dairy cattle. Furthermore, 50% of samples contained zearalenone (ZON) at 700 ng/g, a mycotoxin linked to reproductive disorders. These findings reveal the widespread nature of mycotoxin contamination in TMR, particularly in regions with suboptimal feed storage conditions. The implications for animal health are profound. AFB1 is metabolized in the liver to aflatoxin M1 (AFM1), which is excreted in milk and poses risks to both livestock and consumers Mycotoxins like DON disrupt gut integrity and immune function, while ZEN interferes with reproductive hormones, leading to infertility and reduced milk yield. Even at subclinical levels, chronic exposure can impair growth, feed efficiency, and overall herd health [112]. The quality of TMR is further compromised by microbial and endotoxin contamination.

Table 7. Mycotoxin Contamination in Total Mixed Ration for Ruminants.

Species/Breed	TMR Type/Intervention	Summary of Results	Reference
Dairy cattle	Mycotoxin-contaminated TMR	100% AFB1 (30 ng/g, exceeding EU limits); 100% OTA (48.5 ng/g); 50% ZON (700 ng/g)	[112]
Holstein cows	Grass silage-based TMR	No mycotoxins detected	[109]
Holstein cows	Maize silage-based TMR	90% of samples had DON/ZON	[109]
Dairy cows	Maize silage TMR	Fumonisins (74%), DON (42.5%), ZEN (39.1%) prevalent	[110]
Dairy cows	Fusarium-contaminated TMR	58% samples contaminated (FBs 34%, DON 17%, ZEN 16%)	[111]
Dairy cattle (Pakistan)	Commercial TMR	33.3% AFB1 positive (mean 21.97 ppb)	[113]
Holstein cows	Grass silage TMR	Higher enterobacteria in TMR vs. silage; linked to SCC	[109]
Dairy cows	Maize/grass silage TMR	Highest endotoxins in TMR (293.44 EU/mL)	[114]
Holstein cows	Maize silage TMR	Enterobacteriaceae: 3.93 log10 CFU/g in maize silage	[114]
Dairy cows	Maize silage-based TMR	Maize silage = primary source of DON/ZEA; co-occurrence increases health risks	[25]

Table 7. *Cont.*

Aflatoxin Carryover to Milk			
Holstein cows	AFB1-contaminated TMR + sequestering agent	47% reduction in AFM1 with optimal SA inclusion	[115]
High-yielding cows	AFB1-contaminated TMR	75% milk samples exceeded EU AFM1 limits (50 ng/L)	[116]
Dairy cows (Portugal)	Mycotoxin-contaminated TMR	RC/BEA/enniatins showed 2–10% carryover to milk	[110]
Mitigation Strategies			
Holstein cows	Pelletized SA in TMR	Most effective (0.013 AFM1 excretion)	[115]
Dairy cows	Mycotoxin adsorbents in TMR	Reduced toxin bioavailability	[117]
Dairy cows	Inoculant additives in grass silage TMR	Improved ME content and milk yield	[109]

AFB1 = aflatoxin B1, OTA = ochratoxin A, ZON = zearalenone, DON = deoxynivalenol, ZEN = zearalenone, FBs = fumonisins, SCC = somatic cell count, SA = sequestering agent, AFM1 = aflatoxin M1, RC = roquefortine C, BEA = beauvericin, ME = metabolizable energy. ng/g = nanograms per gram, ppb = parts per billion, EU/mL = endotoxin units per milliliter, log10 CFU/g = logarithmic colony-forming units per gram, ng/L = nanograms per liter.

Vaičiulienė et al. [114] reported that TMR had the highest endotoxin levels (293.44 EU/mL) compared to silages, likely due to bacterial proliferation during feed mixing and storage. Enterobacteriaceae counts were elevated in maize silage (3.93 log10 CFU/g), suggesting poor hygienic conditions that exacerbate mycotoxin risks. These findings indicate that TMR not only introduces mycotoxins but also creates an environment conducive to secondary microbial challenges.

Mitigation strategies are crucial for reducing mycotoxin exposure and ensuring feed safety in TMR. Effective mitigation requires a combination of rigorous feed quality monitoring, improved storage practices, and the strategic use of mycotoxin binders. One key strategy involves the use of sequestering agents, such as clay-based binders, which have been demonstrated to reduce Aflatoxin M1 excretion in milk by 47% when incorporated into TMR [115]. Pelletizing these agents further enhances their efficacy, underscoring the importance of feed processing in mitigation efforts. Furthermore, improving silage management is vital to minimizing fungal growth [109]. This includes ensuring anaerobic conditions during ensiling and using inoculants, which can indirectly reduce mycotoxin risks by improving the overall hygienic quality of the feed. For instance, farms that utilized silage additives reported higher metabolizable energy content and milk yields, indicating better feed preservation and reduced spoilage [109]. In conclusion, mycotoxin contamination in TMR is a multifaceted issue driven by the quality of feed ingredients, storage practices, and environmental factors. Therefore, adopting integrated approaches, including routine mycotoxin screening, enhanced silage management, and the utilization of binding agents, is essential to ensure animal health and food safety. Addressing these challenges will ultimately enhance the sustainability and productivity of ruminant production systems, as farmers and nutritionists must prioritize mycotoxin management to maintain productive and sustainable livestock systems.

6.2. Mycotoxin Mitigation Costs

A significant economic concern in TMR is mycotoxin contamination, which can reduce feed quality and animal health. Maize silage, a common ingredient in TMR, is particularly susceptible to aflatoxins and Fusarium toxins [111]. Contaminated feed leads to decreased productivity, increased disease incidence, and higher veterinary costs [25]. Although effective mitigation strategies, such as mycotoxin binders and improved silage storage, add to operational expenses but are necessary to prevent losses [115]. Routine monitoring and sourcing high-quality ingredients are essential to minimize these risks. In conclusion, the economic benefits of TMR are well-documented, including improved feed efficiency, labor savings, and enhanced livestock productivity. However, these advantages depend on proper formulation, management practices, and investment in technology. Farmers must balance cost-effectiveness with animal health considerations, such as maintaining adequate fiber levels and preventing mycotoxin contamination. For small-scale operations, partial TMR systems or cooperative feed mixing may offer viable alternatives. Ultimately, TMR represents a scientifically validated approach for optimizing livestock production, but its economic viability must be assessed in the context of farm-specific conditions.

7. Implications of Total Mixed Ration for Animal Health

The formulation and delivery of TMR have profound implications for the metabolic and physiological health of ruminants. Blood metabolite profiles serve as critical indicators of this status, reflecting dietary composition, rumen fermentation dynamics, and subsequent metabolic adaptations.

7.1. Blood Metabolites as Health Indicators

7.1.1. Glucose Metabolism

Blood glucose levels are primarily influenced by dietary energy content and rumen fermentation patterns. TMR formulations rich in fermentable carbohydrates (e.g., maize silage, cereal grains) promote increased propionate production in the rumen, which is a primary precursor for hepatic gluconeogenesis [67]. For instance, Hu et al. [118] observed elevated blood glucose in dairy heifers fed a high-straw fermented TMR (HSF), a response attributed to enhanced volatile fatty acid production. However, excessive concentrate inclusion can lead to subacute ruminal acidosis, which may impair glucose metabolism due to systemic inflammation [18].

7.1.2. Nitrogen Utilization

Blood urea nitrogen (BUN) levels are a key reflection of nitrogen metabolism and protein utilization efficiency. Elevated BUN in TMR-fed animals often indicates an imbalance, such as excessive rumen-degradable protein or inadequate energy synchronization, leading to increased ammonia absorption and subsequent hepatic urea synthesis [63]. Conversely, optimized TMR formulations with balanced protein-to-energy ratios (e.g., combining alfalfa silage and maize silage) can lower BUN by improving microbial protein synthesis and nitrogen capture in the rumen [66]. Furthermore, fermented TMR has been shown to reduce BUN, likely by enhancing microbial efficiency and nitrogen retention [64].

7.1.3. Lipid Metabolism

The TMR composition significantly influences lipid metabolism, as evidenced by alterations in serum triglycerides and cholesterol levels. High-concentrate TMR can increase hepatic lipid mobilization, a consequence of elevated insulin resistance and altered VFA profiles [18]. In contrast, TMR formulations incorporating adequate forage (e.g., alfalfa hay) promote healthier fatty acid profiles and can reduce the saturation of intramuscular fat [10]. The strategic inclusion of specific additives, such as oilseeds (e.g., rapeseed), further modulates lipid metabolism, increasing the proportion of beneficial unsaturated fatty acids in animal products [42].

7.1.4. Oxidative Stress

Markers of antioxidant capacity, such as superoxide dismutase and total antioxidant capacity, indicate the level of oxidative stress. High-grain TMR diets can induce oxidative stress due to the increased production of reactive oxygen species associated with rapid rumen fermentation and inflammatory states [63]. Conversely, FTMR inoculated with lactic acid bacteria has been demonstrated to enhance antioxidant capacity by stabilizing rumen pH and reducing systemic inflammatory responses [64].

7.1.5. Synthesis for Animal Health

Collectively, blood metabolite profiles underscore the necessity of balancing fermentable carbohydrates, physically effective fiber, and protein sources in TMR to maintain metabolic health. For lactating dairy cows, this typically means a TMR with moderate starch (25–30% DM) and adequate fiber (≥30% NDF) to stabilize glucose and minimize SARA risk [67]. In growing cattle, FTMR can enhance nutrient digestibility and improve energy retention [46], while for goats and sheep, TMR particle size and fiber content must be carefully optimized to prevent suboptimal fermentation [76]. Future research should explore precision feeding strategies using real-time metabolite monitoring to dynamically tailor TMR formulations for different production stages and individual animal needs.

7.2. Economic Sustainability and Operational Costs of Total Mixed Ration

The adoption of TMR is a significant management decision with major economic implications, impacting feed efficiency, labor requirements, capital investment, and overall farm profitability.

7.2.1. Feed Efficiency and Economic Gains

A primary economic advantage of TMR is its ability to improve feed efficiency. By blending roughages and concentrates into a homogeneous mixture, TMR prevents selective feeding by animals and ensures a consistent and balanced intake of carbohydrates, proteins, and minerals [69]. This nutrient synchronization optimizes rumen microbial activity, leading to enhanced volatile fatty acid production and superior energy utilization [59]. Studies, such as one by Zhong et al. [65], have demonstrated that pelleted TMR increased dry matter intake and average daily gain in lambs due to improved digestibility. These improvements in performance directly translate into economic benefits, as shown in Table 8, where TMR feeding led to lower feed costs per kilogram of gain [22] and higher net income [45]. It is critical, however, that TMR is properly formulated to maintain adequate physically effective fiber (peNDF) to avoid metabolic disorders like SARA, which can erode these economic gains [76].

Table 8. Economic Impact of Feeding Total Mixed Ration to Ruminants.

Species/Breed	TMR Type/Intervention	Findings vs. Control	Reference
Simmental bulls	Dry TMR (straw-based)	Higher feed costs but similar carcass yield vs. conventional TMR	[81]
Crossbred lambs	FTMR (AH-300)	Improved net income due to better feed efficiency	[45]
Dorper Lambs	16% CP Growing TMR	Lowest feed cost/kg gain (RM 8.94 vs. RM 22.92 control)	[22]
Holstein (Dairy)	Confinement TMR	Highest net return ($738/cow) but greatest risk	[91]
Holstein (Dairy)	Automatic TMR (AFS)	75% lower labor costs, 91% energy reduction	[108]
Hanwoo Steers	TMR + Medicinal Plants	Reduced feed costs, increased carcass price	[27]
Sindhi Crossbred	Alkaline-treated TMR	17.75% higher economic benefit/kg gain	[14]

TMR = total mixed ration, FTMR = fermented total mixed ration, CP = crude protein, AFS = automated feeding system. RM = Malaysian ringgit, $ = US dollar, % = percent, kg = kilogram.

7.2.2. Labor and Operational Cost Savings

The operational efficiency of TMR systems, particularly when automated, offers substantial labor savings. Tangorra and Calcante [108] reported that automatic feeding systems (AFS) used for TMR delivery lowered labor requirements by 75% and reduced energy consumption by 91% compared to conventional mixer wagons. Furthermore, AFS improves feeding accuracy and consistency, minimizing nutrient variation and feed waste. While the initial investment in TMR equipment (mixer wagons, AFS) can be substantial, the long-term savings in labor and improved feed utilization contribute significantly to economic sustainability, especially in large-scale operations.

7.2.3. Productivity and Quality Premiums

TMR feeding is consistently associated with enhanced productivity, which directly boosts income. In dairy systems, TMR-fed herds typically produce more milk with higher fat and protein content compared to those on less controlled diets [93]. In beef production, cattle fed TMR often exhibit improved carcass quality metrics, such as better marbling and meat tenderness, which can command premium market prices [29]. Strategies like incorporating specific additives (e.g., medicinal plants) into TMR have also been shown to reduce feed costs while simultaneously increasing carcass value [27].

7.2.4. Economic Considerations and Barriers

Despite the clear benefits, economic barriers to adoption exist. The high capital investment required for TMR equipment can be prohibitive for smallholder farmers [108]. To mitigate this, partial TMR strategies, which combine TMR with grazing or forage supplementation, have been proposed as a cost-effective alternative for smaller operations [12,91]. Furthermore, the pursuit of maximum productivity via high-concentrate TMR formulations must be balanced against potential externalities, such as increased methane emissions [9], which may have future economic implications through regulatory costs or market preferences. Therefore, the economic sustainability of TMR depends on optimizing formulations for both profitability and long-term operational resilience.

8. Challenges in Feeding Total Mixed Ration to Ruminants

TMR is widely adopted in ruminant production due to its ability to provide a nutritionally balanced diet, improve feed efficiency, and enhance productivity. However, several challenges arise in its implementation, affecting both animal health and farm profitability (Figure 5). One major challenge is ruminal acidosis, particularly with high-concentrate TMR formulations that reduce rumen buffering capacity. Bo Trabi et al. [26] found that pelleted TMR (PTMR) reduces particle size, thereby increasing starch digestibility; however, it lowers rumen pH, which increases the risk of SARA. Similarly, Arbaoui et al. [17] reported that high-concentrate TMR resulted in a lower rumen pH (5.58 vs. 5.87 in separate feeding), thereby increasing the risk of acidosis. This implies that while TMR enhances nutrient utilization, improper forage-to-concentrate ratios can disrupt rumen homeostasis, negatively impacting microbial activity and fiber digestion.

Feed sorting is another concern, where animals selectively consume preferred ingredients, leading to nutrient imbalances. Although TMR is designed to minimize this, studies indicate that compact TMR or FTMR may better prevent sorting by improving homogeneity [60,114]. Furthermore, mycotoxin contamination in TMR, particularly from maize silage, poses health risks. Martins et al. [25] highlighted that mycotoxins, such as aflatoxin B1 and deoxynivalenol, are prevalent in TMR, which can impair liver function and reduce feed intake. This emphasizes the importance of rigorous feed quality control and the use of mycotoxin binders to mitigate these effects.

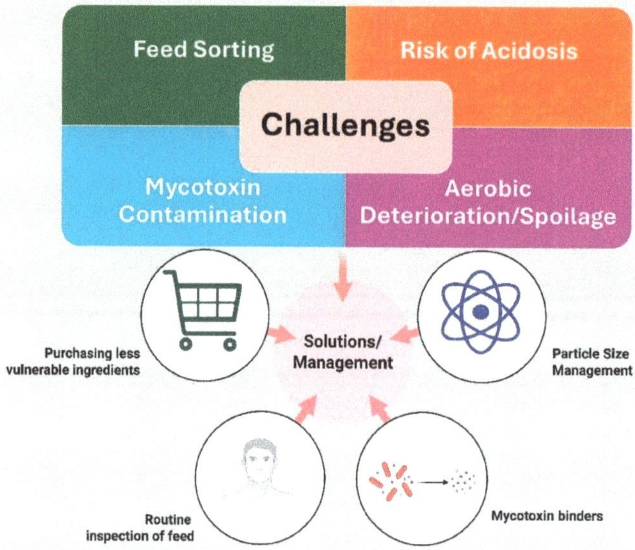

Figure 5. Some challenges of feeding total mixed ration to ruminants and potential solutions.

Methane emissions from TMR-fed ruminants also present environmental challenges. O'Neill et al. [93] observed that TMR-fed dairy cows emitted more methane (397 g/day) than pasture-fed cows (251 g/day), despite higher milk yields. This suggests that while TMR improves productivity, its environmental footprint must be addressed through strategies like FTMR or methane-inhibiting additives [46]. In conclusion, the challenges of feeding TMR include ruminal acidosis, feed sorting, mycotoxin contamination, and methane emissions, highlighting the need for precise formulation, quality control, and innovative strategies such as FTMR or yeast supplementation. Addressing these issues ensures sustainable ruminant production, striking a balance between productivity, animal health, and environmental impact. Future research should focus on optimizing TMR composition and mitigating its associated drawbacks to enhance farm profitability and global food security (Figure 5).

9. Conclusions and Gaps Identified

While TMR offers a precision feeding strategy that significantly enhances ruminant productivity and health, the current body of research identifies several limitations and points towards crucial directions for future investigation. Early studies, such as the one conducted by Lakhani et al. [78], involved a relatively small sample size of Murrah buffaloes ($n = 18$ for the feeding trial, $n = 9$ for in vitro and in vivo comparison) which, while demonstrating the reliability of in vitro techniques for predicting metabolizable energy, suggests that broader validation with more extensive data is needed. Moreover, Tafaj et al. [69], when examining the impact of TMR particle size on rumen fermentation, also noted that their findings were based on limited number of observations (n = 3 cows for the main experiment) and emphasized the need for further studies to verify observed trends and better understand the complex relationships between in vitro gas production and in vivo measurements across different digesta compartments.

A significant challenge in TMR research lies in the translation of in vitro findings to in vivo animal performance and livestock productivity as milk and meat. Researchers, such as Mansfield et al. [119], highlighted that rumen simulation systems, like Rusitec, may not perfectly replicate the natural rumen environment due to factors including reduced protozoa numbers, potential shifts in bacterial populations, and limitations imposed by feed enclosures in nylon bags, which can affect microbial accessibility and nutrient recycling. This discrepancy hinders direct comparisons and the drawing of definitive conclusions for the live animal feeding complex, despite some short-term in vitro studies showing similar fermentation patterns to those of the natural rumen. Furthermore, Mendoza et al. [98] observed that the spot sampling technique used for assessing microbial protein flow might lack sufficient sensitivity to detect small differences between dietary treatments, indicating a methodological limitation that could obscure subtle biological responses. Similarly, Zhong et al. [65] cautioned that their findings on rumen pH and risk of acidosis in dairy cows fed pelleted TMR were based on a single sampling over 42 days, emphasizing the need for longer-term evaluations consisting multiple sample collection to establish the sustained effects on lactating performance and health status. Sun et al. [72] also noted that their study's short duration might have limited the detection of significant differences in carcass traits in fattening lambs supplemented with live yeast.

A notable gap in the current literature, which merits further investigation as detailed below in Section 10 (Future Research Directions), is the predominant focus of studies on dairy and beef cattle. While some research exists on sheep, there is limited research on other ruminants, such as goats. The physical characteristics of TMR ingredients also present limitations. Zhang et al. [48] pointed out that while pelleted TMR can improve growth performance, the fine grinding of roughage often reduces physically effective neutral detergent fiber (peNDF) content, potentially leading to lower rumen pH and compromised fiber degradation. This can sometimes result in minimal or no substantial difference in feed conversion ratio despite increases in average daily gain. Jang et al. [76] found that for goats, traditional peNDF metrics based on particle size distribution might not be as effective in predicting chewing activity or digestibility, suggesting that goats, due to their unique digestive characteristics, may respond differently to feeding pattern and particle size variations, necessitating further research with a broader range of peNDF levels and particle sizes to better identify these effects. Confounding factors can also complicate interpretations of finding in terms of practical on farm applicability. Kronqvist et al. [31] reported that their study design was unable to distinguish between the effects of particle size and dry matter concentration on feed intake, highlighting the need for studies specifically designed to isolate these variables.

Variability in environment and raw material in feed pose additional challenges. Mohammadi Shad et al. [116] reported that aflatoxin B1 (AFB1) contamination in TMR and feed ingredients, particularly cottonseed cake and corn gluten meal, was higher during the rainy season, emphasizing the influence of climatic conditions on feed quality and the need for improved storage and feeding procedures to mitigate mycotoxin occurrence and carryover effect into milk. Furthermore, high-concentration TMR, especially when finely pelleted, can pose a higher risk of acidosis due to rapid fermentation and reduced buffering from saliva if not carefully managed, which may result in reduced feed intake and lower animal performance. An unusually low protein content in the control TMR for one study also highlights the importance of consistent feed quality in comparative trials. These limitations directly inform future research directions.

10. Future Research Directions

This review underpins a need for long-term studies to fully assess the sustained effects of various TMR formulations, including pelleted and fermented TMR (FTMR), on livestock health, animal productivity, and milk composition, particularly at different stages of lactation. Research should focus on optimizing TMR composition, including ideal forage to concentrate ratios, and exploring alternative ingredients, such as sweet sorghum and alfalfa silage mixtures, or novel agricultural by-products, while comprehensively evaluating their impact on rumen function, microbial populations, and digestive enzyme activity. The efficacy of processing methods, such as fermentation, which potentially incorporates lactic acid bacteria and fibrolytic enzymes, in improving nutrient digestibility, reducing methane production, enhancing feed storage quality, and improving aerobic stability, warrants further in vivo validation. Moreover, future investigations should delve deeper into the impact of feed additives, such as yeast cultures for ameliorating rumen pH drops and enhancing fibre degradation, or essential oils and volatile fatty acid solutions as alternative energy sources, with a focus on optimal dosage and long-term effects.

Understanding the intricate interplay between diet, rumen microbiota, and metabolic profiles is crucial for developing precision feeding strategies and identifying markers related to feed efficiency and growth performance. Economic analyses are also important to determine the feasibility of incorporating novel ingredients or feeding strategies, such as the use of wheat middlings in lamb diets or total mixed ration silage in dairy and beef cattle production in tropical countries, to ensure sustainable and profitable livestock farming. Ultimately, continuous research is essential for developing more resilient TMR formulations that can cope with variations in raw material quality and environmental conditions, thereby ensuring optimal animal performance and contributing to global food security in an environmentally responsible manner.

Author Contributions: M.C.D.: conceptualization, writing—original draft. G.J.: writing—review & editing. I.U.G.: conceptualization, methodology, writing—original draft, table/figure visualization, writing—review & editing. S.P.A.: constructive feedback, writing—review & editing. E.N.P.: conceptualization, writing—original draft, methodology, critical evaluation of the content, writing—review & editing. All authors have read and agreed to the published version of the manuscript.

Funding: This research received no external funding.

Institutional Review Board Statement: Not applicable.

Informed Consent Statement: Not applicable.

Data Availability Statement: No new data were created or analyzed in this study. Data sharing does not apply to this article.

Acknowledgments: The authors extend their sincere gratitude to *Ruminants* for the invitation to contribute this invited feature paper. The authors also acknowledge and appreciate the support of their respective institutions. The graphical illustrations were designed using BioRender.com.

Conflicts of Interest: The authors declare no conflicts of interest.

References

1. Ponnampalam, E.N.; Kerr, M.G.; Butler, K.L.; Cottrell, J.J.; Dunshea, F.R.; Jacobs, J.L. Filling the out of Season Gaps for Lamb and Hogget Production: Diet and Genetic Influence on Carcass Yield, Carcass Composition and Retail Value of Meat. *Meat Sci.* **2019**, *148*, 156–163. [CrossRef]
2. Gadzama, I.U. Evaluation of Fresh Microalgae in Ruminant Nutrition: Impact on Rumen Fermentation, Productive Performance and Meat Quality. Master's Thesis, The University of Queensland, Brisbane, Australia, 2024. [CrossRef]

3. Gadzama, I.U.; Ray, S.; Méité, R.; Mugweru, I.M.; Gondo, T.; Rahman, M.A.; Redoy, M.R.A.; Rohani, M.F.; Kholif, A.E.; Salahuddin, M.; et al. *Chlorella vulgaris* as a Livestock Supplement and Animal Feed: A Comprehensive Review. *Animals* **2025**, *15*, 879. [CrossRef]
4. Ponnampalam, E.N.; Butler, K.L.; McDonagh, M.B.; Jacobs, J.L.; Hopkins, D.L. Relationship between Muscle Antioxidant Status, Forms of Iron, Polyunsaturated Fatty Acids and Functionality (Retail Colour) of Meat in Lambs. *Meat Sci.* **2012**, *90*, 297–303. [CrossRef]
5. Jing, H.; Liu, P.; Zheng, S.; Cao, K.; Li, X. Total mixed ration enhances nutrient digestibility, blood biochemical parameters and faecal microbial diversity in horses. *BMC Vet. Res.* **2025**, *21*, 483. [CrossRef] [PubMed]
6. Liu, Y.-X.; Ma, X.-M.; Xiong, L.; Wu, X.-Y.; Liang, C.-N.; Bao, P.-J.; Yu, Q.-L.; Yan, P. Effects of Intensive Fattening With Total Mixed Rations on Carcass Characteristics, Meat Quality, and Meat Chemical Composition of Yak and Mechanism Based on Serum and Transcriptomic Profiles. *Front. Vet. Sci.* **2021**, *7*, 599418. [CrossRef]
7. Pereira, D.M.; De Oliveira, J.S.; Ramos, J.P.D.F.; Cavalcante, I.T.R.; Santos, F.N.D.S.; Da Silva, E.D.S.; Perazzo, A.F.; Macêdo, A.J.D.S.; Tôrres Júnior, P.D.C.; Santos, E.M. Total Mixed Ration Silage Based on Cactus Pear and Cottonseed Cake in the Feeding of Feedlot Finished Lambs. *Trop. Anim. Health Prod.* **2025**, *57*, 50. [CrossRef] [PubMed]
8. Rooney, M.; O'Connor, A.; Gollogly, C.; Mills, A.; Dunne, S.; Chonnacháin, C.N.; Timlin, M.; Brodkorb, A.; O'Callaghan, T.F.; Magan, J.B.; et al. The Impact of Pasture-Fed versus Total Mixed Ration Derived Whole Milk Powder (WMP) on Circulating Fatty Acid Concentrations and Cardiometabolic Health in Healthy Adults: Results from a Randomized Controlled Crossover Trial. *J. Nutr. Biochem.* **2025**, *145*, 110004. [CrossRef] [PubMed]
9. Bharanidharan, R.; Arokiyaraj, S.; Kim, E.B.; Lee, C.H.; Woo, Y.W.; Na, Y.; Kim, D.; Kim, K.H. Ruminal Methane Emissions, Metabolic, and Microbial Profile of Holstein Steers Fed Forage and Concentrate, Separately or as a Total Mixed Ration. *PLoS ONE* **2018**, *13*, e0202446. [CrossRef]
10. Alhidary, I.A.; Abdelrahman, M.M.; Alyemni, A.H.; Khan, R.U.; Al-Saiady, M.Y.; Amran, R.A.; Alshamiry, F.A. Effect of Alfalfa Hay on Growth Performance, Carcass Characteristics, and Meat Quality of Growing Lambs with Ad Libitum Access to Total Mixed Rations. *Rev. Bras. Zootec.* **2016**, *45*, 302–308. [CrossRef]
11. Schären, M.; Seyfang, G.M.; Steingass, H.; Dieho, K.; Dijkstra, J.; Hüther, L.; Frahm, J.; Beineke, A.; Von Soosten, D.; Meyer, U.; et al. The Effects of a Ration Change from a Total Mixed Ration to Pasture on Rumen Fermentation, Volatile Fatty Acid Absorption Characteristics, and Morphology of Dairy Cows. *J. Dairy Sci.* **2016**, *99*, 3549–3565. [CrossRef]
12. Tozer, P.R.; Bargo, F.; Muller, L.D. Economic Analyses of Feeding Systems Combining Pasture and Total Mixed Ration. *J. Dairy Sci.* **2003**, *86*, 808–818. [CrossRef]
13. Kim, S.H.; Alam, M.J.; Gu, M.J.; Park, K.W.; Jeon, C.O.; Ha, J.K.; Cho, K.K.; Lee, S.S. Effect of Total Mixed Ration with Fermented Feed on Ruminal In Vitro Fermentation, Growth Performance and Blood Characteristics of Hanwoo Steers. *Asian Australas. J. Anim. Sci.* **2011**, *25*, 213–223. [CrossRef]
14. Nguyen, H.T.; Nguyen, T.Q.; Le, P.B.T.; Do, D.T.; Nguyen, T.N.; Duong, K.N. Total Mixed Ration (TMR) Based on Alkaline-Treated Rice Straw Improves Growth Rate and Feed Intake of Sindhi Crossbred Cattle. *IOP Conf. Ser. Earth Environ. Sci.* **2025**, *1465*, 012023. [CrossRef]
15. Bae, D.; Shin, C.; Ko, K. Effect of total mixed ration including apple pomace for lactating cows. *Korean J. Dairy Sci.* **1994**, *16*, 295–302.
16. Sarker, N.R.; Yeasmin, D.; Habib, M.A.; Tabassum, F. Feeding Effect of Total Mixed Ration on Milk Yield, Nutrient Intake, Digestibility and Rumen Environment in Red Chittagong Cows. *Asian J. Med. Biol. Res.* **2019**, *5*, 71–77. [CrossRef]
17. Arbaoui, A.; Gonzalo, G.; Belanche, A.; De Vega, A. Unmixed Diet Versus Total Mixed Ration in Beef Cattle Fed High-Concentrate Diets: Effects on Methane Emissions, Animal Performance, and Rumen Fermentation. *Animals* **2025**, *15*, 723. [CrossRef] [PubMed]
18. Li, W.; Ye, B.; Wu, B.; Yi, X.; Li, X.; A, R.; Cui, X.; Zhou, Z.; Cheng, Y.; Zhu, X.; et al. Effect of Total Mixed Ration on Growth Performance, Rumen Fermentation, Nutrient Digestion, and Rumen Microbiome in Angus Beef Cattle during the Growing and Fattening Phases. *Fermentation* **2024**, *10*, 205. [CrossRef]
19. Iraira, S.P.; Madruga, A.; Pérez-Juan, M.; Ruíz-de-la-Torre, J.L.; Rodríguez-Prado, M.; Calsamiglia, S.; Manteca, X.; Ferret, A. Performance, Behaviour and Meat Quality of Beef Heifers Fed Concentrate and Straw Offered as Total Mixed Ration or Free-Choice. *Span. J. Agric. Res.* **2015**, *13*, e0610. [CrossRef]
20. Horcada, A.; Polvillo, O.; Juárez, M.; Avilés, C.; Martínez, A.L.; Peña, F. Influence of Feeding System (Concentrate and Total Mixed Ration) on Fatty Acid Profiles of Beef from Three Lean Cattle Breeds. *J. Food Compos. Anal.* **2016**, *49*, 110–116. [CrossRef]
21. Ferrari, V.B.; Bondurant, R.G.; Erickson, G.E.; Wilke, K.H.; MacDonald, J.C. Effects of Displacing Modified Distillers Grains with Dry-Rolled Corn in a Totally Mixed Ration on Performance of Growing Beef Steers Consuming Cornstalk Residue from Different Harvest Methods. *Appl. Anim. Sci.* **2024**, *40*, 717–724. [CrossRef]

22. Panadi, M.; Mat, K.; Rusli, N.D.; Darmansah, N.F. Cost-Effectiveness of Total Mixed Ration with Varying Crude Protein Levels for Feeding Dorper Lambs for Human Consumption. *HumEnTech* **2022**, *1*, 17–25. [CrossRef]
23. Chen, Y.-H.; Chen, Y.-M.; Tu, P.-A.; Chen, C.-Y.; Wang, H.-T. Stabilizing Total Mixed Ration Dry Matter to Mitigate Environmental-Relative-Humidity Effects on Lactating Cow Performance. *Animals* **2025**, *15*, 1137. [CrossRef]
24. Wang, X.; Li, F.; Zhang, N.; Ungerfeld, E.; Guo, L.; Zhang, X.; Wang, M.; Ma, Z. Effects of Supplementing a Yeast Culture in a Pelleted Total Mixed Ration on Fiber Degradation, Fermentation Parameters, and the Bacterial Community in the Rumen of Sheep. *Anim. Feed Sci. Technol.* **2023**, *296*, 115565. [CrossRef]
25. Martins, D.; Lemos, A.; Silva, J.; Rodrigues, M.; Simões, J. Mycotoxins Evaluation of Total Mixed Ration (TMR) in Bovine Dairy Farms: An Update. *Heliyon* **2024**, *10*, e25693. [CrossRef] [PubMed]
26. Bo Trabi, E.; Seddik, H.; Xie, F.; Lin, L.; Mao, S. Comparison of the Rumen Bacterial Community, Rumen Fermentation and Growth Performance of Fattening Lambs Fed Low-Grain, Pelleted or Non-Pelleted High Grain Total Mixed Ration. *Anim. Feed Sci. Technol.* **2019**, *253*, 1–12. [CrossRef]
27. Lee, S.J.; Kim, D.H.; Guan, L.L.; Ahn, S.K.; Cho, K.W.; Lee, S.S. Effect of Medicinal Plant By-Products Supplementation to Total Mixed Ration on Growth Performance, Carcass Characteristics and Economic Efficacy in the Late Fattening Period of Hanwoo Steers. *Asian Australas. J. Anim. Sci.* **2015**, *28*, 1729–1735. [CrossRef]
28. Cheng, Y.; Wu, Z.; Bi, K.; Wang, X.; Yu, H.; Tian, P. Impact of Inoculating Various Lactic Acid Bacteria on Vitamin A Levels in Total Mixed Ration Silage. *Sci. Rep.* **2025**, *15*, 7343. [CrossRef]
29. Cooke, D.W.I.; Monahan, F.J.; Brophy, P.O.; Boland, M.R. Comparison of Concentrates or Concentrates Plus Forages in a Total Mixed Ration or Discrete Ingredient Format: Effects on Beef Production Parameters and on Beef Composition, Colour, Texture and Fatty Acid Profile. *Ir. J. Agric. Food Res.* **2004**, *43*, 201–216.
30. Cao, Y.; Takahashi, T.; Horiguchi, K.; Yoshida, N.; Cai, Y. Methane Emissions from Sheep Fed Fermented or Non-Fermented Total Mixed Ration Containing Whole-Crop Rice and Rice Bran. *Anim. Feed Sci. Technol.* **2010**, *157*, 72–78. [CrossRef]
31. Kronqvist, C.; Petters, F.; Robertsson, U.; Lindberg, M. Evaluation of Production Parameters, Feed Sorting Behaviour and Social Interactions in Dairy Cows: Comparison of Two Total Mixed Rations with Different Particle Size and Water Content. *Livest. Sci.* **2021**, *251*, 104662. [CrossRef]
32. Liu, W.; Li, E.; Xu, C.; Chen, L.; Wang, X. Nutritional Strategies for Nile Tilapia: Protein and Carbohydrate Balances in Saline-Alkaline Aquaculture. *J. Anim. Sci. Biotechnol.* **2025**, *16*, 86. [CrossRef]
33. Soder, K.J.; Brito, A.F.; Rubano, M.D. Effect of Supplementing Orchardgrass Herbage with a Total Mixed Ration or Flaxseed on Fermentation Profile and Bacterial Protein Synthesis in Continuous Culture. *J. Dairy Sci.* **2013**, *96*, 3228–3237. [CrossRef]
34. Huuskonen, A.; Hietala, S.; Hyvönen, J.; Leinonen, I.; Manni, K. Environmental Impacts and Animal Performance of Finishing Bulls Fed Different Silage-Based Total Mixed Rations. *Livest. Sci.* **2023**, *268*, 105166. [CrossRef]
35. Sant'Anna, M.A.C.; Carlis, M.S.D.P.; Polizel, D.M.; Ferreira, E.M.; Andrighetto, C.; Luz, P.A.D.; Araujo, L.C.D. Megathyrsus Maximus Addition in Total Mixed Ration on Performance, Metabolism, and Quality of Lamb Carcasses. *Cienc. Rural* **2025**, *55*, e20240594. [CrossRef]
36. Schingoethe, D.J. A 100-Year Review: Total Mixed Ration Feeding of Dairy Cows. *J. Dairy Sci.* **2017**, *100*, 10143–10150. [CrossRef] [PubMed]
37. Zhang, G.; Li, Y.; Fang, X.; Cai, Y.; Zhang, Y. Lactation Performance, Nitrogen Utilization, and Profitability in Dairy Cows Fed Fermented Total Mixed Ration Containing Wet Corn Gluten Feed and Corn Stover in Combination Replacing a Portion of Alfalfa Hay. *Anim. Feed Sci. Technol.* **2020**, *269*, 114687. [CrossRef]
38. Du, E.; Mao, N.; Liu, S.; Zhang, H.; Fan, M.; Sun, H.; Zheng, Y.; Cheng, Q.; Wang, C.; Li, F.; et al. Effects of Different Wet Distillers' Grains Ratios on Fermentation Quality, Nitrogen Fractions and Bacterial Communities of Total Mixed Ration Silage. *BMC Microbiol.* **2025**, *25*, 31. [CrossRef]
39. Kondo, M.; Shimizu, K.; Jayanegara, A.; Mishima, T.; Matsui, H.; Karita, S.; Goto, M.; Fujihara, T. Changes in Nutrient Composition and in Vitro Ruminal Fermentation of Total Mixed Ration Silage Stored at Different Temperatures and Periods. *J. Sci. Food Agric.* **2016**, *96*, 1175–1180. [CrossRef]
40. Brambilla, M.; Lazzari, A.; Giovinazzo, S.; Tangorra, F.; Calcante, A.; Bisaglia, C. Operational Evaluation of an Optical Sensor for the Automatic In-Line Estimation of Total Mixed Ration Fibre Length and Particle Size in a Mixing Wagon. *J. Agric. Eng.* **2025**, *56*, 1730.1–1730.6. [CrossRef]
41. Gheller, L.S.; Ghizzi, L.G.; Marques, J.A.; Takiya, C.S.; Grigoletto, N.T.S.; Dias, M.S.S.; Silva, T.B.P.; Nunes, A.T.; Da Silva, G.G.; Fernandes, L.G.X.; et al. Effects of Organic Acid-Based Products Added to Total Mixed Ration on Performance and Ruminal Fermentation of Dairy Cows. *Anim. Feed Sci. Technol.* **2020**, *261*, 114406. [CrossRef]

42. Chen, F.; Wei, J.; Yang, X.; Zhao, N.; Zhang, W.; Huang, S.; Yan, N.; Guo, W. Effect of Pelleted Total Mixed Rations with Different Levels of Intact Rapeseed on Performance, Carcass Traits, Serum Biochemical Indices and Meat Quality of Boer Goats. *Anim. Prod. Sci.* **2019**, *59*, 82. [CrossRef]
43. Miguel, M.; Mamuad, L.; Ramos, S.; Ku, M.J.; Jeong, C.D.; Kim, S.H.; Cho, Y.I.; Lee, S.S. Effects of Using Different Roughages in the Total Mixed Ration Inoculated with or without Coculture of Lactobacillus Acidophilus and Bacillus Subtilis on in Vitro Rumen Fermentation and Microbial Population. *Anim. Biosci.* **2020**, *34*, 642. [CrossRef]
44. Cao, Y.; Takahashi, T.; Horiguchi, K.; Yoshida, N.; Zhou, D. In Vitro Ruminal Dry Matter Digestibility and Methane Production of Fermented Total Mixed Ration Containing Whole-crop Rice and Rice Bran. *Grassl. Sci.* **2012**, *58*, 133–139. [CrossRef]
45. Liu, M.J.; Wang, Y.; Li, Y.Y.; Si, Q.; Bao, J.; Ge, G.T.; Wang, Z.J.; Jia, Y.S.; Du, S. Effects of Alfalfa and Oat Supplementation in Fermented Total Mixed Rations on Growth Performances, Carcass Characteristics, and Meat Quality in Lambs. *Small Rumin. Res.* **2023**, *218*, 106877. [CrossRef]
46. Li, Y.; Lv, J.; Wang, J.; Zhou, S.; Zhang, G.; Wei, B.; Sun, Y.; Lan, Y.; Dou, X.; Zhang, Y. Changes in Carbohydrate Composition in Fermented Total Mixed Ration and Its Effects on in Vitro Methane Production and Microbiome. *Front. Microbiol.* **2021**, *12*, 738334. [CrossRef] [PubMed]
47. Supapong, C.; Cherdthong, A. Can Dietary Fermented Total Mixed Ration Additives Biological and Chemical Improve Digestibility, Performance, and Rumen Fermentation in Ruminants? *Anim. Biotechnol.* **2023**, *34*, 5113–5123. [CrossRef]
48. Zhang, C.; Li, M.M.; Al-Marashdeh, O.; Gan, L.P.; Zhang, C.Y.; Zhang, G.G. Performance, Rumen Fermentation, and Gastrointestinal Microflora of Lambs Fed Pelleted or Unpelleted Total Mixed Ration. *Anim. Feed Sci. Technol.* **2019**, *253*, 22–31. [CrossRef]
49. Ahammed, B.; Alam, J.; Shubho, A.; Ali Biswas, A. Effect of the Fermented Total Mixed Ration on In-Vitro Gas Production and Digestibility in Cattle. *Bangladesh J. Vet. Anim. Sci.* **2021**, *9*, 23–32. [CrossRef]
50. Wang, H.; ur Rehman, K.; Feng, W.; Yang, D.; ur Rehman, R.; Cai, M.; Zhang, J.; Yu, Z.; Zheng, L. Physicochemical Structure of Chitin in the Developing Stages of Black Soldier Fly. *Int. J. Biol. Macromol.* **2020**, *149*, 901–907. [CrossRef]
51. Jairath, G.; Verma, A.K.; Rani, D.; Marappan, G.; Bs, Y.; Singh, B.; Mal, G.; Gopinath, D.; Sharma, R.; Katoch, S.; et al. Self-Fermented Agro-Wastes as Antioxidant Enriched Maize Grain Replacer for Sustainable Animal Feeding. *J. Clean. Prod.* **2023**, *427*, 139223. [CrossRef]
52. Paul, O.; Urmi, S.; Biswas, M. Effect of TMR and Fermented TMR on Ruminal in Vitro Digestion and Gas Production Adv. *Anim. Vet. Sci.* **2023**, *11*, 586–594.
53. Hamidan, M.F.R.; Nadzir, M.N.H.M.; Bakar, M.F.A.; Shohaimi, S.; Bidin, H.; Samat, N. Effect of Azolla Filiculoides Meal Inclusion in the Napier Silage Total Mixed Ration on the in Vitro Cumulative Gas Production and Digestibility. *Pertanika J. Trop. Agric. Sci.* **2022**, *45*, 452–467. [CrossRef]
54. Marchesini, G.; Cortese, M.; Ughelini, N.; Ricci, R.; Chinello, M.; Contiero, B.; Andrighetto, I. Effect of Total Mixed Ration Processing Time on Ration Consistency and Beef Cattle Performance during the Early Fattening Period. *Anim. Feed Sci. Technol.* **2020**, *262*, 114421. [CrossRef]
55. Gadzama, I.U.; Hoffman, L.C.; Holman, B.W.B.; Chaves, A.V.; Meale, S.J. Effects of Supplementing a Feedlot Diet with Microalgae (*Chlorella vulgaris*) on the Performance, Carcass Traits and Meat Quality of Lambs. *Livest. Sci.* **2024**, *288*, 105552. [CrossRef]
56. Zhang, J.; Shang, J.; Hao, Y.; Wang, Y.; Cao, Z.; Yang, H.; Wang, W.; Li, S. Growth Performance, Blood Metabolites, Ruminal Fermentation, and Bacterial Community in Preweaning Dairy Calves Fed Corn Silage-Included Starter and Total Mixed Ration. *J. Dairy Sci.* **2023**, *106*, 4545–4558. [CrossRef] [PubMed]
57. Vibart, R.E.; Burns, J.C.; Fellner, V. Effect of Replacing Total Mixed Ration with Pasture on Ruminal Fermentation. *Prof. Anim. Sci.* **2010**, *26*, 435–442. [CrossRef]
58. Pastorini, M.; Pomiés, N.; Repetto, J.L.; Mendoza, A.; Cajarville, C. Productive Performance and Digestive Response of Dairy Cows Fed Different Diets Combining a Total Mixed Ration and Fresh Forage. *J. Dairy Sci.* **2019**, *102*, 4118–4130. [CrossRef] [PubMed]
59. Boguhn, J.; Kluth, H.; Rodehutscord, M. Effect of Total Mixed Ration Composition on Fermentation and Efficiency of Ruminal Microbial Crude Protein Synthesis In Vitro. *J. Dairy Sci.* **2006**, *89*, 1580–1591. [CrossRef]
60. Bakshi, M.P.S.; Singh, A.S.; Wadhwa, M. Impact of Type and Level of Herbs Supplemented to Total Mixed Ration on the Fermentation Pattern and In Vitro Methane Emission. *Indian J. Anim. Nutr.* **2022**, *39*, 264–271. [CrossRef]
61. Meenongyai, W.; Pattarajinda, V.; Stelzleni, A.M.; Sethakul, J.; Duangjinda, M. Effects of Forage Ensiling and Ration Fermentation on Total Mixed Ration pH, Ruminal Fermentation and Performance of Growing Holstein-Zebu Cross Steers. *Anim. Sci. J.* **2017**, *88*, 1372–1379. [CrossRef]
62. Jahan, R.; Amin, M.; Sarker, N.; Kamal, M. Feeding Effects of Total Mixed Ration on Rumen Metabolic Profile in Cattle. *Bang. J. Anim. Sci.* **2018**, *47*, 35–39. [CrossRef]

63. Zhang, X.; Wang, H.; Guo, X. Effects of Total Mixed Ration with Various Silage on Growth Performance, Serum Parameters, Ruminal Fermentation, and Bacteria Community Profile in Beef Cattle. *Food Sci. Nutr.* **2021**, *9*, 5959–5970. [CrossRef] [PubMed]
64. Song, J.; Ma, Y.; Zhang, H.; Wang, L.; Zhang, Y.; Zhang, G. Fermented Total Mixed Ration Alters Rumen Fermentation Parameters and Microbiota in Dairy Cows. *Animals* **2023**, *13*, 1062. [CrossRef] [PubMed]
65. Zhong, R.; Zhao, C.; Feng, P.; Wang, Y.; Zhao, X.; Luo, D.; Cheng, L.; Liu, D.; Fang, Y. Effects of Feeding Ground versus Pelleted Total Mixed Ration on Digestion, Rumen Function and Milk Production Performance of Dairy Cows. *Int. J. Dairy Tech.* **2020**, *73*, 22–30. [CrossRef]
66. Ku, M.-J.; Miguel, M.A.; Kim, S.-H.; Jeong, C.-D.; Ramos, S.C.; Son, A.-R.; Cho, Y.-I.; Lee, S.-S.; Lee, S.-S. Effects of Italian Ryegrass Silage-Based Total Mixed Ration on Rumen Fermentation, Growth Performance, Blood Metabolites, and Bacterial Communities of Growing Hanwoo Heifers. *J. Anim. Sci. Technol.* **2023**, *65*, 951–970. [CrossRef]
67. Metzler-Zebeli, B.U.; Scherr, C.; Sallaku, E.; Drochner, W.; Zebeli, Q. Evaluation of Associative Effects of Total Mixed Ration for Dairy Cattle Using in Vitro Gas Production and Different Rumen Inocula. *J. Sci. Food Agric.* **2012**, *92*, 2479–2485. [CrossRef]
68. Rodriguez, M.P.; Mariezcurrena, M.D.; Mariezcurrena, M.A.; Lagunas, B.C.; Elghandour, M.M.M.Y.; Kholif, A.M.; Kholif, A.E.; Almaráz, E.M.; Salem, A.Z.M. Influence of live cells or cells extract of *Saccharomyces Cerevisiae* on in vitro gas production of a total mixed ration. *Ital. J. Anim. Sci.* **2015**, *14*, 3713. [CrossRef]
69. Tafaj, M.; Zebeli, Q.; Junck, B.; Steingass, H.; Drochner, W. Effects of Particle Size of a Total Mixed Ration on *in Vivo* Ruminal Fermentation Patterns and Inocula Characteristics Used for in Vitro Gas Production. *Anim. Feed Sci. Technol.* **2005**, *123–124*, 139–154. [CrossRef]
70. Alhadas, H.; Valadares, S.; Souza, C.; Rocha, L.; Pacheco, M.; Silva, B.; Costa, D.; Mantovani, H. PSX-18 Inclusion of Physically Effective Fiber from Forage on Whole Corn Based Diets for Beef Cattle and Its Effects on the Intake and Digestibilities of Dry Matter, Starch and Neutral Detergent Fiber. *J. Anim. Sci.* **2018**, *96* (Suppl. 3), 414. [CrossRef]
71. Felizari, L.; Millen, D.; Perina Bertoldi, G.; Squizatti, M.; Silvestre, A.; Pavão Coffani Nunes, A.; Garcia Mesquita, L.; Millen, D. PSX-26 Ruminal Bacteria and Protozoa Profile during Adaptation and Finishing Periods of Nellore Cattle Previously Submitted to Either Nutritional Restriction or Intake Concentrate Feedstuffs. *J. Anim. Sci.* **2018**, *96* (Suppl. 3), 415–416. [CrossRef]
72. Sun, X.; Wang, H.; You, P.; Pacheco, D.; Wang, M.; Wu, T.; Song, B.; Kang, K.; Li, Y.; Li, B.; et al. Agglomerated Live Yeast (*Saccharomyces cerevisiae*) Supplemented to Pelleted Total Mixed Rations Improves the Growth Performance of Fattening Lambs. *Livest. Sci.* **2022**, *258*, 104855. [CrossRef]
73. Tufarelli, V.; Khan, R.U.; Laudadio, V. Feeding of Wheat Middlings in Lamb Total Mixed Rations: Effects on Growth Performance and Carcass Traits. *Anim. Feed Sci. Technol.* **2011**, *170*, 130–135. [CrossRef]
74. Nkosi, B.D.; Meeske, R. Effects of Ensiling Totally Mixed Potato Hash Ration with or without a Heterofermentative Bacterial Inoculant on Silage Fermentation, Aerobic Stability, Growth Performance and Digestibility in Lambs. *Anim. Feed Sci. Technol.* **2010**, *161*, 38–48. [CrossRef]
75. Nha, B.; Pattarajinda, V. Effect of Physically Effective Neutral Detergent Fibre and Moisture Content in Fermented Total Mixed Ration on Lactating Cow Performance. *Indian J. Anim. Res.* **2019**, *53*, 913–917.
76. Jang, S.Y.; Kim, E.K.; Park, J.H.; Oh, M.R.; Tang, Y.J.; Ding, Y.L.; Seong, H.J.; Kim, W.H.; Yun, Y.S.; Moon, S.H. Effects of Physically Effective Neutral Detergent Fiber Content on Dry Matter Intake, Digestibility, and Chewing Activity in Korean Native Goats (*Capra hircus coreanae*) Fed with Total Mixed Ration. *Asian-Australas J. Anim. Sci.* **2017**, *30*, 1405–1409. [CrossRef]
77. Kumar, V.; Tyagi, A.; Thakur, S.; Singh, N.P.; Chaudhary, J. Effect of Different Feeding Systems on Performance of Lactating Murrah Buffaloes. *Indian J. Dairy Sci.* **2015**, *68*, 61–64.
78. Lakhani, N.; Tyagi, N. Comparative Evaluation of in Vitro Techniques for Predicting Metabolizable Energy Content of Total Mixed Ration for Murrah Buffaloes. *Indian J. Anim. Sci.* **2021**, *91*, 220–224. [CrossRef]
79. Sunarso; Christiyanto, M.; Nuswantara, L.K. Feeding of Total Mixed Ration on the Productivity of Friesian Holstein Cross-Grade Cattle. *IOP Conf. Ser. Earth Environ. Sci.* **2018**, *119*, 012024. [CrossRef]
80. Lailer, P.; Dahiya, S.; Chauhan, T. Complete Feed for Livestock Concept, Present Status and Future Trend: A Review. *Indian J. Anim. Sci.* **2005**, *75*, 84–91.
81. Koch, C.; Schönleben, M.; Mentschel, J.; Göres, N.; Fissore, P.; Cohrs, I.; Sauerwein, H.; Ghaffari, M.H. Growth Performance and Economic Impact of Simmental Fattening Bulls Fed Dry or Corn Silage-Based Total Mixed Rations. *Animal* **2023**, *17*, 100762. [CrossRef]
82. Teshome, D.; Fita, L.; Feyissa, F.; Kitaw, G.; Wondatir, Z. Effect of Total Mixed Ration on Dry Matter Intake, Milk Yield and Composition of Early Lactating Jersey Cows. *J. Biol.* **2017**, *7*, 19–24.
83. Hernández-Ortega, M.; Martínez-Fernández, A.; Soldado, A.; González, A.; Arriaga-Jordán, C.M.; Argamentería, A.; De La Roza-Delgado, B.; Vicente, F. Effect of Total Mixed Ration Composition and Daily Grazing Pattern on Milk Production, Composition and Fatty Acids Profile of Dairy Cows. *J. Dairy Res.* **2014**, *81*, 471–478. [CrossRef]

84. Mäntysaari, P.; Khalili, H.; Sariola, J. Effect of Feeding Frequency of a Total Mixed Ration on the Performance of High-Yielding Dairy Cows. *J. Dairy Sci.* **2006**, *89*, 4312–4320. [CrossRef] [PubMed]
85. Caccamo, M.; Veerkamp, R.F.; Licitra, G.; Petriglieri, R.; La Terra, F.; Pozzebon, A.; Ferguson, J.D. Association of Total-Mixed-Ration Chemical Composition with Milk, Fat, and Protein Yield Lactation Curves at the Individual Level. *J. Dairy Sci.* **2012**, *95*, 6171–6183. [CrossRef] [PubMed]
86. Khy, V.; Prucsasri, P.; Kanthapanit, C.; Chtwachirawong, P. A Comparison of Growth, Feed Efficiency and Carcass Characteristics of Kamphaengsaen Steers Fed Two TMR Fiber Sources During Two Different Feeding Period. *Agric. Nat. Resour.* **2000**, *34*, 216–226.
87. Kim, J.G.; Cheong, E.C.; Li, Y.F.; Kim, H.J.; Farhad, A.; Kim, M.J. Effect of TMR Feed Mixed with Whole Crop Rice on Growth Performance and Meat Quality of Hanwoo Steers. *J. Korean Soc. Grassl. Forage Sci.* **2021**, *41*, 267–272. [CrossRef]
88. Yun, Y.-S.; Seong, H.-J.; Zhang, Q.-M.; Chung, S.-U.; Lee, G.-E.; Jang, S.-Y.; Lee, J.; Lee, S.-H.; Moon, S.-H. Changes in Goat Productivity and Economical Efficiency at Feeding Systems by Castrated Growing Korean Native Goat (*Capra hircus coreanae*). *J. Korean Soc. Grassl. Forage Sci.* **2018**, *38*, 150–155. [CrossRef]
89. Kim, T.; Mayakrishnan, V.; Lim, D.; Yeon, J.; Baek, K. Effect of Fermented Total Mixed Rations on the Growth Performance, Carcass and Meat Quality Characteristics of Hanwoo Steers. *Anim. Sci. J.* **2018**, *89*, 606–615. [CrossRef]
90. Wang, J.; Yang, B.Y.; Zhang, S.J.; Amar, A.; Chaudhry, A.S.; Cheng, L.; Abbasi, I.H.R.; Al-Mamun, M.; Guo, X.F.; Shan, A.S. Using Mixed Silages of Sweet Sorghum and Alfalfa in Total Mixed Rations to Improve Growth Performance, Nutrient Digestibility, Carcass Traits and Meat Quality of Sheep. *Animal* **2021**, *15*, 100246. [CrossRef]
91. Soder, K.J.; Rotz, C.A. Economic and Environmental Impact of Utilizing a Total Mixed Ration in Pennsylvania Grazing Dairy Herds. *Prof. Anim. Sci.* **2003**, *19*, 304–311. [CrossRef]
92. Pankaj, P.K.; Ramana, D.; Nirmala, G.; Maruthi, V.; Shankar, K.S.; Prasad, T.; Singh, V.K. Enhancing Dryland Sheep Productivity with Densified Crop Residue Total Mixed Rations. *Indian J. Dryland Agric. Res. Dev.* **2023**, *38*, 49–53. [CrossRef]
93. O'Neill, B.F.; Deighton, M.H.; O'Loughlin, B.M.; Mulligan, F.J.; Boland, T.M.; O'Donovan, M.; Lewis, E. Effects of a Perennial Ryegrass Diet or Total Mixed Ration Diet Offered to Spring-Calving Holstein-Friesian Dairy Cows on Methane Emissions, Dry Matter Intake, and Milk Production. *J. Dairy Sci.* **2011**, *94*, 1941–1951. [CrossRef]
94. Griinari, J.M.; Dwyer, D.A.; McGuire, M.A.; Bauman, D.E.; Palmquist, D.L.; Nurmela, K.V. Trans-Octadecenoic Acids and Milk Fat Depression in Lactating Dairy Cows. *J. Dairy Sci.* **1998**, *81*, 1251–1261. [CrossRef]
95. Mohammad, M.; Gorgulu, M.; Goncu, S. The Effects of Total Mixed Ration and Separate Feeding on Lactational Performance of Dairy Cows. *Asian Res. J. Agric.* **2017**, *5*, 1–7. [CrossRef]
96. Østergaard, S.; Sørensen, J.T.; Hindhede, J.; Kristensen, A.R. Technical and Economic Effects of Feeding One vs. Multiple Total Mixed Rations Estimated by Stochastic Simulation under Different Dairy Herd and Management Characteristics. *Livest. Prod. Sci.* **1996**, *45*, 23–33. [CrossRef]
97. Renna, M.; Cornale, P.; Lussiana, C.; Battaglini, L.M.; Turille, G.; Mimosi, A. Milk Yield, Gross Composition and Fatty Acid Profile of Dual-purpose AOsta REd PIed Cows Fed Separate Concentrate-forage versus Total Mixed Ration. *Anim. Sci. J.* **2014**, *85*, 37–45. [CrossRef] [PubMed]
98. Mendoza, A.; Cajarville, C.; Repetto, J.L. Digestive Response of Dairy Cows Fed Diets Combining Fresh Forage with a Total Mixed Ration. *J. Dairy Sci.* **2016**, *99*, 8779–8789. [CrossRef]
99. Wongnen, C.; Wachirapakorn, C.; Patipan, C.; Panpong, D.; Kongweha, K.; Namsaen, N.; Gunun, P.; Yuangklang, C. Effects of Fermented Total Mixed Ration and Cracked Cottonseed on Milk Yield and Milk Composition in Dairy Cows. *Asian Australas. J. Anim. Sci.* **2009**, *22*, 1625–1632. [CrossRef]
100. Kolver, E.; Muller, L. Performance and Nutrient Intake of High Producing Holstein Cows Consuming Pasture or a Total Mixed Ration. *J. Dairy Sci.* **1998**, *81*, 1403–1411. [CrossRef]
101. Sakai, T.; Devkota, N.R.; Oishi, K.; Hirooka, H.; Kumagai, H. Evaluation of Total Mixed Ration Silage with Brewers Grains for Dairy Buffalo in Tarai, Nepal. *Anim. Sci. J.* **2015**, *86*, 884–890. [CrossRef] [PubMed]
102. Ponnampalam, E.; Priyashantha, H.; Vidanarachchi, J.; Kiani, A.; Holman, B. Effects of Nutritional Factors on Fat Content, Fatty Acid Composition, and Sensorial Properties of Meat and Milk from Domesticated Ruminants: An Overview. *Animals* **2024**, *14*, 840. [CrossRef] [PubMed]
103. Ponnampalam, E.N.; Norng, S.; Burnett, V.F.; Dunshea, F.R.; Jacobs, J.L.; Hopkins, D.L. The Synergism of Biochemical Components Controlling Lipid Oxidation in Lamb Muscle. *Lipids* **2014**, *49*, 757–766. [CrossRef]
104. Santos-Silva, J.; Alves, S.P.; Francisco, A.; Portugal, A.P.; Dentinho, M.T.; Almeida, J.; da Silva, J.L.R.; Fialho, L.; Cachucho, L.; Jerónimo, E.; et al. Forage Based Diet as an Alternative to a High Concentrate Diet for Finishing Young Bulls—Effects on Growth Performance, Greenhouse Gas Emissions and Meat Quality. *Meat Sci.* **2023**, *198*, 109098. [CrossRef]

105. Alves, S.P.; Vahmani, P.; Mapiye, C.; McAllister, T.A.; Bessa, R.J.B.; Dugan, M.E.R. Trans-10 18:1 in Ruminant Meats: A Review. *Lipids* **2021**, *56*, 539–562. [CrossRef]
106. Johnson, K.A.; Johnson, D.E. Methane Emissions from Cattle. *J. Anim. Sci.* **1995**, *73*, 2483–2492. [CrossRef]
107. Koenig, K.; Beauchemin, K.; McGinn, S. PSX-22 Feeding Condensed and Hydrolysable Tannins in High Protein Finishing Diets to Beef Cattle to Reduce Ammonia Emissions from Manure. *J. Anim. Sci.* **2018**, *96* (Suppl. 3), 415. [CrossRef]
108. Tangorra, F.M.; Calcante, A. *Potential Economic and Environmental Benefits of Automating Milking and Total Mixed Ration (TMR) Feeding on a Dairy Farm*; Springer: Berlin/Heidelberg, Germany, 2024; pp. 935–943.
109. Cogan, T.; Hawkey, R.; Higgie, E.; Lee, M.R.F.; Mee, E.; Parfitt, D.; Raj, J.; Roderick, S.; Walker, N.; Ward, P.; et al. Silage and Total Mixed Ration Hygienic Quality on Commercial Farms: Implications for Animal Production. *Grass Forage Sci.* **2017**, *72*, 601–613. [CrossRef]
110. González-Jartín, J.M.; Rodríguez-Cañás, I.; Alvariño, R.; Alfonso, A.; Sainz, M.J.; Vieytes, M.R.; Gomes, A.; Ramos, I.; Botana, L.M. Occurrence of Mycotoxins in Total Mixed Ration of Dairy Farms in Portugal and Carry-over to Milk. *Food Control* **2024**, *165*, 110682. [CrossRef]
111. Rodríguez-Blanco, M.; Marín, S.; Sanchis, V.; Ramos, A.J. Fusarium Mycotoxins in Total Mixed Rations for Dairy Cows. *Mycotoxin Res.* **2020**, *36*, 277–286. [CrossRef]
112. Sultana, N.; Rashid, A.; Tahira, I.; Hanif, H.; Hanif, N. Distribution of Various Mycotoxins in Compound Feed, Total Mix Ration and Silage. *Pak. Vet. J.* **2013**, *33*, 200–204.
113. Chohan, K.A.; Awan, F.; Ali, M.M.; Iqbal, U.; Ijaz, M. Assessment of Aflatoxin in Dairy Concentrate Feeds, Total Mixed Rations, Silage and Various Feed Ingredients in Pakistan. *Pak. J. Zool.* **2016**, *48*, 277–280. [CrossRef]
114. Vaičiulienė, G.; Bakutis, B.; Jovaišienė, J.; Falkauskas, R.; Gerulis, G.; Kerzienė, S.; Baliukonienė, V. Prevalence of Mycotoxins and Endotoxins in Total Mixed Rations and Different Types of Ensiled Forages for Dairy Cows in Lithuania. *Toxins* **2021**, *13*, 890. [CrossRef] [PubMed]
115. Masoero, F.; Gallo, A.; Diaz, D.; Piva, G.; Moschini, M. Effects of the Procedure of Inclusion of a Sequestering Agent in the Total Mixed Ration on Proportional Aflatoxin M1 Excretion into Milk of Lactating Dairy Cows. *Anim. Feed Sci. Technol.* **2009**, *150*, 34–45. [CrossRef]
116. Mohammadi Shad, Z.; Ghavami, M.; Atungulu, G.G. Occurrence of Aflatoxin in Dairy Cow Feed Ingredients and Total Mixed Ration. *Appl. Eng. Agric.* **2019**, *35*, 679–686. [CrossRef]
117. Gallo, A.; Minuti, A.; Bani, P.; Bertuzzi, T.; Cappelli, F.P.; Doupovec, B.; Faas, J.; Schatzmayr, D.; Trevisi, E. A Mycotoxin-Deactivating Feed Additive Counteracts the Adverse Effects of Regular Levels of Fusarium Mycotoxins in Dairy Cows. *J. Dairy Sci.* **2020**, *103*, 11314–11331. [CrossRef]
118. Hu, Z.; Zuo, M.; Ding, S.; Zhong, Y.; Xue, M.; Zheng, H. Integrating Metabolomics and Genomics to Uncover the Impact of Fermented Total Mixed Ration on Heifer Growth Performance Through Host-Dependent Metabolic Pathways. *Animals* **2025**, *15*, 173. [CrossRef]
119. Mansfield, H.; Endres, M.; Stern, M. Comparison of Microbial Fermentation in the Rumen of Dairy Cows and Dual Flow Continuous Culture. *Anim. Feed Sci. Technol.* **1995**, *55*, 47–66. [CrossRef]

ruminants

MDPI

Review

Influence of Virtual Fencing Technology in Cattle Management and Animal Welfare

Ishaya Usman Gadzama [1,*], Homa Asadi [2], Qazal Hina [3] and Saraswati Ray [4]

1 School of Agriculture and Food Sustainability, University of Queensland, Gatton, QLD 4343, Australia
2 Department of Plant Protection, Faculty of Agriculture, Vali-e-Asr University of Rafsanjan, Rafsanjan 7718897111, Iran; homaasadi9@gmail.com
3 Department of Animal Nutrition, University of Veterinary and Animal Sciences, Lahore 54000, Pakistan; qazalhina5@gmail.com
4 School of Environmental and Rural Science, Faculty of Science, Agriculture, Business and Law (SABL), University of New England, Armidale, NSW 2351, Australia; raiisonu3225@gmail.com
* Correspondence: ishayausman@gmail.com

Simple Summary: Virtual fencing (VF) technology offers a modern approach to cattle management by using GPS-enabled collars to create invisible boundaries instead of physical fences. Studies indicate that cattle quickly learn to associate auditory cues with mild electrical pulses, achieving high containment rates (≥90%) within days. While welfare impacts are generally minimal, with cortisol levels comparable to traditional fencing, some short-term behavioral disruptions and occasional collar-related abrasions have been reported. This technology enhances pasture management flexibility but faces challenges such as high costs, connectivity issues, and individual animal variability. Further research is needed to optimize training protocols, assess long-term welfare effects, and improve scalability for diverse farming systems.

Abstract: Virtual fencing (VF) technology represents an innovative approach to livestock management, utilizing GPS-enabled collars to establish invisible boundaries through auditory and mild electrical stimuli. While VF offers potential benefits such as enhanced pasture management flexibility and reduced labor costs, its widespread adoption faces challenges including high initial investment costs, connectivity issues, GPS accuracy limitations, potential device durability concerns, and individual animal variability in learning and response. Furthermore, despite studies showing rapid learning and generally minimal long-term welfare impacts, questions remain regarding optimizing training protocols, addressing occasional short-term behavioral disruptions and collar abrasions, assessing long-term welfare effects across diverse systems (especially intensive and dairy), and improving scalability. To comprehensively assess the potential and limitations of this technology and guide its future development and implementation, a review integrating existing knowledge on the efficacy, welfare implications, and practical applications of VF in cattle production systems is essential. This review examines the efficacy, welfare implications, and practical applications of VF in cattle production systems. Studies demonstrate that cattle rapidly learn to associate auditory cues with electrical pulses, achieving high containment rates (≥90%) within days, with minimal long-term welfare impacts as indicated by stable cortisol levels. However, short-term behavioral disruptions and occasional collar-related abrasions have been reported, particularly in dairy cattle. While VF enhances pasture management flexibility and reduces labor costs, challenges such as connectivity issues, individual animal variability, and high initial investment costs limit its widespread adoption. The findings suggest that VF is a promising tool for precision livestock farming, though further research is needed to optimize training protocols, assess long-term welfare effects, and improve scalability across diverse farming systems.

Keywords: precision livestock farming; animal welfare; stress indicators; GPS-enabled collars; animal behavior; associative learning; grazing management

1. Introduction

Grazing livestock has been a fundamental agricultural practice for centuries, shaping landscapes and providing essential food resources [1]. Historically, controlling the movement of grazing animals relied primarily on herding and physical barriers such as stone walls, hedges, and wire fences [2]. While these traditional methods were effective, they required significant labor and infrastructure investment. The past century has seen substantial changes in temperate grazing dairy systems, particularly due to the increasing demand for animal-derived protein, which is projected to nearly double by 2050 [3]. This growing demand has intensified pressure on grazing land, necessitating more efficient and flexible management strategies [4].

In response, technological advancements have led to the development of precision livestock farming (PLF), which aims to enhance animal management through automated monitoring [5,6]. PLF technologies, including electronic identification systems, on-animal sensors (e.g., accelerometers and GPS), and stationary management systems, are increasingly being adopted in both intensive and extensive farming operations [7,8]. Within this technological progression, virtual fencing (VF) has emerged as an innovative solution to manage animal movement without physical barriers [2,6]. VF systems typically involve wearable collars that deliver sensory cues, such as audio warnings followed by mild electrical pulses, to contain animals within predefined virtual boundaries [9]. This technology represents a significant advancement in flexible, data-driven grazing management, offering potential benefits for pasture utilization and environmental conservation [10].

Virtual fencing is defined as an enclosure or boundary system without physical structures, relying instead on wearable devices such as collars [2,4,6]. When an animal approaches a virtual boundary, it first receives an audio cue, followed by a low-intensity electrical pulse if it continues moving forward [9]. This operant conditioning approach trains animals to retreat upon hearing the initial warning, minimizing the need for electrical corrections over time [7]. Research has demonstrated that cattle can successfully adapt to these systems, with prior exposure to physical electric fences accelerating learning [11,12]. Companies such as Halter, Vence, Nofence, eShepherd/Gallaghers, and Boviguard have commercialized VF technology, leveraging GPS tracking and wireless communication to dynamically adjust grazing areas [13].

A key advantage of VF is its ability to facilitate rotational and strip-grazing systems, which traditionally require labor-intensive fence adjustments [10]. By enabling farmers to modify grazing zones remotely via smartphones or computers, VF enhances pasture utilization efficiency while reducing labor costs [14]. Furthermore, VF can protect environmentally sensitive areas, such as riparian zones, by temporarily excluding livestock to prevent soil erosion and water contamination [9,15]. Studies have shown that VF can effectively restrict cattle from vulnerable ecosystems while maintaining pasture productivity [16].

A critical aspect of VF is the animal's ability to learn and adapt to virtual boundaries through associative learning [7]. Research on beef heifers in rotational grazing systems has shown a decline in electric-to-acoustic signal ratios over time, indicating successful behavioral adaptation [12]. Similarly, nursing cows undergoing VF training exhibited reduced electrical stimuli as they learned to respond to audio cues alone [14]. Furthermore, some research found no significant difference in FCM levels between virtually and physically fenced cattle, indicating comparable stress levels [17], while others similarly reported no significant differences in welfare indicators such as lying time and weight gain [13]. These findings highlight the need for further investigation into the long-term welfare implications of virtual fencing across different livestock species. This study characterizes the state of the art regarding the use of VF as an innovative tool for modern grazing management. Specifically, it (1) explores the technological foundations of VF systems, including their reliance on wearable collars with GPS and auditory–electrical stimuli to control livestock movement without physical barriers; (2) reviews research on animal learning and behavioral adaptation, assessing how cattle respond to virtual boundaries over time; and (3) evaluates animal welfare implications, particularly stress responses associated with electrical stimuli, through analyses of physiological indicators (e.g., fecal cortisol metabolites, FCM) and behavioral metrics.

Beyond animal behavior, this paper discusses the practical applications of VF in grazing systems, including its role in rotational grazing, pasture optimization, and environmental protection (e.g., preventing overgrazing near riparian zones) [9,15]. It also situates VF within the broader framework of PLF, highlighting how sensor-based monitoring enhances livestock management efficiency. Finally, the paper considers barriers to adoption, such as cost and farmer acceptance, while emphasizing the technology's potential to support sustainable livestock production in the face of growing global food demand and environmental challenges.

2. Materials and Methods

2.1. Literature Search

To comprehensively assess the potential of virtual fencing (VF) for cattle management, a comprehensive literature search was conducted across multiple academic databases, including Google Scholar, ScienceDirect, Scopus, PubMed, and Web of Science. The goal was to gather peer-reviewed studies evaluating the efficacy, animal welfare implications, and practical applications of VF in grazing systems. Both controlled experiments and field trials were included to ensure a balanced understanding of real-world applicability. While peer-reviewed journal articles formed the core of the review, relevant conference papers and technical reports were also considered to capture emerging trends in the field.

2.2. Searching Criteria

The search strategy employed targeted keywords such as "virtual fencing cattle behavior", "precision livestock farming grazing", "animal welfare virtual fencing", and "GPS collars grazing management". Boolean operators (AND/OR) were used to refine the results, focusing primarily on studies involving cattle. Additional filters were applied to prioritize studies examining animal learning, stress responses (e.g., cortisol levels), and pasture utilization efficiency. No restrictions were placed on publication years to ensure the inclusion of both foundational and recent research. To maintain scientific rigor, non-peer-reviewed sources, commercial advertisements, and non-English publications without verified translations were excluded. Reference lists of key articles were manually screened to identify further relevant studies.

3. The Concept and Mechanism of Virtual Fencing for Cattle

Virtual fencing represents an innovative advancement in precision livestock farming, offering a boundary control system that eliminates the need for physical barriers [18]. This technology has gained attention for its potential to revolutionize grazing management through targeted grazing approaches, labor reduction, and application in environmentally sensitive areas where traditional fencing proves impractical [19]. Unlike conventional fencing methods, VF systems utilize wearable technology, typically GPS-enabled collars, to establish and maintain virtual boundaries (Figure 1).

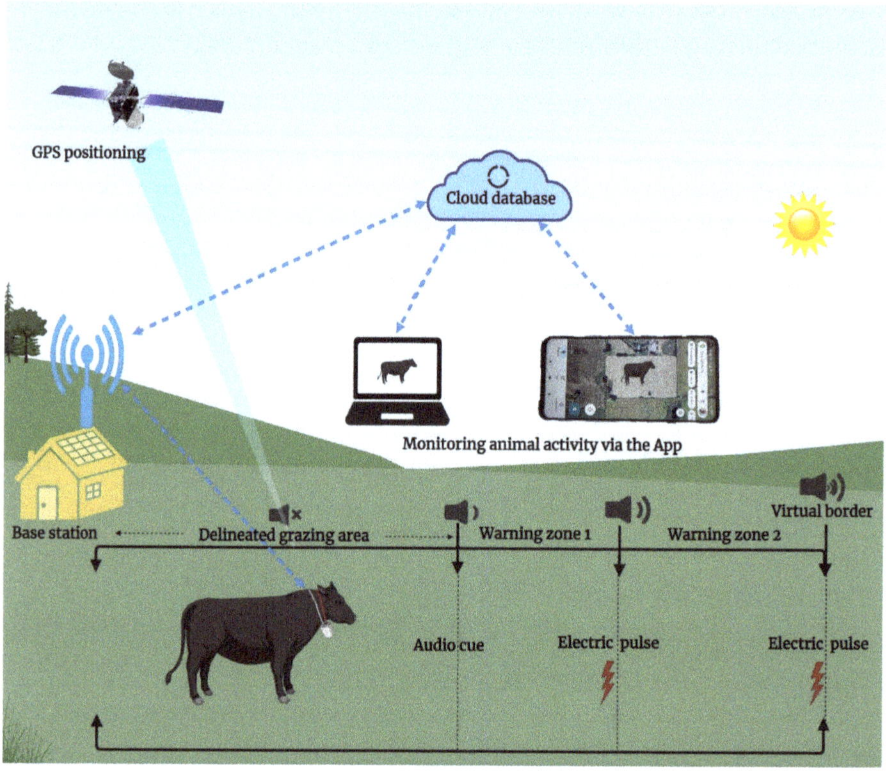

Figure 1. Detailed schematic representation of the virtual fencing concept, including the components used. GPS—Global Positioning System.

The operational mechanism of virtual fencing relies on the sophisticated integration of positioning technology and behavioral conditioning [20]. As the authors of [21] explain, these systems continuously monitor an animal's position through GPS tracking, comparing it against predefined virtual boundaries established by farmers using digital mapping tools. When cattle approach these boundaries, the system initiates a two-stage deterrent process: first emitting an audible warning signal, followed by a mild electrical stimulus if the animal continues moving toward the boundary [22]. This design is rooted in operant conditioning principles, where animals learn to associate the initial auditory cue with the subsequent aversive stimulus [7].

Research demonstrates that cattle can effectively learn this association, with studies showing a significant reduction in electrical stimuli required over time as animals learn to respond to audio cues alone [9]. Verdon et al. [23] reported that experienced cattle showed containment rates comparable to physical fencing, with electrical stimuli decreasing by up to 80% after the initial learning period. This learning capacity is crucial for both the effectiveness and ethical justification of virtual fencing systems [24]. Current commercial systems like eShepherd and NoFence have demonstrated impressive containment capabilities, with some studies reporting ≥99% effectiveness in keeping cattle within designated areas [25]. However, as Goliński et al. [4] note, these systems may not completely eliminate boundary breaches to the same degree as physical barriers. Despite this limitation, the technology offers unique advantages, including the ability to create exclusion zones within pastures and precisely track animal movements [10]. These features enable more dynamic grazing management strategies, such as rotational grazing systems that can be adjusted remotely in response to pasture conditions [16].

Several virtual fencing systems and manufacturers have been evaluated in research (Table 1). For instance, the Halter® system, developed for intensive pastoral dairy farming, utilizes sound and vibration as primary cues and a low-energy electrical pulse as a secondary aversive cue [26]. Research has shown that cows can learn to respond to sound cues within a day and can be remotely herded to the milking parlor [27]. Another widely studied system, Nofence®, has demonstrated success in keeping cattle within virtual enclosures, with animals learning to respond to auditory warnings over time during a 4-week evaluation period [28,29]. The Halter® system was evaluated over a 6-week period, while the eShepherd® system by Gallagher has been tested on pasture-raised Angus and Jersey cows, in a 4-week study following a 1-week acclimation period (Table 1), showing effective boundary compliance through proximity-based beep and pulse responses [24–29]. Previous research by Anderson and Hale [30] and Umstätter [2] laid the groundwork by defining virtual fencing and reviewing its evolution, highlighting its potential in rangeland settings.

A wide range of parameters have been measured to assess efficacy, animal welfare, and behavioral responses to virtual fencing (Table 1). Technology efficacy, often measured by the percentage of time animals remain within virtual boundaries, has generally shown high containment rates (≥99%) for various cattle types, including heifers, steers, dry cows, and lactating dairy cows [31–34]. Langworthy et al. [10] studied lactating dairy cows in intensive grazing systems and found that while virtual front-fences did not entirely eliminate entry into exclusion zones, cows were contained within inclusion zones ≥ 99% of the time. These findings align with earlier studies on beef and dry dairy cattle [35,36]. However, Verdon et al. [23] reported occasional containment failures, suggesting that effectiveness may vary depending on environmental conditions and herd dynamics. The number and type of cues delivered (audio vs. electrical pulses) are frequently recorded to assess animal learning. A common finding is that electrical pulses decrease over time as cattle learn to associate auditory warnings with boundary limits [37].

Animal welfare is a critical aspect investigated in virtual fencing studies. Researchers have measured physiological stress indicators such as cortisol levels in milk, hair, and feces [29,38]. While Langworthy et al. [10] noted transient increases in milk cortisol concentrations during the initial week of virtual fencing, Sonne et al. [17] found no significant changes in manure cortisol levels. A systematic review by Wilms et al. [13] concluded that most welfare indicators, including fecal cortisol metabolites (FCMs), showed no significant differences between virtually and physically fenced cattle, with some studies even reporting lower stress levels in VF-managed herds. Behavioral responses, such as time spent near virtual fences, lying duration, and activity levels, have also been analyzed [39,40]. Studies indicate that cattle may initially avoid fence boundaries but adapt over time [41].

Individual differences in behavior suggest that temperament influences adaptation, with more exploratory animals requiring more corrective stimuli [42].

3.1. Associative Learning Is Crucial in the Context of Virtual Fencing and Animal Welfare

Virtual fencing technology is used to manage livestock movement using wearable devices that emit sensory cues, typically pairing a non-aversive audio signal with an electrical stimulus if the animal continues toward the boundary. The success of this system relies on associative learning, where animals learn to associate the audio cue with the subsequent aversive stimulus, allowing them to predict and control their interactions with the virtual boundary [20,22]. This predictability and controllability are critical for animal welfare, as they enable animals to avoid the electrical pulse by responding to the benign audio cue alone, reducing stress and improving welfare outcomes [24]. Studies have shown that species like cattle, sheep, and goats can learn this association, with most animals reducing their reliance on electrical stimuli over time [4,34]. However, individual variation exists, with some animals requiring more training or failing to learn, which may compromise welfare [41,42].

While acute stress may occur during initial learning, successful associative learning minimizes long-term welfare impacts, as evidenced by declining electrical stimuli in trained animals [38]. Social learning and herd behavior also influence individual responses, with group training enhancing learning efficiency [9,23,39]. However, further long-term studies are needed to assess welfare impacts in intensive systems and address individual differences.

3.2. Learning Behavior and Social Adaptation

Learning behavior is a key focus, with studies examining how quickly cattle respond to audio cues and avoid electrical pulses. Naïve heifers have been shown to comply with virtual fence boundaries within 3–7 days, with learning retention observed over extended periods [43]. The decreasing ratio of electrical to acoustic signals over time indicates successful associative learning [44]. Social learning also plays a role, as cattle react to herd mates receiving stimuli, suggesting that group training may enhance compliance [45]. Training protocols, whether individual or group-based, have been explored to optimize learning efficiency, with some studies recommending phased training to minimize stress [46].

3.3. Impact on Livestock Performance and Pasture Management

The impact of virtual fencing on livestock performance, including weight gain and milk production, has been evaluated. Hamidi et al. [34] found no negative effects on beef heifer weight gain compared to traditional electric fencing. Similarly, Verdon et al. [23] reported no adverse effects on dairy cow milk yield or live weight. Pasture utilization studies present mixed findings: some suggest lower grazing efficiency with VF, while others highlight its potential to prevent overgrazing and protect sensitive ecosystems [44]. Virtual fencing is applied in rotational grazing, strip-grazing, and conservation grazing, offering labor and cost savings over physical fences [28]. It also enables dynamic pasture management, such as creating firebreaks or protecting riparian zones [47]. However, limitations remain, including the need for further research on large-scale deployments, suckler cows with calves, and long-term welfare impacts [48]. Future studies should explore individual animal variability, optimal training methods, and integration with precision livestock farming technologies [44]. While challenges remain, virtual fencing represents a significant advancement in precision livestock farming, providing a flexible, welfare-compatible tool for pasture-based management. Ongoing innovations and research will continue to enhance its applicability, making it a promising solution for sustainable cattle production.

Table 1. Overview of virtual fencing studies in cattle: devices, manufacturers, and measured parameters (2001–2025).

Year	Animal Species/Breed	Virtual Fencing Device	Device Manufacturer	Location on Animal	Behavior/Parameters Measured	Reference
2025	Kinsella Composite (KC) crossbred beef cattle (heifers and cows)	Nofence®	Nofence AS, Batnfjordsøra (Norway)	Neck collar with adjustable strap	Electrical pulses (EPs) and audio cues (ACs) received EP-to-AC ratio (E:A) Inclusion zone frequency (IZF, % time within boundaries) Escape frequency/duration Step counts (via leg sensor) Lying/standing time	[42]
2025	Kinsella Composite (KC) crossbred beef cattle (heifers and cows)	IceQube+ activity sensor	Peacock Technologies (Stirling, UK)	Lower left rear leg	Activity patterns (step counts) Lying vs. standing time	[42]
2025	Fleckvieh heifers	Nofence®	Nofence AS, Batnfjordsøra (Norway)	Neck	Success ratio (auditory/electrical), escape alerts, GNSS, fecal cortisol, body weight gain	[36]
2024	Fleckvieh heifers	Nofence®	Nofence AS, Batnfjordsøra (Norway)	Neck collar	Behavior metrics, herbage intake, stress indicators (fecal cortisol)	[43]
2024	Angus cows	Vence® VF system	Vence (vence.io; Merck & Co., Inc., Rahway, NJ, USA)	GPS-enabled VF collar	Percentage of cow locations in different management zones (riparian exclusion, ridge exclusion, grazing area, large-area exclusions), noncompliance	[44]
2024	Lactating Holstein-Friesian	Nofence®	Nofence AS, Batnfjordsøra (Norway)	Neck collar	Acoustic warnings, electrical pulses, step count, milk yield, hair cortisol	[28]
2024	Pasture-raised Angus beef, Jersey dry cows	eShepherd®	Gallagher, Hamilton (New Zealand)	Collar	Proximity to boundary (beep/pulse response)	[29]
2023	Dairy cows	Nofence®	Nofence AS, Batnfjordsøra (Norway)	Neck collar	Cow positions, audio tones (ATs), electric pulses (EPs), activity	[45]
2023	Free-ranging cattle	Nofence®	Nofence AS, Batnfjordsøra (Norway)	Below the neck	Accelerometry-inferred behaviors (feeding, resting, scratching)	[38]
2023	Fleckvieh heifers	Nofence® (Model: C2.1)	Nofence, AS, Batnfjordsøra (Norway)	Attached to the neck	GPS data (walking distance, lying time, spatial pattern of movement), lying time (validated with observational data), active time, spatial distribution (Camargo's Index of Evenness)	[46]
2023	Cattle	Nofence®	Nofence AS, Batnfjordsøra (Norway)	Neck collar	Distance to boundary, acoustic warnings, aversive stimuli	[4]
2022	Cattle	Vence®	Vence Corporation, San Diego, CA (USA)	GPS collar	Location, welfare, animal distribution	[41]
2022	Pregnant Limousin cows	Nofence®	Nofence AS, Batnfjordsøra (Norway)	Neck collar	Hair cortisol, signal responses	[47]

Table 1. *Cont.*

Year	Animal Species/Breed	Virtual Fencing Device	Device Manufacturer	Location on Animal	Behavior/Parameters Measured	Reference
2022	12 pregnant Angus cows	Nofence®	Nofence AS, Batnfjordsøra (Norway)	Neck collar	Activity levels	[35]
2021	Dairy cattle (Holstein-Friesian)	Pre-commercial prototype (eShepherd®)	Agersens, Melbourne (Australia)	Neckband	Location, stimuli count, time per zone, speed	[48]
2021	Lactating Dairy Cows (Friesian/Jersey)	eShepherd pre-commercial prototype	Agersens, Melbourne (Australia)	Neckband	Time in exclusion zone, stimuli ratio	[10]
2020	Non-lactating Holstein Friesian	GPS/DGPS collar with stimuli unit	MediaTek (Hsinchu, Taiwan), Trimble (Sunnyvale, CA, USA)	Neck collar	Boundary challenges, behavior (grazing/walking/drinking)	[27]
2019	Holstein-Friesian dairy cows	eShepherd™ collar (automated prototype)	Agersens, Melbourne (Australia)	Top of the neck	Grazing time, GPS position, audio/electric pulses	[49]
2018	Cattle	Halter®	Halter (Auckland, New Zealand)	Neck-collar + head-halter	Health (body temperature), response to audio/tactile/visual/electrical stimuli	[26]
2015	Dairy cattle	Cowbell collar (audible alarm + shock)	Cambridge Industrial Design (Cambridge, UK), Teagasc (Oak Park, Carlow, Ireland)	Around the neck	Grazing, socializing, lying, milk yields	[32]
2001	Cattle	Directional Virtual Fence (DVFTM)	Anderson & Hale (Patent)	GPS collar	Location relative to boundary	[30]

VF = virtual fencing; EP = electrical pulse; AC = audio cue; E:A = electrical-to-audio ratio; IZF = inclusion zone frequency (% time within boundaries); GNSS = Global Navigation Satellite System; GPS = Global Positioning System; DGPS = differential GPS; AT = audio tone.

4. Commercial Cattle Monitoring Systems in Pasture-Based Systems

Diverse commercial Precision Livestock Farming (PLF) technologies are designed for pasture-based livestock systems (Table 2). These systems vary in terms of sensor types (e.g., accelerometers, GPS, thermometers), target species, and geographic origins, reflecting the adaptability of PLF solutions to different agricultural contexts [50]. Devices such as smaXtec (Graz, Austria/Ainring, Germany) and Allflex SenseHub (Netanya, Israel) utilize accelerometers and thermometers to monitor rumination, body temperature, and calving events, aligning with research indicating that accelerometers effectively detect behavioral changes linked to health and reproduction, while thermometers enable early disease detection [50]. In contrast, Ceres Tag (Lutwyche, QLD, Australia) and digitanimal (Madrid, Spain) integrate GPS with accelerometers to track activity and geofencing, addressing challenges like theft prevention and pasture utilization, particularly valuable in large-scale grazing systems where manual monitoring is impractical [50].

Table 2. Some devices used in commercial cattle monitoring systems and their sensors, and functionalities.

Device	Sensors	Outputs/Functionalities	Country
Datamuster	Walk-over-weigh (weighing crate)	Maternal parentage, reproductive efficiency, growth rates, calving, property mapping	Australia
smaXtec (GmbH)	Accelerometer, thermometer	pH, body temperature, calving, heat detection, health, rumination	Austria, Germany
Ceres Tag	GPS, Accelerometer	Activity, geofencing, health monitoring	Australia
Allflex SenseHub	Accelerometer	Health, rumination, feed intake, heat detection, calving, activity, heat stress	Israel
Moomonitor+	Accelerometer	Activity, resting, feeding, rumination, heat detection	Ireland
IceTag/IceQube	Accelerometer	Lameness, activity, resting, heat detection	UK
Moocall	Accelerometer	Calving, heat detection	Ireland
CalveSense	Accelerometer	Calving monitoring	Israel
eShepherd®	GPS, Accelerometer	Virtual fencing, activity monitoring, pasture management	Australia
Halter®	GPS, Accelerometer	Virtual fencing, activity monitoring, pasture management	New Zealand
Vence®	GPS, Accelerometer	Virtual fencing, activity monitoring, pasture management	USA
Nofence®	GPS, Accelerometer	Virtual fencing, activity monitoring	Norway
digitanimal	GPS, Accelerometer, thermometer	Activity, geofencing, body temperature	Spain

Abbreviations and symbols: GPS = Global Positioning System; GmbH = Gesellschaft mit beschränkter Haftung (German for "company with limited liability"); pH = potential of hydrogen (measure of acidity/alkalinity); UK = United Kingdom; USA = United States of America. Adapted from Aquilani et al. [50].

Virtual fencing technologies like eShepherd (Australia), Halter (New Zealand), and Vence (USA) rely on GPS and accelerometers to guide livestock via auditory cues and electric stimuli, reducing labor costs and enabling dynamic pasture allocation, though their success depends on consistent animal training and ethical welfare considerations [50]. Reproductive efficiency is a key focus for devices like Datamuster (Australia) and Moocall (Ireland), which use walk-over-weigh systems and accelerometers to predict calving and maternal parentage, supporting automated weight monitoring for genetic selection and reduced postpartum losses [50]. Similarly, IceTag (UK) and Moomonitor+ (Ireland) prioritize lameness and heat detection (Table 2), addressing welfare concerns in pasture-based systems.

Despite these advancements, challenges such as cost, data transmission reliability, and farmer adoption persist. GPS collars and multi-sensor systems face limitations in battery life, connectivity, and affordability for small-scale farmers [50]. This highlights the need for user-friendly, energy-efficient designs to enhance scalability. Emphasizing multi-sensor integration while acknowledging economic and operational barriers is crucial for broader implementation.

It is important to note that the virtual front-fence does not eliminate cow entry into the exclusion zone, unlike a conventional temporary electric front-fence. While cattle typically learn to associate acoustic cues with boundaries within 12 days [31], occasional entry into exclusion zones still occurs. Strategies to guide cattle back to inclusion zones include dynamic fencing adjustments, incentives like food, and leveraging their gregarious nature—collared individuals often return to rejoin herds [9]. However, standardized protocols for managing exclusion zone entries remain understudied. Informal observations suggest calves without collars may wander into restricted areas, while stakeholder insights propose combining movement incentives with gradual virtual boundary shifts to encourage voluntary returns [9,13,22,47].

Effective management of exclusion zone entries is vital for animal welfare, as frequent boundary breaches increase exposure to aversive stimuli and stress. While initial learning phases may temporarily elevate stress markers like milk cortisol [10,17], studies on beef cattle show no long-term welfare impacts [11,12,22,47]. Maintaining environmental predictability is critical; frequent boundary changes in intensive systems may undermine controllability, prolonging stress even after cue-to-pulse associations are learned [11,12]. Individual variability in learning efficiency also necessitates tailored approaches to minimize welfare risks, particularly in dairy systems with high stocking densities [51].

5. Impact of Virtual Fencing Devices on Animal Behavior and Welfare

Virtual fencing systems have been increasingly studied for their effectiveness in livestock containment and their implications for animal behavior and welfare [52]. The findings from multiple studies suggest that while these devices generally achieve successful containment, their impact on animal welfare varies depending on factors such as device design, electric pulse intensity, acoustic signals, and individual animal responses. Most collars used in VF systems weigh between 1.4 and 1.45 kg, with Nofence® and eShepherd® being the most commonly tested [53–56]. The lightweight design appears to have minimal physical impact, though some studies have reported occasional jaw abrasions, particularly with prolonged use [34,36]. This suggests that while collar weight itself may not be a major welfare concern, improper fit or extended wear could lead to minor physical discomfort in some animals. The electric pulses delivered by these devices typically range from 0.2 to 3 kV, often preceded by an 82 dB acoustic warning lasting 5–20 s [31,39,45,57]. The use of an auditory cue before the electric pulse is critical in promoting associative learning, allowing animals to avoid the stimulus over time. Studies have found that cattle generally adapt within 5–7 days, with a decreasing ratio of electric to acoustic stimuli [9,56,57]. However, individual variability exists, with some animals requiring more stimuli than others [46,57].

5.1. Impact of Device Weight and Electric Pulses on Animal Welfare

The design and weight of virtual fencing (VF) collars play a significant role in animal welfare, with variations in materials and construction influencing comfort and long-term usability. Fisher and Cornish [53] reported that device weight and materials directly affect animal welfare, noting that the eShepherd collar (~1.5 kg; Figure 2) incorporates a nylon strap and counterweight, while lighter alternatives such as the Halter system (1.42 kg) and Nofence cattle collar (1.465 kg for cattle) use materials like polyester, foam, and rubber to reduce strain. In contrast, the Vence collar (1.13 kg) employs stainless steel and plastic chain links, which may require careful fitting to avoid pressure sores [53]. While cattle generally tolerate heavier collars well, smaller livestock like sheep may experience fatigue, highlighting the need for species-specific designs [53]. These findings align with studies on cattle, where collars weighing 1.4–1.45 kg (e.g., Nofence® and eShepherd®) showed minimal adverse behavioral effects [58–62]. However, a study involving lactating dairy cows fitted with pre-commercial prototype eShepherd neckband-mounted virtual fence devices reported the development of abrasions on the lower jaw of some cows after 23 days of use [34,36]. This led to the experiment being terminated 11 days earlier than planned. This observation is particularly important as abrasions have not been reported in other research or commercial trials utilizing beef breed or non-lactating dairy cattle [9], and previous research on beef cattle using virtual fencing found no effect on skin abrasions [9]. Some animals in the study also showed indications of increasing milk cortisol concentrations during the first week, which could suggest stress [36]. Wilms et al. [13] also noted this finding by Verdon et al. [12] and Langworthy et al. [10], mentioning the possible connection between abrasions and decreased grazing time, although higher milk cortisol was not proven in injured animals [36]. The authors suggested that the abrasions were likely caused by the design of the prototype device, which was initially intended for extensively grazed cattle and may need modifications for intensive pastoral production systems like those used for dairy cows [36]. They hypothesized that the abrasions could be linked to differences in the morphology, grazing behavior, and management practices of dairy cattle compared to extensive beef systems [34,36]. Specifically, the frequent pasture allocations common in dairy systems may require cows to repeatedly rediscover the virtual boundary each day, potentially reducing their perception of environmental predictability and controllability [36]. The premature end to the experiment limited the ability to draw firm conclusions about the long-term effects of virtual fencing on dairy cow welfare. This highlights the clear need for longer-term research, especially within intensive pastoral systems.

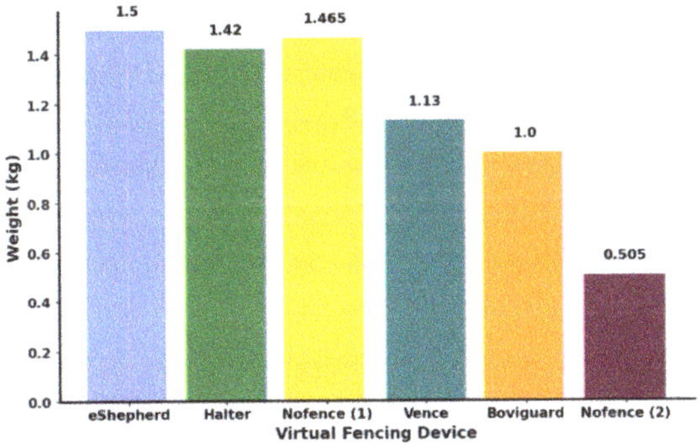

Figure 2. Weight of virtual fencing devices. Nofence® (1) for cattle; Nofence® (2) for sheep/goats. Adapted from Colusso et al. [48], Lomax et al. [49], and Fisher and Cornish [53].

Electric pulses, typically ranging from 0.2 J at 3 kV to low-voltage 800 V stimuli, are often preceded by acoustic warnings to facilitate associative learning. Harland et al. [33,42] found that cattle progressively reduced their reliance on electrical stimuli, with heifers decreasing their electrical-to-audio (E:A) response ratio from 17.9% during training to 5.2% during grazing. Similarly, Confessore et al. [47] reported that Limousin cows learned to avoid restricted zones with minimal stress, as indicated by stable cortisol levels. However, Lomax et al. [49] observed that some individuals received repeated pulses, raising concerns about potential aversion and welfare costs in less responsive animals. This variability suggests that while most cattle adapt well, individual differences in temperament and learning speed must be considered to minimize unnecessary stress. The implications of these findings highlight a balance between effective containment and animal welfare. While current VF systems are generally well-tolerated by cattle, improvements in collar materials, such as softer straps or adjustable fits, could mitigate abrasion risks [36]. Additionally, optimizing pulse intensity and ensuring consistent acoustic warnings may further reduce reliance on electrical stimuli, particularly for slower-learning individuals [53,57].

5.2. Role of Acoustic Signals in Animal Training

Acoustic cues play a crucial role in conditioning cattle to virtual boundaries. Most studies reported using tones between 82 dB and 2700 Hz, with rising pitch signals proving effective in eliciting avoidance behavior [39,53,58]. Verdon et al. [11,12] noted that dairy cows achieved 90% containment within four days, with decreasing reliance on pulses as they learned to respond to auditory warnings. However, individual variability exists, as Sonne et al. [17] found that some heifers required more time to adapt, while Fisher and Cornish [53] reported that shorter audio tones improved responsiveness. These findings suggest that while acoustic signals are generally effective, their duration and frequency should be optimized to minimize stress and enhance learning efficiency.

5.3. Cortisol Data and Stress Implications

Cortisol measurements across studies indicate that VF systems do not generally induce significant stress compared to traditional electric fencing (EF). Hamidi et al. [43,46] and Harland et al. [33,42] found no significant differences in fecal cortisol metabolites (FCMs) between the VF and EF groups, while [17] reported stable manure cortisol levels throughout their study. Confessore et al. [28] further supported this, showing no significant changes in hair cortisol concentrations (HCCs) before and after VF exposure. However, Verdon et al. [12] recorded elevated milk cortisol levels in dairy cows during early VF exposure, though these stabilized over time. These discrepancies may reflect differences in study duration, animal type (beef vs. dairy), or individual stress susceptibility. Overall, the cortisol data suggest that once animals adapt, VF does not impose chronic stress, but initial exposure may require careful management to mitigate short-term welfare impacts.

5.4. Behavioral Adaptations and Welfare Outcomes

Behavioral observations indicate that cattle generally adapt well to VF, with most studies reporting high containment rates (>90%) and reduced escape attempts over time (Table 3) [9,46]. Harland et al. [33] found that cattle learned to avoid virtual boundaries with minimal welfare concerns, even in winter conditions, while Fuchs et al. [31] reported successful herd conditioning in dairy cows, albeit with slightly more vocalizations and displacements compared to EF groups. However, some studies highlight potential welfare trade-offs, such as reduced lying time [9] and altered grazing distribution [44]. Additionally, Langworthy et al. [10] noted that pasture depletion reduced VF efficacy, requiring more auditory cues, which could increase stress if not managed properly.

Table 3. Comparative analysis of virtual fencing systems: behavioral responses, welfare impacts, and technical specifications across cattle studies.

Study Reference	Animal Type (n)	Device (Weight)	Electric Pulse	Acoustic Signal	Cortisol Data	Key Behavioral and Welfare Findings
[36]	Beef heifers (n = 32)	Nofence® (1.45 kg)	0.2 J at 3 kV (max 3)	82 dB (increasing pitch)	No significant FCM differences	Comparable to EF, adaptation over time, slower initial transitions
[33]	Yearling beef heifers, bulls, cows + calves	Nofence® (1.45 kg)	Mild electrical pulse (1.5–3 kv)	Audio cues (frequency not specified)	Not measured	Cattle learned to avoid virtual boundaries with minimal welfare concerns. The system provided effective containment, despite some collar losses, without major welfare issues. Animals adapted well, showing no significant behavioral stress. It was also effective in winter conditions, with no adverse welfare effects.

Table 3. *Cont.*

Study Reference	Animal Type (n)	Device (Weight)	Electric Pulse	Acoustic Signal	Cortisol Data	Key Behavioral and Welfare Findings
[42]	Yearling heifers (n = 49)	Nofence® (1.45 kg)	1.5–3 kV	82 dB (5–20 s)	Not measured	Heifers adapted to VF boundaries in 5–7 days; E:A ratio decreased from 17.9% (training) to 5.2% (grazing). High individual variability in response.
[42]	First-calf cows (n = 39, same animals as heifers in Year 1)	Nofence® (1.45 kg)	1.5–3 kV	82 dB (5–20 s)	Not measured	Cows retained prior learning (E:A ratio: 1.6% during re-training, 2.2% during grazing). Presence of uncollared calves did not significantly reduce containment (>99% compliance).
[42]	Bulls (n = 2 in Year 1; n = 3 in Year 2)	Nofence® (1.45 kg)	1.5–3 kV	82 dB (5–20 s)	Not measured	Excluded from analysis due to small sample size.
[43]	Fleckvieh heifers	1.45 kg	0.2 J at 3 kV	82 dB at 1 m (rising pitch)	No effect on fecal cortisol compared to a physical fence	Minor and inconsistent differences in activity budget
[55]	Dairy cows (n = 80)	Halter® (1.4 kg)	Up to 0.18 J (20 µs)	2700 Hz crescendo	Not reported	90% containment (<1.7 min beyond fence), autonomous herding by day 4, decreasing pulse ratio
[54]	Angus cattle (n = 17)	Nofence®	0.2 J (1 s)	82 dB rising tone (5–20 s)	Not measured	The cattle effectively responded to audio alerts and remained securely contained within the enclosure. No spatial issues or welfare concerns were observed.
[44]	Angus cows (n = 12)	Vence®	800 V (0.5 s)	0.5 s auditory cue	Not reported	Reduced exclusion use, grazing distribution changes, some noncompliance
[31]	Dairy cows (n = 20)	Nofence®	25× weaker than EF; 0.1 ± 0.7/day	Audio tone (1.9 ± 3.3/day)	Milk cortisol: no difference vs. EF	Successful herd conditioning No welfare differences vs. EF More vocalizations/displacements (VF)
[14]	Nursing Brangus cows (n = 28)	Nofence® (1.45 kg)	Mild electric pulse	-	Not reported	Cows learned to avoid restricted zones and spent more time in containment areas. They needed fewer audio-electric cues, relied more on auditory signals, traveled less daily, explored smaller areas, and showed reduced overall activity. Training minimized negative impacts on their comfort, well-being, and welfare.
[17]	Angus cows (n = 5)	Nofence®	0.2 J at 3 kV for one second	82 dB tones increasing in pitch for 5–20 s	Manure cortisol concentrations ranged from 12 to 42 ng/g *w*/*w*, with a mean of 20.6 ± 5.23 ng/g *w*/*w*. No significant differences among individuals. Levels remained stable throughout the study	No negative effects on cattle behavior and welfare VF was comparable to traditional electric fencing. There was no evidence of stress from VF, as indicated by manure cortisol levels. The study suggests using manure cortisol analysis as a noninvasive stress measure for large grazers during VF.

Table 3. *Cont.*

Study Reference	Animal Type (n)	Device (Weight)	Electric Pulse	Acoustic Signal	Cortisol Data	Key Behavioral and Welfare Findings
[47]	Limousin cows (n = 20)	Nofence®	Low electric pulse	Rise tone scale of sounds lasting from 5 s to 20 s	Hair cortisol concentration (HCC) in pg/mg. Mean Ti (beginning): 1.14 pg/mg; mean Tf (end): 0.82 pg/mg. No significant differences were observed in HCC before and after the VF treatment	Cows learned with fewer stimuli over time and a lower ratio of sounds to pulses. They stayed in designated zones more consistently, with minimal escapes. Stress levels remained stable, and shorter audio tones showed better responsiveness.
[12]	Lactating dairy cows (mixed breed) (n = 30)	eShepherd® prototype (~1.4 kg)	Short, sharp (kV range; confidential) Short, sharp	Non-aversive audio tone	Increased milk cortisol levels: higher at d5 (VF) vs. d8 (EF)	Lower activity/grazing (d4–6) No aversion behaviors Jaw abrasions influenced grazing behavior (study halted) Cows were successfully contained behind the virtual fence for 10 days.
[48]	Non-lactating dry cows (Holstein-Friesian) (n = 34)	eShepherd® prototype ~1.4 kg (~725 g unit)	Short, sharp (kV range; confidential) Short	Distinctive audio tone	Not measured	Cows learned to remain within the VF over time, even with restricted feed. Restricted cows received more stimuli Learned retention (≤17% AT→EP) Feed motivation influenced behavior
[10]	Dairy cattle	eShepherd® (0.73 kg)	Short, sharp electrical pulse sequence in the kilovolt range	Distinctive but non-aversive audio cue within the animal's hearing range	Not reported	Cows learned to respond to audio cues to avoid electrical stimuli, with a daily ratio of 0.18 ± 0.27. Stayed within the inclusion zone ≥ 99% of the time. Pasture depletion slightly reduced efficacy, and more audio cues were needed when entering new paddocks. VF effectively contained the herd without affecting production metrics but did not fully prevent entry into the exclusion zone. Some cows developed jaw abrasions from the neckband, and social learning may have occurred.
[9]	Beef steers (n = 64)	eShepherd® (~1.4 kg)	Low voltage 800 V (<1 s)	785 Hz	Fecal cortisol metabolite (FCM) concentrations (ng/g of dry feces). No differences between fence types overall. Concentrations decreased across time for all cattle. Cohort 1 had significantly higher overall concentrations than cohort 2	VF successfully contained animals over 4 weeks with minimal welfare impacts. Animals spent slightly less time lying (<20 min/day). They all learned to respond to audio cues, showing no avoidance behaviors or significant stress differences (measured via FCM). Individual learning rates varied. Further research is needed for pastured dairy and beef cattle.

Table 3. *Cont.*

Study Reference	Animal Type (n)	Device (Weight)	Electric Pulse	Acoustic Signal	Cortisol Data	Key Behavioral and Welfare Findings
[49]	Dairy cows (n = 12)	eShepherd® (~1.4 kg)	Short, sharp pulse in the kilovolt range	Distinctive audio tone (within the animal's hearing range)	Not measured	Cows were successfully contained 99% of the time, but there was significant variation in the stimuli received and paddock usage among individuals. Some cows experienced potential welfare costs due to repeated electrical pulses and possible aversion to the fence line. Further research is needed on the long-term welfare impacts.

Abbreviations and Symbols: VF = virtual fencing; EF = electric fencing; EP = electrical pulse; E:A = electrical-to-audio ratio; FCMs = fecal cortisol metabolites; HCC = hair cortisol concentration; dB = decibel; J = Joule; kV = kilovolt; µs = microsecond; Hz = Hertz; s = second; min = minute; ng/g = nanogram per gram; pg/mg = picogram per milligram; *w*/*w* = wet weight; n = sample size; vs. = versus; ~ = approximately; ± = standard deviation.

Beef cattle appear to adapt more readily to VF than dairy cattle, possibly due to differences in temperament and prior handling ([36] vs. [12]). Additionally, younger animals and previously trained individuals (e.g., first-calf cows in [33,42]) show faster learning, suggesting that prior experience enhances VF efficacy. The presence of uncollared companions (e.g., calves) did not significantly disrupt containment [33,42], supporting VF use in mixed groups (Table 3). However, device-related injuries [12] and individual variability in stimulus response [49] highlight the need for ergonomic collar designs and tailored training protocols.

Virtual fencing technology has demonstrated significant potential in livestock management, with studies reporting high containment rates and rapid animal learning (Table 4). Grinnell et al. [36] found that VF achieved a success ratio of 94.6% to 97.9%, comparable to electric fencing, with no adverse effects on welfare or weight gain. Similarly, Harland et al. [49] observed that cattle learned to associate audio cues with electrical pulses within 5–7 days, reducing the EP-to-AC ratio from 17.9% during training to just 5.2% during grazing. These findings suggest that VF is an effective containment method, though individual variability in learning rates was noted [49,50].

Table 4. Summary of key findings, limitations, and future directions in virtual fencing studies.

Reference	Summary of Main Findings	Limitations of the Study	Future Directions/Recommendations
[36]	VF success ratio improved (94.6% to 97.9%); no welfare/weight gain differences vs. EF; proposed 90% success threshold benchmark.	Short duration (8 weeks); double-fence setup; infrequent cortisol sampling; high forage availability; human intervention.	Longer-term studies; open-range testing; frequent stress sampling; varied forage conditions; minimal human intervention.
[33]	Naïve cattle learned to associate audio cues (ACs) with electrical pulses (EPs) within 5–7 days, reducing the EP-to-AC ratio from 17.9% (training) to 5.2% (grazing). Heifers retained learning over a 300-day hiatus; first-calf cows with prior VF experience had even lower EP rates (1.6% re-training, 2.2% grazing). Cattle stayed within VF boundaries > 99% of the time, though cows with uncollared calves had slightly more escapes. Individual variability observed (high-, moderate-, low-stimuli cohorts), but all achieved similar containment rates. Stocking density influenced cow behavior (higher density = more ACs/EPs).	No non-VF control group for direct comparisons. Confounding factors (uncollared calves, prior VF experience) complicated behavioral interpretations. Small sample sizes for high-stimuli cohorts (n = 5–7). Limited to temperate grasslands; may not generalize to arid/tropical systems. Ethical concerns about EPs, though no adverse welfare impacts noted.	Further research on long-term welfare effects. Investigate scalability for commercial herds. Optimize training durations. Explore alternatives to EPs for welfare-sensitive applications. Test VF in diverse climates and production systems.

Table 4. *Cont.*

Reference	Summary of Main Findings	Limitations of the Study	Future Directions/Recommendations
[42]	Network connectivity was generally good, with mean connection intervals within the optimal 15 min window. Poor connections occurred < 1% of the time. Collars primarily used 4G-LTE, but rural areas in Canada may have limited network availability. Most collar failures were due to connectivity issues, with older (v2.1) collars showing increased failures in the second year. Physical collar loss occurred, especially with bulls. Battery charge remained high (>96%) in all trials, including winter, though solar charging was lower in winter. Habitat and animal behavior influenced charging. Collar battery capacity was sufficient for cattle in both summer and winter.	Limited generalizability due to study being conducted in only two Alberta locations; broader rural Canada has poor cellular access. Study duration may have been insufficient to assess long-term collar performance. Unmeasured factors (cloud cover, precipitation, animal behavior, wildfires) could affect solar charging. Focused on technical performance; impacts on animal behavior, welfare, and grazing management need further study.	Longer-term studies to assess collar durability and performance over multiple years. Research in areas with poorer network coverage to evaluate VF feasibility. Further investigation into environmental and behavioral factors affecting solar charging. Studies on animal welfare, behavior, and grazing efficiency under VF systems. Improved collar design (e.g., better fit, reduced loss risk, enhanced connectivity in remote areas).
[29]	Solar-powered eShepherd collars reduced fencing costs and enabled dynamic grazing.	Cellular dependency, small sample (n = 60), and no cost–benefit comparisons.	Off-grid collar solutions, larger trials, and diversified trade strategies.
[28]	No age-related differences in VF learning; younger cows received more ATs later, older cows responded faster. No long-term stress (milk yield, hair cortisol stable).	Short (31-day) trial; small sample (n = 20); only Holstein-Friesians; heat waves may have confounded results.	Longer studies; multi-breed trials; forage availability effects.
[55]	Dairy cows using the Halter VF system showed rapid learning, with electrical pulse (EP) rates dropping from 60% (Day 1) to 2.6% post-training. Also, 90% of cows spent ≤1.7 min/day beyond boundaries, and 50% received no paddock pulses by the final week. Superior to earlier VF systems due to bidirectional sound, vibration, and machine learning.	Short 28-day period; no long-term welfare metrics; sample limited to mid-lactation cows in one region; operational mismanagement caused pulse spikes.	Longer-term studies; welfare assessments (e.g., cortisol); economic feasibility analysis; testing in diverse environments.
[13]	No welfare differences between VF and PF in weight gain/lying behavior; lower FCM in VF cattle ($p = 0.0165$). Learning evidenced by reduced electric/acoustic signal ratio ($p = 0.0014$).	Small sample (n = 13 studies); beef cattle focus; short trial durations; variable protocols; FCM may miss acute stress.	Long-term dairy cow studies; standardized protocols; assess individual learning differences; multimodal stress indicators.
[54]	No consistent leadership hierarchy found in Angus cows using Nofence© (W = 0.15, $p < 0.001$). Daily rank variability suggests dynamic social interactions rather than stable leadership.	GPS inaccuracies (3.5–10 m); short observation period (45 days); focused on spatial metrics rather than direct social interactions.	Longer observation periods; integrate direct interaction data; improve GPS precision; assess age/experience effects.
[43]	Heifers successfully learned virtual fencing (VF) over 12 days, showing fewer strong reactions (scores 2 and 3) and more mild reactions (score 1). Increased acoustic signals and decreased electric pulses (higher success ratio). Collar data showed a 91.3% success ratio by trial end. Improved confidence ratio (phases 2→3). Faster mode switching (teach→operate) in Round 2 vs. Round 1. Cattle learn to associate audio cues with shocks, adjusting behavior to avoid pulses.	Only 37% of collar cues were observed (group dynamics hindered tracking). Bias risk: observers focused on the first reacting heifer. Technical issues: inconsistent acoustic signals before mode activation across collars. Training variables: unclear if 12 days is optimal; physical fence (PF) proximity may influence learning. Single escape incident after PF removal. Small pasture size (3000 m^2) and visual support from physical fences may limit real-world applicability	Refine observation methods for group settings. Improve collar algorithms for consistency. Study longer training periods and PF-VF distance effects. Test VF reliability in PF-free environments. Explore individual vs. herd learning differences. Single escape incident suggests further research is needed on fence distance effects. Confidence ratio requires refinement.

Table 4. *Cont.*

Reference	Summary of Main Findings	Limitations of the Study	Future Directions/Recommendations
[38]	Tri-axial accelerometer data from collars effectively classified cattle behaviors (accuracy: 0.998). Orientation correction did not improve performance. A 20 s smoothing window optimized results. Rare behaviors were classified but with lower accuracy.	Short-duration behaviors excluded; individual variability not assessed; smoothing may obscure rapid transitions; rare behaviors underrepresented.	Include short-duration behaviors; account for individual differences; optimize smoothing parameters; collect more rare behavior data.
[45]	Cows adapted to VF within 3 days; EP/AT ratio declined. No lasting welfare impacts (milk yield, cortisol stable). More vocalizations/displacements initially.	Small sample (n = 20); short duration; group-based training; controlled conditions.	Individual training protocols; on-farm trials; larger/longer welfare assessments.
[4]	Virtual fencing (eShepherd, Nofence, etc.) reduces labor/costs and benefits ecologically sensitive areas. Auditory cues minimize welfare concerns, but individual learning varies.	High costs; variable efficacy in large/diverse herds; GPS reliability issues; underdeveloped regulations.	Cost reduction; improve GPS reliability; welfare optimization; stakeholder engagement for regulatory frameworks.
[46]	VF collars showed 92% precision for lying time detection Strong correlation between UAV RGBVI and herbage changes Behavior changes: ↓ lying time, ↑ walking distance over days Improved spatial distribution (Camargo's Index) Random forest model showed moderate correlation (R^2 = 0.43) between UAV data and animal activity Demonstrated potential for "grid grazing" precision management	Small-scale case study: only 2 pastures/treatment for 15 days Limited imagery: RGB-only (no NIR/NDVI data) Environmental variables not fully controlled Potential animal acclimatization bias	Larger-scale trials needed for validation Incorporate NIR/NDVI sensors for better pasture assessment Study forage scarcity scenarios Develop decision support systems for practical farming use Explore long-term effects of grid grazing
[59]	Overall, 98% containment; 76% fewer electric cues; rapid learning; reduced movement/exploration.	Individual variability; artificial pen setting; no calf data; short duration (12 days); audio design questions.	Pasture testing; include calves; longer trials; optimize auditory signals; assess social learning.
[58]	Pasture systems provide ecosystem services (carbon sequestration, biodiversity) and nutritionally enhanced milk (higher CLA/omega-3). Rotational grazing and virtual fencing improve sustainability, but continuous grazing risks degradation. Multi-species/silvopastoral systems enhance resilience.	Focused on temperate regions; limited long-term data on innovations; economic feasibility of technologies understudied; climate change impacts not addressed.	Context-specific research; long-term studies of grazing innovations; cost–benefit analyses; tropical/arid system adaptations.
[60]	Wearables (e.g., SCR Heatime Pro+) showed moderate–high accuracy for rumination/eating. Satellite NDVI correlated well with forage biomass (r = 0.74–0.94). Virtual fencing (e.g., Nofence) contained cattle without welfare harm.	Wearables less accurate in grazing vs. confinement. Satellite limitations: weather dependency, multi-species pastures. Virtual fencing lacked long-term welfare/economic data. High costs hinder small-scale adoption.	Pasture-specific algorithm refinements, cost reduction, and longitudinal studies on scalability. Autonomous tools (e.g., CowBot) need further validation.
[41]	Personality matters: consistent activity differences (walking distance) between calves Active calves grow faster: positive correlation between movement and weight gain Predictable plasticity: personality type influences environmental adaptation	Controlled environment: may not reflect all farm conditions Limited behaviors: focused mainly on locomotion Small sample: 64 calves total	Expand to production traits Develop personality-based management Include more behavior metrics

Table 4. *Cont.*

Reference	Summary of Main Findings	Limitations of the Study	Future Directions/Recommendations
[41]	A 42% reduction in fine fuel biomass within VF boundaries A rate of 50% forage utilization inside vs. 5% outside fuel breaks High containment: dry cows (100%) vs. cow/calf pairs (75%) Low shock rates: 2.3/day (dry cows), 10.1/day (cow/calf pairs) Effective learning: cattle associated audio cues (0.5 s tone) with boundaries Water access improved containment	Case study design limits generalizability Mixed groups may have influenced behavior No cortisol measurements for welfare assessment Short 30-day trial may not show long-term effects No visual cues with virtual boundaries	Longer-term studies on habituation Welfare assessments (e.g., cortisol levels) Separate trials for dry cows vs. cow/calf pairs Visual cue integration to aid learning Testing in diverse landscapes
[61]	No welfare difference: Similar stress (fecal cortisol) and behavior between VF and EF Equal growth rates: no difference in daily weight gain Learning curve: shock frequency decreased after initial period Individual variation: different learning speeds among calves	Short duration: 21-day grazing period Limited population: dairy-origin calves only Missing specs: no VF pulse details provided Single stress measure: only cortisol metabolites analyzed	Longer-term studies Breed/age comparisons Standardized VF parameters Multi-measure welfare assessment
[17]	No cortisol changes (20.6 ± 5.23 ng/g) over 18 days; stable individual levels (12–42 ng/g); supports noninvasive monitoring.	Small sample (n = 5); short duration; no control group; methodological cortisol variability; pregnant cows only.	Larger samples; longer studies; include controls; standardize cortisol methods; diverse physiological states.
[57]	Electric shocks in farm management (fencing, trainers, prods) inherently cause pain, raising ethical concerns unless justified by welfare benefits. Virtual fencing allows controlled shocks but may disproportionately affect slow learners. Technologies like prods/poultry wires were deemed ethically indefensible due to stress.	Reliance on manufacturer data may underreport risks. Long-term behavioral impacts (especially in sheep) were insufficiently explored. Gaps in validated non-aversive alternatives (e.g., tactile collars). Societal desensitization to animal pain was theorized but not tested.	Longitudinal welfare studies, interdisciplinary research (ethics/tech/economics), and development of non-aversive alternatives.
[50]	PLF tools (RFID, GPS, accelerometers) enable monitoring of health, behavior, and pasture use. Virtual fencing reduces infrastructure but has welfare/response variability. Remote sensing (UAVs/satellites) aids biomass assessment, though integration with sensors is limited.	Battery life/transmission range constraints, high costs, variable animal responses to virtual fencing, remote sensing accuracy affected by vegetation/weather.	Improve battery/connectivity solutions; cost reduction; standardize virtual fencing protocols; integrate sensor/remote sensing data.
[47]	Rapid learning: 89% reduction in electric pulses over 4 trials Effective containment: 92% success rate in final trial No chronic stress: HCC levels remained stable (Δ = 0.07 µg/g). No hair cortisol changes; breed-specific learning Behavior adaptation: audio response time improved by 65%	Limited to adult Limousins Short duration (21 days) Single-breed study; group training effects; No pasture utilization data Small cortisol sample (n = 16); 2G dependency.	Long-term welfare studies (6+ months) Multi-breed trials (Angus, Hereford) Individual training Calf-specific protocols Grazing efficiency metrics Standardize virtual fencing protocols; integrate sensor/remote sensing data.
[35]	Significant learning; herd behavior influenced responses; no activity changes post-stimulus.	Small sample (n = 12); no control; short-term focus; sheep-calibrated accelerometers; single breed.	Larger, more diverse samples; controls; long-term welfare; cattle-specific sensors; multiple breeds/demographics.
[10]	The virtual front-fence did not entirely prevent cow entry into the exclusion zone; individual cows were generally contained within the inclusion zone for ≥99% of the time. EP/audio tone (AT) ratio dropped to 0.18 ± 0.27. Pasture depletion had minimal impact (≤28 s/h in exclusion zones). Uniform grazing near VF, unlike dry cows (Lomax et al., 2019 [49]). Comparable welfare/energy intake to electric fencing.	Short duration (10 days/treatment); no herd replication; neckband abrasions in dairy cows; excluded grazing-function audio cues.	Longer studies with herd replication; device design improvements; testing in varied terrains.

Table 4. *Cont.*

Reference	Summary of Main Findings	Limitations of the Study	Future Directions/Recommendations
[12]	VF initially matched electric fencing in welfare metrics, but days 4–6 showed reduced activity/grazing and higher cortisol. Early termination due to neckband abrasions.	Small sample (n = 30); early termination; cortisol variability; no visual cue testing.	Improved collar design; long-term welfare studies; visual cue integration.
[48]	Dairy cattle contained successfully (89% time in inclusion zone). No difference in stimuli between fresh/residual pasture days. Diurnal activity patterns observed near boundaries.	Non-lactating cows used; short restriction period; GPS inaccuracies (8.4 m buffer); small sample (n = 10); social dynamics not measured.	Include lactating cows; longer restriction periods; improve GPS accuracy; larger samples; assess social interactions.
[62]	Digital tools (GPS collars, drones, virtual fences) improved health/behavior monitoring. Virtual fences had mixed welfare outcomes due to individual learning variability. Drones risked stress if flown too close.	Sensor accuracy lacked standardization. GPS errors in dense vegetation. Small/long-term virtual fencing trials. Drone stress data was limited. Connectivity/battery issues in remote areas.	Standardized validation protocols, ethical frameworks for shock use, and adaptive tech for remote pastures.
[27]	Cows learned VF boundaries but relied on visual cues (e.g., white tape). Removal of visual cues increased boundary challenges. Reduced grazing/rumination during training suggested stress.	Small sample (n = 9); short training; directional ambiguity of audio cues; no social learning analysis; unquantified stress.	Larger/longer studies; improved audio cue directionality; integration of social learning; physiological stress measures.
[63]	Group-trained cows relied more on social cues; reduced paired stimuli (45%→14%).	Technical failures; short tests (10 min); artificial setting; feed motivation declined.	Improve collar reliability; longer tests; pasture applications; sustained motivation methods.
[49]	Virtual fencing (VF) collars effectively contained dairy cows (99% success rate), with reduced stimuli over time (EP:AT ratio dropped from 20% to 12%), indicating learning. However, individual variability was high, and some cows avoided fence zones, suggesting welfare concerns.	Short duration (6 days), no physiological stress measures (e.g., cortisol), non-lactating cows, pasture quality not measured, and proprietary collar algorithms limited transparency.	Longer trials with larger herds, physiological welfare assessments with lactating cows, quantify pasture, and transparent collar specs for reproducibility.
[9]	Virtual fencing contained cattle as effectively as electric tape over 4 weeks, with 71.51% of interactions using only audio cues. No difference in fecal cortisol metabolites (stress indicator) between groups. Slightly reduced lying time (<20 min/day) in virtual fence groups (p = 0.02) but within normal range. Weight gain differences between cohorts suggest environmental factors influence results.	Short duration (4 weeks); technical malfunctions in neckbands; pasture quality not quantified; small sample size (n = 8/group); individual temperament not assessed.	Longer-term studies; improved device reliability; quantify pasture quality; larger sample sizes; assess individual temperament effects.
[64]	Fog-enabled WSN with edge mining (IEM) achieved 83–95% activity classification accuracy, reducing cloud dependency. Energy-efficient but performance depended on parameter tuning (γ, ε). Jerky motions reduced accuracy.	Used human (not cattle) acceleration data. Parameter sensitivity required cloud optimization. Limited to binary activity states (e.g., standing/walking). Scalability in large herds untested.	Species-specific algorithm training, on-farm parameter optimization, and broader behavioral classification (e.g., grazing/lameness).
[65]	Precision tools (e.g., Grasshopper rising plate meter, PastureBase Ireland) improved pasture management accuracy (r^2 = 0.99). Virtual fencing enabled dynamic grazing but required GPS reliability.	Limited validation in commercial pastures, GPS dependency, and lack of long-term welfare data.	Wider validation in diverse climates, improved GPS robustness, and cost-effective scaling for small farms.
[32]	CID's cowbell-shaped tracking collar provided real-time grazing and milk yield data, integrating GPS and virtual fencing.	Small trial (n = 50), durability claims unverified, reliance on mobile networks.	Larger-scale trials, peer-reviewed durability testing, and alternative data transmission (e.g., LoRaWAN).

Table 4. *Cont.*

Reference	Summary of Main Findings	Limitations of the Study	Future Directions/Recommendations
[56]	Broadcast audio cues (8 kHz tones, dog barks) reduced cattle presence near speakers but habituation risks and inconsistent sound levels limited efficacy. Mobile sound delivery may improve consistency.	Stationary speakers caused variable exposure. Small sample (n = 24), artificial paddock setup. Uncontrolled variables: age, breed, hearing sensitivity. No long-term habituation data.	Mobile sound systems, field testing in diverse terrains, and integration with wearable devices.

Abbreviations and Symbols: VF = virtual fencing; EF = electric fencing; ACs = audio cues; EPs = electrical pulses; AT = audio tone; PF = physical fencing; UAV = unmanned aerial vehicle; NDVI = normalized difference vegetation index; RGBVI = red–green–blue vegetation index; FCMs = fecal cortisol metabolites; HCC = hair cortisol concentration; W = Kendall's W (coefficient of concordance); Δ = change; ↑ = increase; ↓ = decrease; n = sample size; r = correlation coefficient.

Despite these successes, several limitations were identified across studies (Table 4). Short trial durations (e.g., 8 weeks in Grinnell et al. [36]; 28 days in Verdon et al. [55]) and small sample sizes (e.g., n = 20 in Confessore et al. [28]; n = 13 studies in Wilms et al. [13]) restrict the generalizability of the findings. Additionally, reliance on controlled environments, such as double-fence setups [36] or artificial paddocks [62], may not reflect real-world conditions. Technical challenges, including GPS inaccuracies [54] and collar connectivity issues [42], further complicate implementation. Ethical concerns regarding EPs were also raised, though no direct welfare harm was reported [33,57].

Future research should prioritize longer-term studies to assess habituation and welfare impacts, as suggested by multiple authors [9,55]. Improved collar designs, better GPS accuracy, and alternative stimuli (e.g., vibration or bidirectional sound) could enhance reliability [4,55]. Testing in diverse environments, such as arid or tropical systems [33], and integrating visual cues [27] may further optimize VF efficacy. Economic feasibility remains a barrier, with high costs limiting small-scale adoption [60]. The implications of these findings are twofold: VF offers a promising tool for precision grazing and ecological management [48,58], but scalability depends on addressing technical, ethical, and economic challenges. Future studies should adopt standardized protocols, larger sample sizes, and multi-breed comparisons to ensure broader applicability [13,54]. Refining technology and validating long-term impacts could revolutionize sustainable livestock management through virtual fencing.

5.5. Challenges in the Durability and Reliability of Virtual Fencing

Like any technology, the long-term adoption and success of virtual fencing technology depend heavily on the reliability and durability of the equipment, particularly the collars worn by the animals [2,18]. Studying potential problems with VF collars and equipment over time is crucial for ensuring the technology is a practical and welfare-friendly alternative to traditional fencing methods. Several studies and stakeholder discussions highlight concerns regarding the technical performance and reliability of virtual fencing systems. Prior research has documented connection issues or unspecified technical problems that can interfere with collar function or data storage and retrieval [4,41]. For instance, a study involving an earlier prototype of VF collars for dairy cows reported malfunction issues that led to the removal of animals from the experiment [23].

A specific technical challenge reported in the literature is GPS drift. GPS drift refers to the discrepancy between an animal's actual location and the location recorded by the collar's GPS receiver. This can be influenced by environmental factors such as the quality of the GPS receiver and antenna, the number of satellites visible, proximity to buildings, heavy tree cover, steep slopes, hilly terrain, and even thunderstorms [4,25]. Issues of GPS drift can lead to areas intended for exclusion becoming inadvertently accessible to animals. Stakeholders in one study highlighted that collars were not fit to hold animals within small paddocks in some cases, leading farmers to abandon their use for precision grazing [54]. Furthermore, GPS drift can be ambiguous and stressful for animals, potentially causing them to receive unnecessary electric pulses [63]. Careful planning of virtual boundaries is suggested to minimize issues related to drift, particularly avoiding tight corridors and corners [25,48].

The reliability of VF systems is considered a key barrier to their widespread implementation [2]. Uncertainties regarding technical reliability, coupled with challenges in training animals to use the system, can hinder its adoption by farmers [21]. Concerns about potential system malfunctions also raise safety issues, such as animals escaping the virtual boundary, which can lead to questions about liability [9]. While advancements in electronic communication and device design hold the potential to enhance effectiveness and reduce costs, studies addressing the detailed technical performance of VF collars are still limited [11,12,63]. Placing studies on collar durability and reliability in a broader context is important because these issues directly impact the ability of farmers to effectively manage livestock, protect sensitive environmental areas, and maintain animal welfare [9,24]. Ensuring the reliability of the technology is essential not only for practical on-farm applications but also for gaining public acceptance and social license for its use [6,16,18]. Therefore, continued research and development focused on enhancing the technical robustness and durability of virtual fencing collars and equipment are vital for the successful and widespread adoption of this innovative technology in livestock grazing systems [4,18,21].

5.6. Comparison Between Virtual and Physical Fencing Systems

Virtual fencing eliminates the need for physical infrastructure, reducing long-term labor and material costs while maintaining 99% effectiveness in cattle containment through associative learning (Table 5) [19]. This system enables rapid boundary adjustments [6,9,22,24,47], supports biodiversity by excluding sensitive areas from grazing [1,6,13,16,37,40], and avoids landscape fragmentation [41]. Furthermore, virtual fences can be integrated with monitoring technologies, allowing for precise tracking of individual animals [38] and dynamic grazing patterns, which may enhance pasture management [13,38,40]. Importantly, Sonne et al. [17] emphasize that virtual systems reduce injury risks to wildlife and livestock compared to traditional barriers (Table 5). In contrast, physical fences remain a widely understood and reliable method, particularly in harsh weather or rugged terrains where GPS signals may falter [4,21]. Their visual clarity provides immediate deterrence, avoiding the need for animal training [20]. However, physical barriers pose significant ecological drawbacks, including habitat fragmentation and restricted wildlife movement [66–68]. Additionally, their high installation and maintenance costs [44] and labor-intensive upkeep [69] limit scalability, particularly in remote regions.

Table 5. Comparative analysis of virtual and physical fencing systems in modern livestock management.

Aspect	Virtual Fences	Physical Fences	References
Advantages			
Cost and Labor	Reduces long-term material costs and labor.	No technological dependency.	[4,38,68]
Flexibility	Rapid boundary modification. Dynamic grazing patterns.	Reliable in all weather/terrain. Clear visual boundaries.	[2,9]
Environmental Impact	No landscape fragmentation. Supports biodiversity and wildlife migration.	Effective physical containment.	[17,35]
Effectiveness	99% effective for cattle containment. Precise tracking in difficult terrains.	Established, widely understood technology.	[9,19,69]
Animal Welfare	No significant stress differences vs. physical fences. Eliminates injury risks.	No ethical concerns (no electric stimuli).	[17,19,20]
Integration	Integrates with monitoring/research systems. Social–psychological benefits.	No animal training required.	[19,21]

Table 5. *Cont.*

Aspect	Virtual Fences	Physical Fences	References
Disadvantages			
Cost and Labor	High initial costs (collars per animal). Requires ongoing tech maintenance.	High material/labor costs. Expensive installation/maintenance.	[21,44,69]
Technical Issues	GPS accuracy/signal loss in rugged terrain. Dependency on GSM coverage.	Inflexible boundaries. Rigid infrastructure limits adaptability.	[41,45]
Animal Welfare	Ethical concerns (electric impulses). Initial grazing time reduction.	Injury risks for wildlife/livestock. Higher aversive visual stimuli.	[9,20,56]
Environmental Impact	No predator protection. Technical malfunctions.	Fragments landscapes. Restricts wildlife movement/migration.	[19,68,69]
Operational Challenges	Requires animal training. Continuous monitoring needed.	Labor-intensive installation/maintenance. Infeasible in protected habitats.	[4,9]

Despite the promise of virtual fencing, challenges persist. High initial costs, dependency on GPS accuracy [45], and ethical concerns about electrical stimuli [56] require careful consideration. For instance, Campbell et al. [9] reported minor stress differences between virtual and physical systems, but prolonged GPS signal loss in complex terrains remains a hurdle [41]. Meanwhile, physical fences, while reliable, are incompatible with conservation goals in protected habitats [19]. Virtual fencing aligns with global trends toward precision agriculture and ecological preservation, whereas physical fences retain value in predictable, small-scale operations. Future innovations must address cost barriers and technological limitations to ensure equitable access for farmers while safeguarding animal welfare and ecosystems.

6. Conclusions

Virtual fencing technology has emerged as a transformative tool in precision livestock farming, offering a flexible and labor-efficient alternative to traditional fencing. Studies demonstrate that cattle rapidly learn to associate auditory cues with mild electrical stimuli, achieving high containment rates (≥90%) with minimal long-term welfare impacts, as evidenced by stable cortisol levels and normal behavioral patterns. However, short-term behavioral disruptions and occasional collar-related abrasions highlight the need for optimized device design and training protocols. These integrated findings suggest that VF can significantly enhance pasture management by enabling dynamic rotational grazing, protecting sensitive ecosystems, and reducing labor costs. However, challenges such as high initial costs, connectivity limitations, and individual animal variability must be addressed for its broader adoption. Regulatory standards should ensure welfare-compliant designs, while farmer education programs can facilitate smoother transitions to VF systems. The novelty of VF lies in its integration of GPS tracking and behavioral conditioning, providing a scalable solution for sustainable livestock production. Future research should focus on long-term welfare assessments, cost-effective scalability, and the development of non-aversive stimuli. As the technology evolves, VF has the potential to redefine cattle management, aligning productivity with environmental and animal welfare goals.

Author Contributions: I.U.G.: conceptualization, data extraction, table visualization, writing—original draft, project management, coordination, graphical abstract, writing—review and editing, final proofreading before publication. H.A.: writing—review and editing. Q.H.: writing—review and editing. S.R.: writing—original draft, writing—review and editing. All authors have read and agreed to the published version of the manuscript.

Funding: This research received no external funding.

Institutional Review Board Statement: Not applicable.

Informed Consent Statement: Not applicable.

Data Availability Statement: No new data were created or analyzed in this study. Data sharing does not apply to this article.

Acknowledgments: The authors thank Ruminants for the invitation to submit this invited feature article and for waiving the Article Processing Charge. The graphical abstract was created with BioRender.com.

Conflicts of Interest: The authors declare no conflicts of interest.

References

1. Roche, J.R.; Berry, D.P.; Bryant, A.M.; Burke, C.R.; Butler, S.T.; Dillon, P.G.; Macmillan, K.L. A 100-Year Review: A Century of Change in Temperate Grazing Dairy Systems. *J. Dairy Sci.* **2017**, *100*, 10189–10233. [CrossRef] [PubMed]
2. Umstätter, C. The Evolution of Virtual Fences: A Review. *Comput. Electron. Agric.* **2011**, *75*, 10–22. [CrossRef]
3. Gadzama, I.U.; Ray, S.; Méité, R.; Mugweru, I.M.; Gondo, T.; Rahman, M.A.; Redoy, M.R.A.; Rohani, M.F.; Kholif, A.E.; Salahuddin, M.; et al. Chlorella vulgaris as a Livestock Supplement and Animal Feed: A Comprehensive Review. *Animals* **2025**, *15*, 879. [CrossRef] [PubMed]
4. Goliński, P.; Sobolewska, P.; Stefańska, B.; Golińska, B. Virtual Fencing Technology for Cattle Management in the Pasture Feeding System—A Review. *Agriculture* **2022**, *13*, 91. [CrossRef]
5. Gadzama, I.U.; Ray, S. Precision Livestock Farming in Pasture-Based Dairy Systems: Monitoring Grazing Behavior. Wikifarmer. 2024. Available online: https://www.researchgate.net/publication/385053381 (accessed on 12 March 2025).
6. Gadzama, I.U. Virtual Fencing for Livestock Management: Effects on Cattle Behavior, Welfare, and Productivity. ResearchGate. 2025. Available online: https://www.researchgate.net/publication/388036239 (accessed on 12 March 2025).
7. Lee, C.; Colditz, I.G.; Campbell, D.L.M. A Framework to Assess the Impact of New Animal Management Technologies on Welfare: A Case Study of Virtual Fencing. *Front. Vet. Sci.* **2018**, *5*, 187. [CrossRef] [PubMed]
8. Gadzama, I.U. How AI is Revolutionizing Animal Farming: Benefits, Applications, and Challenges. Wikifarmer. 2025. Available online: https://www.researchgate.net/publication/387828770 (accessed on 10 April 2025).
9. Campbell, D.L.M.; Lea, J.M.; Keshavarzi, H.; Lee, C. Virtual Fencing Is Comparable to Electric Tape Fencing for Cattle Behavior and Welfare. *Front. Vet. Sci.* **2019**, *6*, 445. [CrossRef]
10. Langworthy, A.D.; Verdon, M.; Freeman, M.J.; Corkrey, R.; Hills, J.L.; Rawnsley, R.P. Virtual Fencing Technology to Intensively Graze Lactating Dairy Cattle. I: Technology Efficacy and Pasture Utilization. *J. Dairy Sci.* **2021**, *104*, 7071–7083. [CrossRef]
11. Verdon, M.; Horton, B.; Rawnsley, R. A Case Study on the Use of Virtual Fencing to Intensively Graze Angus Heifers Using Moving Front and Back-Fences. *Front. Anim. Sci.* **2021**, *2*, 663963. [CrossRef]
12. Verdon, M.; Langworthy, A.; Rawnsley, R. Virtual Fencing Technology to Intensively Graze Lactating Dairy Cattle. II: Effects on Cow Welfare and Behavior. *J. Dairy Sci.* **2021**, *104*, 7084–7094. [CrossRef]
13. Wilms, L.; Komainda, M.; Hamidi, D.; Riesch, F.; Horn, J.; Isselstein, J. How Do Grazing Beef and Dairy Cattle Respond to Virtual Fences? A Review. *J. Anim. Sci.* **2024**, *102*, skae108. [CrossRef]
14. Nyamuryekung'e, S.; Cox, A.; Perea, A.; Estell, R.; Cibils, A.F.; Holland, J.P.; Waterhouse, T.; Duff, G.; Funk, M.; McIntosh, M.M.; et al. Behavioral Adaptations of Nursing Brangus Cows to Virtual Fencing: Insights from a Training Deployment Phase. *Animals* **2023**, *13*, 3558. [CrossRef] [PubMed]
15. Grudzinski, B.; Fritz, K.; Dodds, W. Does Riparian Fencing Protect Stream Water Quality in Cattle-Grazed Lands? *Environ. Manag.* **2020**, *66*, 121–135. [CrossRef]
16. Schillings, J.; Holohan, C.; Lively, F.; Arnott, G.; Russell, T. The Potential of Virtual Fencing Technology to Facilitate Sustainable Livestock Grazing Management. *Animal* **2024**, *18*, 101231. [CrossRef] [PubMed]
17. Sonne, C.; Alstrup, A.K.O.; Pertoldi, C.; Frikke, J.; Linder, A.C.; Styrishave, B. Cortisol in Manure from Cattle Enclosed with Nofence Virtual Fencing. *Animals* **2022**, *12*, 3017. [CrossRef]
18. Anderson, D.M. Virtual Fencing—Past, Present and Future. *Rangel. J.* **2007**, *29*, 65–78. [CrossRef]
19. Jachowski, D.S.; Slotow, R.; Millspaugh, J.J. Good Virtual Fences Make Good Neighbors: Opportunities for Conservation. *Anim. Conserv.* **2014**, *17*, 187–196. [CrossRef]
20. Lee, C.; Henshall, J.M.; Wark, T.J.; Crossman, C.C.; Reed, M.T.; Brewer, H.G.; O'Grady, J.; Fisher, A.D. Associative Learning by Cattle to Enable Effective and Ethical Virtual Fences. *Appl. Anim. Behav. Sci.* **2009**, *119*, 12–22. [CrossRef]
21. Anderson, D.M.; Estell, R.E.; Holechek, J.L.; Ivey, S.; Smith, G.B. Virtual Herding for Flexible Livestock Management–A Review. *Rangel. J.* **2014**, *36*, 205–221. [CrossRef]
22. Campbell, D.L.; Lea, J.M.; Haynes, S.J.; Farrer, W.J.; Leigh-Lancaster, C.J.; Lee, C. Virtual Fencing of Cattle Using an Automated Collar in a Feed Attractant Trial. *Appl. Anim. Behav. Sci.* **2018**, *200*, 71–77. [CrossRef]

23. Verdon, M.; Lee, C.; Marini, D.; Rawnsley, R. Pre-Exposure to an Electrical Stimulus Primes Associative Pairing of Audio and Electrical Stimuli for Dairy Heifers in a Virtual Fencing Feed Attractant Trial. *Animals* **2020**, *10*, 217. [CrossRef]
24. Lee, C.; Campbell, D.L.M. A Multi-Disciplinary Approach to Assess the Welfare Impacts of a New Virtual Fencing Technology. *Front. Vet. Sci.* **2021**, *8*, 637709. [CrossRef]
25. Boyd, C.S.; O'Connor, R.; Ranches, J.; Bohnert, D.W.; Bates, J.D.; Johnson, D.D.; Davies, K.W.; Parker, T.; Doherty, K.E. Virtual Fencing Effectively Excludes Cattle from Burned Sagebrush Steppe. *Rangel. Ecol. Manag.* **2022**, *81*, 55–62. [CrossRef]
26. Rahman, A.; Smith, D.V.; Little, B.; Ingham, A.B.; Greenwood, P.L.; Bishop-Hurley, G.J. Cattle Behavior Classification from the Collar, Halter, and Ear Tag Sensors. *Inf. Process. Agric.* **2018**, *5*, 124–133. [CrossRef]
27. McSweeney, D.; O'Brien, B.; Coughlan, N.E.; Férard, A.; Ivanov, S.; Halton, P.; Umstätter, C. Virtual Fencing Without Visual Cues: Design, Difficulties of Implementation, and Associated Dairy Cow Behaviour. *Comput. Electron. Agric.* **2020**, *176*, 105613. [CrossRef]
28. Confessore, A.; Schneider, M.K.; Pauler, C.M.; Aquilani, C.; Fuchs, P.; Pugliese, C.; Dibari, C.; Argenti, G.; Accorsi, P.A.; Probo, M. A Matter of Age? How Age Affects the Adaptation of Lactating Dairy Cows to Virtual Fencing. *J. Anim. Sci.* **2024**, *102*, skae137. [CrossRef]
29. Agar, C. Costly Fencing Renewal Replaced with Virtual Fencing; The Cows Wear a Collar on Which Hangs a Two-Sided Solar Panel Approximately the Same Size as a Swiss Cowbell. *Ont. Farmer* **2024**, *B*, 10. Available online: https://www.proquest.com/magazines/costly-fencing-renewal-replaced-with-virtual-cows/docview/3072070025/se-2?accountid=14723 (accessed on 18 April 2025).
30. Anderson, D.M.; Hale, C.S. Animal Control System Using Global Positioning and Instrumental Animal Conditioning. U.S. Patent 6,232,880 B1, 15 May 2001.
31. Fuchs, P.; Stachowicz, J.; Schneider, M.K.; Probo, M.; Bruckmaier, R.M.; Umstätter, C. Stress Indicators in Dairy Cows Adapting to Virtual Fencing. *J. Anim. Sci.* **2024**, *102*, skae024. [CrossRef]
32. Banks, T. Wearable Tech for Cows Helps Herds Communicate with Farmers. Design Week (Online). 28 July 2015. Available online: https://www.proquest.com/magazines/wearable-tech-cows-helps-herds-communicate-with/docview/1699347294/se-2?accountid=14723 (accessed on 15 April 2025).
33. Harland, A.J.; Novais, F.J.; Fitzsimmons, C.J.; Church, J.S.; da Silva, G.M.; Londono-Mendez, M.C.; Bork, E.W. Evaluating Virtual Fencing as a Tool to Manage Beef Cattle for Rotational Grazing Across Multiple Years. *J. Environ. Manag.* **2025**, *381*, 125166. [CrossRef]
34. Hamidi, D.; Grinnell, N.A.; Komainda, M.; Riesch, F.; Horn, J.; Ammer, S.; Traulsen, I.; Palme, R.; Hamidi, M.; Isselstein, J. Heifers Don't Care: No Evidence of Negative Impact on Animal Welfare of Growing Heifers When Using Virtual Fences Compared to Physical Fences for Grazing. *Animal* **2022**, *16*, 100614. [CrossRef]
35. Aaser, M.F.; Staahltoft, S.K.; Korsgaard, A.H.; Trige-Esbensen, A.; Alstrup, A.K.O.; Sonne, C.; Pertoldi, C.; Bruhn, D.; Frikke, J.; Linder, A.C. Is Virtual Fencing an Effective Way of Enclosing Cattle? Personality, Herd Behaviour and Welfare. *Animals* **2022**, *12*, 842. [CrossRef]
36. Grinnell, N.A.; Hamidi, D.; Komainda, M.; Riesch, F.; Horn, J.; Traulsen, I.; Palme, R.; Isselstein, J. Supporting Rotational Grazing Systems with Virtual Fencing: Paddock Transitions, Beef Heifer Performance, and Stress Response. *Animal* **2025**, *19*, 101416. [CrossRef] [PubMed]
37. Musinska, J.; Skalickova, S.; Nevrkla, P.; Kopec, T.; Horky, P. Unlocking Potential, Facing Challenges: A Review Evaluating Virtual Fencing for Sustainable Cattle Management. *Livest. Sci.* **2025**, *295*, 105693. [CrossRef]
38. Versluijs, E.; Niccolai, L.J.; Spedener, M.; Zimmermann, B.; Hessle, A.; Tofastrud, M.; Devineau, O.; Evans, A.L. Classification of Behaviors of Free-Ranging Cattle Using Accelerometry Signatures Collected by Virtual Fence Collars. *Front. Anim. Sci.* **2023**, *4*, 1083272. [CrossRef]
39. Stenfelt, J.; Yngvesson, J.; Blokhuis, H.J.; Rørvang, M.V. Dairy Cows Did Not Rely on Social Learning Mechanisms When Solving a Spatial Detour Task. *Front. Vet. Sci.* **2022**, *9*, 956559. [CrossRef] [PubMed]
40. Campbell, D.L.M.; Haynes, S.J.; Lea, J.M.; Farrer, W.J.; Leigh-Lancaster, C.J.; Lee, C. Temporary Exclusion of Cattle from a Riparian Zone Using Virtual Fencing Technology. *Animals* **2018**, *9*, 5. [CrossRef]
41. Boyd, C.S.; O'Connor, R.C.; Ranches, J.; Bohnert, D.W.; Bates, J.D.; Johnson, D.D.; Davies, K.W.; Parker, T.; Doherty, K.E. Using Virtual Fencing to Create Fuel Breaks in the Sagebrush Steppe. *Rangel. Ecol. Manag.* **2023**, *89*, 87–93. [CrossRef]
42. Harland, A.J.; Novais, F.J.; Durunna, O.N.; Fitzsimmons, C.J.; Church, J.S.; Bork, E.W. Evaluation of the Technical Performance of the Nofence Virtual Fencing System in Alberta, Canada. *Smart Agric. Technol.* **2025**, *10*, 100713. [CrossRef]
43. Hamidi, D.; Grinnell, N.A.; Komainda, M.; Wilms, L.; Riesch, F.; Horn, J.; Hamidi, M.; Traulsen, I.; Isselstein, J. Training Cattle for Virtual Fencing: Different Approaches to Determine Learning Success. *Appl. Anim. Behav. Sci.* **2024**, *273*, 106220. [CrossRef]
44. Murray, B.D.; Wagner, K.L.; Reuter, R.; Goodman, L.E. Use of Virtual Fencing to Implement Critical Conservation Practices. *Rangelands* **2024**, *47*, 41–49. [CrossRef]
45. Fuchs, P.; Stachowicz, J.; Schneider, M.; Probo, M.; Bruckmaier, R.M.; Umstätter, C. PSVI-16 Conditioning Dairy Cows to a Virtual Fencing System Is Compatible with Animal Welfare. *J. Anim. Sci.* **2023**, *101* (Suppl. 3), 334–335. [CrossRef]
46. Hamidi, D.; Hütt, C.; Komainda, M.; Grinnell, N.A.; Horn, J.; Riesch, F.; Hamidi, M.; Traulsen, I.; Isselstein, J. Grid Grazing: A Case Study on the Potential of Combining Virtual Fencing and Remote Sensing for Innovative Grazing Management on a Grid Base. *Livest. Sci.* **2023**, *278*, 105373. [CrossRef]

47. Confessore, A.; Aquilani, C.; Nannucci, L.; Fabbri, M.C.; Accorsi, P.A.; Dibari, C.; Argenti, G.; Pugliese, C. Application of Virtual Fencing for the Management of Limousin Cows at Pasture. *Livest. Sci.* **2022**, *263*, 105037. [CrossRef]
48. Colusso, P.I.; Clark, C.E.F.; Ingram, L.J.; Thomson, P.C.; Lomax, S. Dairy Cattle Response to a Virtual Fence When Pasture on Offer Is Restricted to the Post-Grazing Residual. *Front. Anim. Sci.* **2021**, *2*, 791228. [CrossRef]
49. Lomax, S.; Colusso, P.; Clark, C.E.F. Does Virtual Fencing Work for Grazing Dairy Cattle? *Animals* **2019**, *9*, 429. [CrossRef]
50. Aquilani, C.; Confessore, A.; Bozzi, R.; Sirtori, F.; Pugliese, C. Review: Precision Livestock Farming Technologies in Pasture-Based Livestock Systems. *Animal* **2022**, *16*, 100429. [CrossRef]
51. Waterhouse, T. Virtual fencing systems: Balancing production and welfare outcomes. *Livestock* **2023**, *28*, 227–234. [CrossRef]
52. Bello, R.W.; Moradeyo, O.M. Monitoring Cattle Grazing Behavior and Intrusion Using Global Positioning System and Virtual Fencing. *Asian J. Math. Sci.* **2019**. Available online: https://www.ajms.in/index.php/ajms/article/view/235 (accessed on 22 April 2025).
53. Fisher, A.; Cornish, A. *Independent Scientific Literature Review on Animal Welfare Considerations for Virtual Fencing*; Department of Agriculture, Fisheries and Forestry: Canberra, ACT, Australia, 2022.
54. Lund, S.M.; Jacobsen, J.H.; Nielsen, M.G.; Friis, M.R.; Nielsen, N.H.; Mortensen, N.Ø.; Skibsted, R.C.; Aaser, M.F.; Staahltoft, S.K.; Bruhn, D.; et al. Spatial Distribution and Hierarchical Behaviour of Cattle Using a Virtual Fence System. *Animals* **2024**, *14*, 2121. [CrossRef]
55. Verdon, M.; Hunt, I.; Rawnsley, R. The Effectiveness of a Virtual Fencing Technology to Allocate Pasture and Herd Cows to the Milking Shed. *J. Dairy Sci.* **2024**, *107*, 6161–6177. [CrossRef]
56. Umstätter, C.; Brocklehurst, S.; Ross, D.W.; Haskell, M.J. Can the Location of Cattle Be Managed Using Broadcast Audio Cues? *Appl. Anim. Behav. Sci.* **2013**, *147*, 34–42. [CrossRef]
57. Grumett, D.; Butterworth, A. Electric Shock Control of Farmed Animals: Welfare Review and Ethical Critique. *Anim. Welf. J.* **2022**, *31*, 373–385. [CrossRef]
58. Heins, B.J.; Pereira, G.M.; Sharpe, K.T. Precision Technologies to Improve Dairy Grazing Systems. *JDS Commun.* **2023**, *4*, 318–323. [CrossRef] [PubMed]
59. Wróbel, B.; Zielewicz, W.; Staniak, M. Challenges of Pasture Feeding Systems—Opportunities and Constraints. *Agriculture* **2023**, *13*, 974. [CrossRef]
60. Occhiuto, F.; Vázquez-Diosdado, J.A.; King, A.J.; Kaler, J. Using Precision Technology to Investigate Personality and Plasticity of Movement in Farmed Calves and Their Associations with Weight Gain. *Anim. Sci. Proc.* **2023**, *14*, 160. [CrossRef]
61. Holohan, C.; Gordon, A.; Palme, R.; Buijs, S.; Lively, F. An Assessment of Young Cattle Behaviour and Welfare in a Virtual Fencing System. In Proceedings of the XXV International Grassland Congress, Covington, KY, USA, 14–19 May 2023; pp. 1569–1571. [CrossRef]
62. Herlin, A.; Brunberg, E.; Hultgren, J.; Högberg, N.; Rydberg, A.; Skarin, A. Animal Welfare Implications of Digital Tools for Monitoring and Management of Cattle and Sheep on Pasture. *Animals* **2021**, *11*, 829. [CrossRef]
63. Colusso, P.I.; Clark, C.E.F.; Lomax, S. Should Dairy Cattle Be Trained to a Virtual Fence System as Individuals or in Groups? *Animals* **2020**, *10*, 1767. [CrossRef]
64. Bhargava, K.; Ivanov, S.; Kulatunga, C.; Donnelly, W. Fog-Enabled WSN System for Animal Behavior Analysis in Precision Dairy. In Proceedings of the 2017 International Conference on Computing, Networking and Communications (ICNC), Silicon Valley, CA, USA, 26–29 January 2017; IEEE: New York, NY, USA, 2017; pp. 504–510.
65. French, P.; O'Brien, B.; Shalloo, L. Development and Adoption of New Technologies to Increase the Efficiency and Sustainability of Pasture-Based Systems. *Anim. Prod. Sci.* **2015**, *55*, 931–935. [CrossRef]
66. Hayward, M.W.; Kerley, G.I. Fencing for Conservation: Restriction of Evolutionary Potential or a Riposte to Threatening Processes? *Biol. Conserv.* **2009**, *142*, 1–13. [CrossRef]
67. Osipova, L.; Okello, M.M.; Njumbi, S.J.; Ngene, S.; Western, D.; Hayward, M.W.; Balkenhol, N. Fencing Solves Human-Wildlife Conflict Locally but Shifts Problems Elsewhere: A Case Study Using Functional Connectivity Modelling of the African Elephant. *J. Appl. Ecol.* **2018**, *55*, 2673–2684. [CrossRef]
68. Hoag, D.; Vorster, A.G.; Ehlert, K.; Evangelista, P.; Edwards-Callaway, L.; Mooney, D.F.; Virene, J. Beef Cattle Producer Perspectives on Virtual Fencing. *Rangel. Ecol. Manag.* **2024**, *96*, 143–151. [CrossRef]
69. Wätzold, F.; Jauker, F.; Komainda, M.; Schöttker, O.; Horn, J.; Sturm, A.; Isselstein, J. Harnessing Virtual Fencing for More Effective and Adaptive Agri-Environment Schemes to Conserve Grassland Biodiversity. *Biol. Conserv.* **2024**, *297*, 110736. [CrossRef]

ruminants

MDPI

Review

Enhanced Animal Welfare and Labeling in Cattle, Sheep, and Goats

Maria Papageorgiou [1,*], Agori Karageorgou [1], Ouranios Tzamaloukas [2] and Panagiotis Simitzis [1,*]

1 Laboratory of Animal Breeding and Husbandry, Department of Animal Science, Agricultural University of Athens, 11855 Athens, Greece
2 Department of Agricultural Sciences, Biotechnology and Food Science, Cyprus University of Technology, Limassol P.O. Box 50329, Cyprus; akarageorgou@aua.gr (A.K.); ouranios.tzamaloukas@cut.ac.cy (O.T.)
* Correspondence: mpapageorgiou@aua.gr (M.P.); pansimitzis@aua.gr (P.S.)

Simple Summary: For a sustainable tomorrow, the welfare of humans, animals, and the environment must be considered as a whole. Nowadays, animal welfare has become a public concern, with consumers becoming increasingly aware of the quality of food products and husbandry practices, and willing to pay more for animal products that prioritize animal welfare. Thus, there is need for transparent communication of the animal welfare level through product labeling. Cattle, sheep, and goats account for nearly all dairy production and are among the primary species raised for meat. In Europe, there is increasing development of animal welfare labeling schemes by private or public initiatives. All these schemes exceed the welfare level of conventionl systems and sometimes even that of organic systems, particularly when they were developed by animal welfare organization initiatives. The few standards developed for evaluating animal welfare in intensive systems, although mainly reinforcing the regulations for conventional husbandry, enhance welfare by analyzing the regulations in detail, reducing ambuigity and ensuring clear guidance. The standards developed for extensive systems promote a higher welfare level by promoting access to pasture, comfort, social bonding, and environmental enrichment, thus promoting positive aspects in the animals' lives by evaluating animal welfare in a more comprehensive way.

Abstract: In 2020, the European Union endorsed its "Farm-to-Fork" strategy, emphasizing the need for transparency in the food production chain and communication of the sustainability level and nutritional value of food products to the consumer through labeling. For animal-based products, this also includes information about the husbandry systems under which the animals are raised. At the same time, people are becoming increasingly concerned both as citizens and as consumers about animal welfare issues in production species, as animal welfare is considered an integral part of sustainability and food security. This has led to the development of various enhanced animal welfare labeling schemes, initiated by public or private entities, or even as a partnership of both. Specifically for cattle, sheep, and goats, various standards have been developed and implemented in Europe, all establishing higher welfare standards compared to conventional farming, and in some cases exceeding the minimum requirements for organic farming as set by Regulation (EU) 2018/848. Most of these standards, especially those developed by NGOs advocating for animal welfare or through public initiative, were developed for semi-intensive to extensive systems. They primarily incorporate animal-based measures, including positive welfare indicators, offering a holistic approach to animal welfare evaluation. Although there is significant heterogeneity in European animal welfare standards, nearly all of them promote access to pasture, comfort, environmental enrichment, and, in some cases, even mother–young bonding.

 https://doi.org/10.3390/ruminants5020019

Keywords: labeling; animal welfare; animal welfare standard; cattle; meat; dairy; sheep; goat; positive animal welfare; mother–young bonding

1. Introduction

Nowadays, animal welfare is a fundamental component of agricultural development, gaining increasing concern on a worldwide basis [1,2]. In 2016, the United Nations Committee on World Food Security formally denoted animal welfare as a key pillar of sustainable agricultural development, recognizing its link to food security and emphasizing that animals, humans, and the environment should be considered interconnected [3]. In other words, a production system cannot be considered sustainable today if the animals are kept below a minimum welfare level deemed acceptable by society [1,4]. Gradually, this level is getting higher, particularly in developed countries, where people as citizens advocate for stricter legislation for animal protection and as consumers for ethical production practices for animal-derived products [5]. Animal welfare is considered a common good that is beneficial for society, and animal-derived products originating from animals raised under an adequate, societally acceptable welfare level are perceived as products of high quality [1,6]. Thus, consumers are willing to pay a premium for products from animals kept under enhanced welfare standards [7,8]. In the EU, 90% of citizens consider animal welfare a key factor that influences their decision to purchase animal-derived products, believing that products that are more friendly and ethical to animals are also healthier and more friendly to the environment [9]. This trend is a result not only of ethical concerns, but also of pressure from animal welfare non-governmental organizations (NGOs) demanding improved animal welfare practices in livestock systems [10].

On the other hand, the pressure for improved animal welfare standards for production animals contrasts with the growing international demand for products of animal origin, especially milk and meat [7,8]. It can be challenging for a farming system to balance animal and environmental sustainability with economic viability, since, although consumers are willing to pay more for animal-derived products from animals kept at an improved welfare level, this willingness is complex and influenced by various factors [10–15]. Additionally, improved animal welfare often results in an increased price of the final product, which can be unaffordable for some consumers [8,11]. Considering the rising global population and the increasing demand for animal-derived products, especially in developing countries, this challenge becomes even more complex [8–12]. As a result, there is still intensification in livestock production, and the welfare of the animals is severely compromised in some farming systems.

Cattle, sheep, and goats are livestock species whose welfare raises high concern among consumers. According to the Food and Agriculture Organization (FAO), 81% percent of the world's milk supply is produced by dairy cattle, 15% by buffalo, and 4% by sheep, goats, and camels [16]. In 2024, the EU's milk output reached approximately 235 million tons, increased by 0.6% compared to 2023 [17]. This growth was primarily driven by yield improvements rather than an increase in animal numbers, which shows the intensification of breeding that could lead to welfare impairment of highly productive animals [16,17]. Moreover, goat and sheep herds are expected to expand at a faster rate. World milk production is expected to rise by 1.7% by 2030 compared to 2019, outpacing the growth of other agricultural commodities. Per-capita consumption of both fresh and processed dairy products is estimated to increase in the EU from 23.6 kg in 2018 to 25.2 kg (milk solids) by 2030, including mainly milk, cheese, and yogurt. The production and consumption of meat from cattle and small ruminants follow the same trend as dairy products [16].

Although meat consumption has been shifting towards poultry, growth in beef and sheep meat consumption is expected to increase by 5.9% and 15.7%, respectively, by 2030 on a global basis, representing 20% and 5%, respectively, as a source of meat protein [18].

These numbers underscore how challenging it is to achieve a sufficient animal welfare level for livestock ruminants in the various existing production systems, from purely extensive to highly intensive, and at the same time to satisfy the various consumer demands and the price that they can afford for the final animal-derived product. In 2020, the European Union emphasized the need for transparency in food labeling through its "Farm-to-Fork" strategy [19]. This strategy highlights the importance of informing consumers about the sustainability level of food production, the nutritional value of food products, and the level of animal welfare in farming systems [19]. This initiative is an effort to inform consumers through product labeling and communicate not only the quality of the products but also the husbandry systems under which the animals are raised [18,20].

In Europe, consumers are increasingly concerned about animal welfare and husbandry practices in cattle [14,15], sheep [10,13,21], and goat [13,21] farming systems; thus, transparent labeling to communicate this information is crucial. Consumers are willing to and do pay a higher price for milk when cattle are raised with access to pasture [14]. Veal meat is considered a product of added value when the welfare of the animals is enhanced, and consumers are particularly concerned about the provision of painkillers during the castration of calves and bulls [15]. Labeling influences consumer decisions for choosing beef, exhibiting a preference to organic or animal welfare labels [14,15,22]. Organic bovine products are considered a socially responsible and ethical consumption choice [23,24]. Positive animal welfare information on labeling affects the emotional responses of consumers towards milk [25,26] and yogurt [27], who perceive them as tastier and healthier than conventional dairy products [25,27] and a more conscious choice [26,27]. Sheep and goat milk consumption is rising in Southern Europe, since in consumers' minds small ruminants are often raised in extensive systems and their husbandry is more environmentally friendly compared to cattle husbandry [13,21]. Additionally, the European public perception of sheep and goat dairy products is affected by the origin and the impact of the product on the environment and the welfare level of the animals [10,13,21]. For example, in Greece, one out of three consumers will pay more for organic-labeled products [21], while in the UK, 93% of consumers would pay more for a product with a label that indicates improved animal welfare [28]. Consumers are also becoming concerned about cow–calf separation and are willing to pay more for meat from calves reared by their mothers or milk by cattle that have suckled the calf, even for a limited period [29].

The above outcomes highlight the importance of informing consumers about the animal welfare practices in farming systems. As part of this effort, new welfare protocols are being developed through the initiative of private, public, and non-profit organizations [20,30–33]. These protocols include animal welfare provisions that exceed the minimum legal requirements of an acceptable animal welfare level [20,30–33]. Also, they establish standards that often go beyond the standards of organic farming as laid down by Regulation (EU) 2018/848 on organic production and labeling [32,33]. Not only do they include standards that ensure the protection of animals from negative welfare aspects, but also often positive welfare indicators that promote the experience of positive emotional states in animals [20,32,33].

This paper aims to identify and review the main European protocols regarding the assessment and labeling of welfare in cattle, sheep, and goats, covering both dairy and meat production systems. We will analyze and compare the retrieved animal welfare standards in terms of their structure, content, geographical origin and extent, initiation, and audit frequency. In addition, we will examine the additional provisions included in these standards that go beyond those of conventional husbandry systems, enhancing animal welfare in relation to husbandry practices, transport, and slaughter. A special focus will be also placed on the environmental enrichment criteria and the positive welfare indicators that are incorporated into the standards and improve animal welfare. After reviewing all these factors, we will discuss whether the retrieved labeling schemes require a higher level of animal welfare not only compared to conventional systems but also in relation to organic husbandry. Finally, we will reflect on current implementations and challenges and propose directions for future research and development of animal welfare labeling schemes.

2. Materials and Methods

All the labeling schemes were retrieved through Google and Google Scholar. Some of them were already known, while the rest were identified by using the following combinations of keywords: "animal welfare" AND "labelling scheme*" OR "labeling scheme", "animal welfare" AND "labelling" OR "labeling", "animal welfare" AND "protocol*" OR "standard*", "dairy" OR "milk" OR "yogurt" OR "cheese" OR "butter" AND "animal welfare" AND "labelling" OR "labeling", "meat" OR "beef" OR "bovine" OR "veal" OR "lamb*" AND "animal welfare" AND "labelling" OR "labeling", "cattle" OR "sheep" OR "goat*" AND "animal welfare" AND "labelling" OR "labeling".

In total, 19 labeling schemes were identified. We selected only labeling schemes that are used in Europe but were not necessarily developed by European countries. Then, we visited each scheme's official website to examine and download their standards and policy for the audits. Sixteen labeling schemes had their full standards publicly available online, which we downloaded, analyzed, and included in the review. Three labeling schemes (Für Mehr Tierschutz, Compromiso Bienestar Animal PAWS, and Animal Welfare Interovic Spain) only had selected criteria available online rather than the full list of standards, so they were not included in our review.

3. Overview of Enhanced Animal Welfare Labeling Schemes in Cattle, Sheep, and Goats in Europe

In Table 1, we present the selected animal welfare labeling schemes. As shown in the table, fifteen standards have been developed for the welfare assessment of dairy cattle, eight for dairy sheep, and seven for dairy goats. Regarding meat husbandry systems, fourteen, eleven, and eight schemes address the welfare evaluation of beef cattle, sheep, and goats, respectively.

Table 1. Labeling schemes with enhanced animal welfare standards in Europe and information about geographical scope, country of origin, and initiation (private or public). The table also indicates the number of tiers of each scheme (single or multiple), the species covered, and the audit frequency.

Labeling Scheme	Species	Country	Territorial Extent	Initiation	Number of Tiers (Single/Multiple)	Frequency of Audits	References
KRAV	Cattle (dairy, beef), sheep (dairy, meat), goats (dairy, meat)	Sweden	National (also Germany, Spain, Latvia, Denmark, Bulgaria)	Private	1	Annual	[34]
Svensk Sigill	Cattle (dairy, beef), sheep (dairy, meat)	Sweden	National	Private	3—IP Sigill Grundcertifiering/naturbeteskött/klimatcertifiering	Annual	[35–37]
Bedre Dyrevelfaerd	Cattle (dairy, beef)	Denmark	National (also Baltic region, Norway, Sweden)	Public–private	3	Annual	[38]

Table 1. *Cont.*

Labeling Scheme	Species	Country	Territorial Extent	Initiation	Number of Tiers (Single/Multiple)	Frequency of Audits	References
Anbefalet af Fyrenrs Beskyttelse	Cattle (dairy, beef), sheep (dairy, meat)	Denmark	National (also Norway, Sweden)	Private	1	At least annual (different times of year)	[39–41]
Beter Leven	Cattle (dairy, beef)	Netherlands	National	Public–private	3	Annual	[42–47]
Meadow Milk	Dairy cattle	Netherlands	National	Private	1	Annual	[48]
EKO	Cattle (dairy), goats (dairy)	Netherlands	National	Public–private	3	Once every two years, maximum 30-month interval	[49,50]
AMA Gütesiegel	Cattle (dairy, beef), sheep (dairy, meat), goats (dairy, meat)	Austria	National	Public–private	3 (dairy: basic, plus, plus outdoors/beef: basic, Mer Tierwohl, bio/Freiland/ Aim/Weidehaltung)	Annual	[51–53]
Demeter	Cattle (dairy, beef), sheep (dairy, meat), goats (dairy, meat)	Germany	National (also international)	Private	1	Annual	[54]
Bio Suisse	Cattle (dairy, beef), sheep (dairy, meat), goats (dairy, meat)	Switzerland	National (also international)	Private	1	Annual	[55]
Red Tractor	Cattle (dairy, beef), sheep (meat), goats (dairy)	UK	National	Private	1	Annual (max 18 months)	[56–58]
RSPCA Approved	Cattle (dairy, beef)	UK	National (also Ireland, Australia)	Private	1	Annual	[59–61]
Scotch Assured	Cattle (beef), sheep (meat)	UK (RSPCA)	National	Private	1	Annual (different times of year)	[62]
Bienestar Animal Welfair	Cattle (dairy, beef), sheep (dairy, meat), goats (dairy, meat)	Spain	National (also Spain, Portugal)	Public (research institutes)	1	Minimum annual	[63–67]
Animal Welfare Certified by GAP	Cattle (dairy, beef), sheep (meat), goats (meat)	Global GAP	International	Private	6 (1–5, 5+)	Every 15 months (to cover all seasons)	[68–71]
Animal Welfare Approved by AGW	Cattle (dairy, beef), sheep (dairy, meat), goats (dairy, meat)	A Greener World	International	Private	1	Minimum annual and on all farms	[72–77]

3.1. Structure, Content, Geography, Initiation, and Audits

3.1.1. Number of Tiers

The labeling schemes can follow either a single- or a multi-tiered structure. The higher the tier, the more provisions and enrichments are required and the more improved the animal welfare level is. As shown in Table 1, the majority of schemes are single-tiered, meaning that a farming system either passes or fails to pass a certification audit. However, some labeling schemes employ a multi-tiered structure, certifying animal welfare at ascending levels. The number of tiers ranges from three [38,42–47] to six [68–71]. In these cases, the tier is indicated on the label either numerically [68–71] or symbolically with a corresponding number of shapes, such as hearts for Bedre Dyrevelfaerd [38] and stars for Beter Leven [42–47].

3.1.2. Country of Origin and Geographical Extent

Almost all animal welfare standards have been developed in Central and Northern Europe. Only one scheme, Welfair, originates from Southern Europe. Countries in Central and Northern Europe often implement national regulations that are stricter than those of the EU regarding animal welfare. This can be explained by the fact that the debate on animal welfare, as well as the concern of consumers about husbandry practices, is more intense in these countries [20].

The geographical scope of most schemes is mainly national [35–37,42–53,56–58,62]. However, it is also common for a standard to be applied beyond its country of origin, in neighboring countries (KRAV [34], Bedre Dyrevelfaerd [38], Anbefalet af Fyrenrs Beskyttelse [39–41], RSPCA Approved [60,61], Welfair [63–67]).

3.1.3. Initiation

Most of the standards were developed through private initiatives. However, the Bedre Dyrevelfaerd [38], Beter Leven [42–47], EKO [49,50], and AMA Gütesiegel [51–53] schemes are a result of public and private collaboration. For these four standards, all audits are conducted and overseen by private control bodies. The only exception is Bedre Dyrevelfaerd [38]. Although inspections under this scheme are carried out by private inspection bodies, the audit oversight is managed by public authorities [20,38]. So far, Denmark remains the only European country that certifies animal welfare at the public level. Nonetheless, efforts are currently underway to develop a public certification system also in Italy and Germany [20]. The Welfair standards were developed by academic institutions. Their criteria are based on the Welfare Quality animal welfare assessment protocols for dairy cows [63], fattening cattle [64], and veal calves [65]. These protocols are the outcome of collaborative research within the Welfare Quality Network, involving multiple European universities [78]. Although the network does not have its own labeling system, it provides standardized certification protocols that can serve as a basis for the development of national-level schemes [78]. The AWIN protocols for sheep [66] and goats [67] were developed on the same philosophy and form the basis of Welfair standards for small ruminants.

Private initiators of improved animal welfare standards are often non-government organizations (NGOs). RSPCA Approved, spread throughout England and Ireland, was established by the non-profit animal protection organization Royal Society for the Prevention of Cruelty to Animals—one of the oldest organizations advocating for animal welfare [79]. Anbefalet af Fyrenrs Beskyttelse is an NGO focusing on animal protection with a territorial scope mainly in the Baltic and Scandinavian regions [80]. Global GAP [81] and A Greener World have an international scope and presence [82]. In some cases, animal welfare labeling schemes are also designed to support product differentiation in the market by associating improved animal welfare with high-quality products. For example, Scotch Assured was developed by the collaboration of Quality Meat Scotland and the Scottish Society for the Prevention of Cruelty to Animals to differentiate Scottish red meat of cattle and sheep in the market. Meat products carrying the Scotch Assured logo originate from animals that have not only been raised under improved animal welfare conditions but have also been born, raised, and slaughtered in Scotland [83].

3.1.4. Frequency of Audits

For all labeling schemes, audits are conducted annually. The only exception is EKO, requiring auditing of farms every two years (with a maximum interval of 30 months). This standard focuses on promoting access to pasture for dairy cattle and goats and is the only scheme that mandates farms to be both organic and non-GMO certified [49,50]. Organic certification requires annual audits according to Regulation (EU) 2018/848 and already sets high animal welfare standards. This may explain the less frequent audits for EKO compared to other standards. Some standards require the audits to be conducted at different times of the year (Anbefalet af Fyrenrs Beskyttelse [39–41], Scotch Assures [62]) or every 15 months to cover all seasons (Animal Welfare Certified by GAP [68–71]). This variation highlights the importance of evaluating animal welfare during different seasonal conditions, particularly those that may pose additional challenges to animal well-being.

3.1.5. Type of Production Systems Covered

As analyzed in Table 2, the standards vary significantly in terms of the type of system that they address, ranging from purely intensive to purely extensive. Most schemes were developed for semi-extensive to extensive systems, since the level of animal welfare is improved when ruminants have access to pasture or the outdoors. Red Tractor [56–58] and Welfair [63–67] are the only schemes developed for evaluating animal welfare in intensive systems.

Table 2. Labeling schemes with information about the conversion period, the main type of production system that they assess (extensive or intensive), whether they focus exclusively on promoting animal welfare and not other sustainable aspects, and whether they require organic certification according to Regulation (EU) 2018/848 on organic production and labeling of organic products.

Labeling Scheme	Production System Types	Focusing Exclusively on Animal Welfare	Mandatory Organic Certification	Conversion Period	Breed	References
KRAV	Extensive	No	No	Cattle: 6 months milk, 12 meat, sheep, goats: 6 months milk/meat	Suitable for organic	[34]
Svensk Sigill	Extensive	No	No	One season on pasture ("naturbeteskött"), 60 days lamn IP, up to 120 days cattle)	All (natural calving)	[35–37]
Bedre Dyrevelfaerd	Semi-intensive–extensive	Yes	No	No	All	[38]
Anbefalet af Fyrenrs Beskyttelse	Extensive	Yes	No	Animals born and raised in the herd	Sheep: all, 75% documented beef cattle breeds and crosses thereof (natural calving)	[39–41]
Beter Leven	Extensive	Yes	No	No	Breeds with a high incidence of caesarean section and stress during transport excluded	[42–47]
Meadow Milk	Extensive	No	No	No	All	[48]
EKO	Extensive	No	Yes	No	All	[50,59]

Table 2. *Cont.*

Labeling Scheme	Production System Types	Focusing Exclusively on Animal Welfare	Mandatory Organic Certification	Conversion Period	Breed	References
AMA Gütesiegel	Extensive	No	No	6 months, 2 months calves (organic time credited)	All (animals purchased from farm's local region)	[51–53]
Demeter	Extensive	No	No	Milk, all: 6 months, meat sheep, lamb: 6 months organic, 12 conv. beef: 1/3 of their lives if organic, 2/3 if conv.	Suitable for organic farming	[54]
Bio Suisse	Extensive	No	No	6 months	Suitable for organic farming	[55]
Red Tractor	Intensive	No	No	No	All	[56–58]
RSPCA Approved	Semi-extensive–extensive	Yes	No	No	All	[59–61]
Scotch Assured	Extensive	No	No	Born, reared, slaughtered in Scotland—entire life under standard	All	[62]
Bienestar Animal Welfair	Intensive	Yes	No	No	All	[63–67]
Animal Welfare Certified by GAP	All—extensive	Yes	No	A mimuim of 2 years in tier 4 is required to reach tier 5 in tier 4 to reach t	All—animals selective to adapt, easy calving, polled	[68–71]
Animal Welfare Approved by AGW	Extensive	Yes	No	No	All—easy calving, closed herd, polled animals	[72–77]

3.1.6. Exclusive Focus on Promoting Animal Welfare

As shown in Table 2, standards developed by NGOs prioritize the promotion of animal welfare, without addressing other aspects of sustainability (Bedre Dyrevelfaerd [38], Anbefalet af Fyrenrs Beskyttelse [39–41], Beter Leven [42–47], RSPCA Approved [59–61], Animal Welfare Certified by GAP [68–71], Animal Welfare Approved by AGW [72–77]). However, it is also common for some standards to promote environmental sustainability alongside animal welfare, especially those that have been developed for the evaluation of animal welfare in extensive systems and require access to pasture (Meadow Milk [48], EKO [49,50], Demeter [54], Bisuisse [55]). In certain cases, social sustainability is also promoted (KRAV [34], Svensk Sigill [35–37]).

3.1.7. Conversion Period

Nine out of sixteen animal welfare schemes do not require any conversion period. When a conversion period is mandatory, it is usually six months for all dairy ruminants and for small ruminants raised for meat and one year for beef cattle (KRAV [34], Demeter [54]). In these cases, the length of the mandated conversion period is the same as the one imposed by the standards of organic agriculture according to Regulation (EU) 2018/848. Some schemes, such as AMA Gütesiegel [51–53] and Demeter [54], also allow organic time to be credited. Anbefalet af Fyrenrs Beskyttelse [39–41] requires a close herd and Scotch Assured specifies that animals must have spent their entire life under the conditions of the standard [62].

3.1.8. Breeds

Regarding the breeds permitted by the standards, almost all schemes impose no specific restrictions. However, they all emphasize that breed selection should prioritize natural calving, excluding breeds or animals with a high incidence of caesarean sections.

Maximum thresholds for caesarean sections and dystocia rates within the herd can also be established (Animal Welfare Approved by GAP [68,69]). Certain standards developed mainly for welfare assessment in extensive systems allow breeds suitable for organic farming (KRAV [34], Demeter [54], BioSuisse [55]), while others allow only animals from local breeds (AMA Gütesiegel [51–53], Scotch Assured [62]).

3.2. *Husbandry Practices*

3.2.1. Tail Docking

Tail docking of sheep is prohibited under most standards (although it is still permitted for veterinary reasons), as shown in Table 3. Demeter [54] states that this practice is permitted but not systematically. However, in labeling schemes created for intensive systems (Red Tractor [55], Welfair [65]) or for meat quality (Scotch Assured [61]), tail docking is allowed. Welfair [65] explains that the length of the tail must cover the anus/vulva.

Table 3. Handling practices regarding tethering, tail docking, castration of males, and dehorning/disbudding under the selected animal welfare standards.

Labeling Scheme	Tail Docking (Sheep)	Castration of Males	Specific Identification/Dehorning/ Disbudding Practices	Tethering (Cattle)	References
KRAV	No	Yes (calves under 8 weeks)	Only hot dehorning methods permitted	Yes (max 50 animals, during the stable period, routine exercise)	[34]
Svensk Sigill	No (yes for veterinary reasons)	Yes	Identification in an acceptable way for animal's welfare	Not mentioned	[35–37]
Bedre Dyrevelfaerd	No	Not mentioned/as conventional	Not mentioned/as conventional	No	[38]
Anbefalet af Fyrenrs Beskyttelse	No	Yes (bulls under 10 months)	Not mentioned/as conventional	No	[39–41]
Beter Leven	No	Yes (anesthetic up to at least e days)	Freeze branding, disbudding paste not permitted	No	[42–47]
Meadow Milk	Not mentioned	Not mentioned	Not mentioned	Not mentioned	[48]
EKO	Not mentioned	Not mentioned	Not mentioned	Not mentioned	[49,50]
AMA Gütesiege	Not mentioned	Not mentioned	Not mentioned	As of 1 January 2024, tethered housing with access to pasture less than 90 days not permitted	[51–53]
Demeter	Yes (not systematically)	Yes (to improve the health, welfare, hygiene of animals)	Dehorning not permitted (annual review)	No	[54]
Bio Suisse	No (yes for veterinary reasons)	Yes (if prescribed by a veterinarian)	Not mentioned/ as conventional	No (bovines yes if regular access to range and/or pasture according to Article 75 of the Direct Payments Ordinance)	[55]
Red Tractor	Yes	Yes	Cutting/sawing horn (must not be used routinely	No	[56–58]
RSPCA Approved	Permitted only if the calf is intended to be raised beyond 12 months of age		Not mentioned/ as conventional	No	[59–61]

Table 3. *Cont.*

Labeling Scheme	Tail Docking (Sheep)	Castration of Males	Specific Identification/Dehorning/ Disbudding Practices	Tethering (Cattle)	References
Scotch Assured	Yes	Yes	Chemical cauterization not recommended	Yes (recommended no, untethered exercise must be provided daily)	[62]
Bienestar Animal Welfair	Yes (must cover the anus/vulva)	Yes	Not mentioned/ as conventional	No	[63–67]
Animal Welfare Certified by GAP	Yes for tiers 1–4, no tiers for 5, 5+	Yes (tiers 2–5+, not only pain relief, also non-steroidal anti-inflammatory drug)	Disbudding is prohibited in tiers 5, 5+, dehorning prohibited in all tiers	No	[68–71]
Animal Welfare Approved by AGW	Not permitted	Yes	Caustic paste may be used to disbud calves by 7 days of age (instructions of use)/hot branding prohibited	Not permitted	[72–77]

3.2.2. Castration

Castration of males is permitted by all schemes, with the exceptions of Animal Welfare Certified by GAP, which forbids the practice in tiers 5, 5+ [68–71], and RSPCA Approved, which allows castration and tail docking only for animals that will be raised beyond the age of twelve months [59–61].

3.2.3. Dehorning

When it comes to dehorning, Beter Leven prohibits the use of caustic paste [42,43]. Scotch Assured [63] and Animal Welfare Approved by AGW [72,73] permit its use but distinctly mention that the practice is not recommended. Also, the scheme provides detailed instructions on its use since it can be dangerous for the animals. Demeter [54] underscores the need for dehorning and genetic hornless breeding of cattle, having also the strategy of "horns belong to the cow" [84].

3.2.4. Tethering

Tethering of cattle is not allowed by most standards, and the few that allow it (KRAV on farms with less than fifty animals [34], BioSuisse [55], Scotch Assured [62]) require that the animals be provided with routine exercise.

3.2.5. Husbandry Practices Concerning Dairy Calves

A common practice in dairy farming is the immediate separation of newborn animals from their mothers after birth. As shown in Table 4, some animal welfare evaluation schemes require that the mother have licked the newborn before the separation (Beter Leven [42–47]), while others mandate that the newborn be allowed to suckle for 6–12 h (Animal Welfare Certified by GAP [73,74,76] tiers 2–4) or up to one day (KRAV [34], Bedre Dyrevelfaerd [38] tier 3, Anbefalet af Fyrenrs Beskyttelse [40,41]). Animal Welfare Certified by GAP in tiers 5, 5+ requires that the newborn stay with the mother until weaning. The duration of individual housing for calves also varies considerably between standards from one week (KRAV [35], Bedre Dyrevelfaerd [38], Anbefalet af Fyrenrs Beskyttelse [40,41], Demeter [54]) to up to eight weeks (Red Tractor [56–58], Scotch Assured [62], BioSuisse [55],

Svesk Sigill [35,36]). Animal Welfare Certified by GAP permits individual housing of calves only in the lowest tiers. Weaning time differs from a minimum of 5 weeks (Red Tractor [56–58], Scotch Assured [62]) to natural weaning (Animal Welfare Certified by GAP [73,74,76] in tiers 5, 5+) for all dairy animals.

Table 4. Handling practices regarding dairy calf welfare in the selected animal welfare labeling schemes.

Labeling Scheme	Do Dairy Newborns Suckle? (Min Time)	Minimum Weaning Age	Housing of Calves in Individual Pens	References
KRAV	Calves: 1 day, lambs/kids: 3 days	Calves: 3 months, lambs: 8 weeks, kids: 45 days	Up to 1 week	[34]
Svensk Sigill	Not mentioned	Calves: 3 months, lambs: 8 weeks (25 kgs min)	Up to 8 weeks	[35–37]
Bedre Dyrevelfaerd	Tier 3: calves for 1 day	Calves tier 1: min 8 weeks, tier 2: 10 weeks, tier 3: 12 weeks	Up to 1 week	[38]
Anbefalet af Fyrenrs Beskyttelse	At least 1 day	Calves: 5 months, lambs: 3 months	Up to 1 week	[39–41]
Beter Leven	Separation after the mother has licked the newborn	Calves: 3 months tier 1, 6 months tier 3	Up to 2 weeks	[42–47]
Meadow Milk	Not mentioned	Not mentioned	Not mentioned	[48]
EKO	Not mentioned	Not mentioned	Not mentioned	[49,50]
AMA Gütesiege	Not mentioned	Calves: 3 months	Up to 8 weeks	[51–53]
Demeter	Not mentioned	Not mentioned/as conventional	Up to 1 week	[54]
Bio Suisse	Not mentioned	Calves: 3 months, lambs/kids: 35 days	Up to 8 weeks	[55]
Red Tractor	No	Calves: 5 weeks, lambs/kids: 5 weeks	Up to 8 weeks	[56–58]
RSPCA Approved	Not mentioned	Calves: 8 weeks	Not mentioned/as conventional	[59–61]
Bienestar Animal Welfair	Not mentioned	Not mentioned	Not mentioned	[63–67]
Animal Welfare Certified by GAP	6 h tier 1, 12 h tiers 2–4, until weaning 5, 5+	Calves: 6 months, tiers 1–4, 8 months tiers 5, 5+ natural weaning step 5, lambs/kids: 12 weeks tiers 1–4, 16 weeks tiers 5, 5+	Up to 2 weeks tier 1, not permitted in other tiers	[68–71]
Animal Welfare Approved by AGW	Recommended: calves should be reared by their mothers	Calves: 6 months/dairy calves 6 weeks from milk	Up to 4 weeks	[72–77]

3.2.6. Access to Pasture and Stocking Density on Pasture

Access to pasture is a mandatory animal welfare provision in all labeling schemes except for Red Tractor [56–58] and Welfair [63–67], as these two were developed for animal welfare assessment in intensive systems. Nevertheless, the Welfare Quality Assessment protocol for fattening cattle includes as a positive welfare indicator access to pasture or to an outdoor loafing area [64]. This is also incorporated in the Welfair standards. Table 5 provides a detailed analysis of the pasture requirements specified by each scheme. Usually, the requirements are defined by the minimum days per year/grazing period and/or the minimum hours of pasture access per day. For calves older than six months and cows, access requirements vary from 120 days per year and 6 h per day (Bedre Dyrevelfaerd [38] tier 1, Meadow Milk [48]) to 180 days per year and 8 h (EKO [49]) or 6 h per day (Anbefalet af Fyrenrs Beskyttelse [40]). Animal Welfare Certified by GAP [73,74,76] in tiers 5, 5+ again sets the highest welfare standard, requiring 200 days per year and unrestricted access in tier 5+. For sheep and goats, continuous grazing is promoted (KRAV [34], Animal Welfare Certified by GAP [70,71]). Standards like Meadow Milk [48] and EKO [49,50] were specifically developed to promote natural grazing. Additionally, Animal Welfare Approved by AGW certifies dairy cattle as grassfed certified if the animals have grazed continuously for one year prior to slaughter [85]. Most labeling schemes also define maximum stocking density on pasture so that the coverage will not be denuded, and the animals will have space to move freely and synchronize their behaviors (Table 5).

Table 5. Access to pasture, stocking density on pasture, daylight provision, and maximum allowed mortality rates in the selected standards.

Labeling Scheme	Pasture	Stocking Density on Pasture	Daylight/Day–Night Rhythm	Mortality	References
KRAV	Cattle over six months: outdoors most of the day during the grazing period (min 12 h), sheep/goats: around the clock in grazing period	Not mentioned	Daylight must be let in via an area equivalent to at least 3% of the floor area	Suckling calf mortality 0–24 h: 11%, calf mortality 1–180 days: 10%, young animal mortality 6–15 months (heifers) 4%, cows that die natural death/euthanized 10%	[34]
Svensk Sigill	"Grundcertifering" or "naturbeteskött" certified: minimum half of the grazing period (for animals older than 6 months), dairy cattle: min 10 h/day, beef: whole day	Low, not harming the pasture	Windows or openings for light to enter	Not specified	[35–37]
Bedre Dyrevelfaerd	Tier 1: no; tier 2: cattle min 150 days, cattle more than 6 months 1 May–1 November, calves less than 6 months, 1 May–1 September; tier 3: all more than 4 months 1 May–1 November, calves less than 4 months 1/5–1/9	Not mentioned	Not mentioned	As of 1 January 2028, max 8% for cattle and 10% for calves under the last 24 months	[38]
Anbefalet af Fyrenrs Beskyttelse	Sheep: min 7 months, beef cattle: min $5\frac{1}{2}$ months, dairy cattle: min 6 h/day from 1 April–1 November	Beef cattle: min 0.2 ha/animal, dairy cattle: min 1 ha/animal	Light area must be at least 5% of the floor area	Not mentioned	[39–41]
Beter Leven	Beef cattle tier 1: 8 h/day for 150 day/year or 24 h/day for 115 days/year, tier 3: 210 days of 12 h/day or 160 days of 24 h/day, dairy cattle tier 1: 6 h/day for 120 days/year, tier 3: 8 h/day for 180 days	All cattle tier 1: 10 cows/ha, tier 3: 6.5 cows/ha	Beef cattle: daylight-permeable area covers at least 6.7%, dairy: 10% of the floor area	Not mentioned	[42–47]
Meadow Milk	Dairy cattle: 6 h/day for 120 days/year or 720 h in total, min 120 days/year	10 cows/ha	Not mentioned	Not mentioned	[48]
EKO	Dairy cattle: 8 h/day for 180 days (15 April – 12 October), dairy goats: 5 h/day for 120 days	6.5 cows/ha, 100 goats/ha	Not mentioned	Not mentioned	[49,50]
AMA Gütesiegel	Dairy cattle: 120 days/year, 2 h/day dairy plus standard and 6 h/day dairy plus outdoors standard	No	Sizes of the surface of the windows min 3% of the floor area	Not mentioned	[51–53]
Demeter	All access to pasture during the summer half-year	2 livestock units/ha (1.4 manure units/ha)	Not mentioned	Not mentioned	[54]
Bio Suisse	Cattle: regular (exception bulls, fattening calves), sheep, goats: every day during the growing season	2.5 livestock units/ha	Natural light	Not mentioned	[55]

Table 5. *Cont.*

Labeling Scheme	Pasture	Stocking Density on Pasture	Daylight/Day–Night Rhythm	Mortality	References
Red Tractor	No	No	Normal daylight hours (natural or artificial), night hours not specified	Not specified	[56–58]
RSPCA Approved	Calves older than 8 weeks must be given continuous access to paddock	Not mentioned	Min 100 lux during daylight hours	Not specified	[59–61]
Scotch Assured	Yes, not specified	Not mentioned	Natural, so that animals can see clearly	Not mentioned	[62]
Bienestar Animal Welfair	No—but in cattle access to outdoor loafing area or pasture (in days per year or hours per day) and access to pasture in expression of other behaviors (appropriate behavior)	No	Not mentioned	Not mentioned—but criterion score lamb mortality (appropriate nutrition), veal mortality	[63–67]
Animal Welfare Certified by GAP	Beef cattle: 2/3 of their lives in tiers 1–2, 3/4 of their lives in tier 4/all their lives on range or pasture in tiers 5, 5+, dairy cattle 6 h/day in tiers 4, 5 and min 200 days/year in tier 5+, sheep/goats: unrestricted daily access	So that vegetation coverage not denuded more than 20%	Not specified, due to continuous access to pasture	Pre-weaning calf mortality max 6% per 12-month period in steps 1–3 and 4% in 4–5+/mortality for weaned heifers to 3-weeks pre-calving max 2% per 12 months all steps, sheep/goats pre-weaning 8%	[68–71]
Animal Welfare Approved by AGW	Beef cattle: continuous (calves only growing season), dairy if 1 year prior to slaughter continuously then grassfed certified	Not mentioned	Maximum day-length 16 h	Not specified	[72–77]

3.2.7. Mortality Rates

In addition to the previously mentioned provisions and practices that enhance animal well-being, another important factor that varies between the retrieved animal welfare standards is mortality. Although mortality rates are influenced by various management and breeding factors, high mortality generally indicates poor welfare conditions. Thus, three of the selective schemes establish maximum acceptable mortality rates for different animal groups. KRAV [34] permits a maximum calf mortality rate of 10% for calves aged 1–180 days and 4% for older animals. Bedre Dyrevelfaerd [38] allows a maximum mortality rate of 8% for cattle and 10% for calves (<180 days of age) based on data from the past two years (this criterion will come into effect on 1 January 2028). Animal Welfare Certified by GAP [68,69] imposes the strictest limits, even at low tiers, allowing a maximum pre-weaning calf mortality rate of 6% at tiers 1–3 and 4% at higher tiers. Notable, it is also the only standard that sets a mortality threshold for sheep and goats, limiting pre-weaning mortality to 8% [70,71].

3.2.8. Day–Night Rhythm and Daylight Provision

As presented in Table 5, almost all schemes promote natural lighting and day–night rhythm, especially those that require access to pasture (Biosuisse [55], Animal Welfare Certified by GAP [68–71]). Although Red Tractor [56–58] was developed for animal welfare assessment in intensive systems, it still requires normal daylight hours, either through natural or artificial light. Additionally, most standards set explicit standards regarding the amount of natural light that must enter the facilities by specifying the minimum size of window surface area as a percentage of the total floor area. This requirement varies from 3% (KRAV [34], AMA Gütesiegel [51–53]) to up to 10% (Beter Leven [44,45]) in the standards.

3.3. Transport and Slaughter

The rearing of the animals is described with details in all standards; however, it is equally important to also assess animal welfare during transport and slaughter. As shown in Table 6, transport and slaughter are not typically analyzed as separate sections or standards. In most cases, even when transport criteria are addressed, the requirements are those included in the conventional Regulation (EC) 1/2005 on the protection of animals during transport. Although all standards recommend keeping transport time to a minimum, ideally less than 8 h if possible, only Animal Welfare Certified by GAP 5+ [68,69] sets a stricter requirement, prohibiting transport and demanding on-farm slaughter. Regarding slaughter conditions, only Red Tractor [86], Animal Welfare Approved by AGW [87], and Welfair [88,89] provide separate detailed standards, with the latter incorporating various animal-based indicators to evaluate welfare.

Table 6. Labeling schemes with information on whether they include a separation section/standard about animal welfare evaluation during transport and slaughter. The table also presents the maximum allowed transport time under each standard.

Labeling Scheme	Separate Section/Standard About:		Maximum Transport Time	References
	Transport	Slaughter		
KRAV	No	No	The ideal minimum, but not specified	[34]
Svensk Sigill	Yes ("naturbeteskött", "klimatcertifiering")	No	Not mentioned (8 h)	[35–38]
Bedre Dyrevelfaerd	No	No	8 h	[38]
Anbefalet af Fyrenrs Beskyttelse	Yes	No	8 h (including loading/unloading)	[39–41]
Beter Leven	Yes	No	8 h	[42–47]
Meadow Milk	No	No	Not mentioned	[48]
EKO	No	No	Not mentioned	[49,50]
AMA Gütesiegel	Yes	No	Not mentioned	[51–53]
Demeter	No	No	Minimized, on-farm slaughter recommended	[54]
Bio Suisse	No	No	8 h	[55]
Red Tractor	Yes	Yes (meat and poultry processing scheme standards section [86])	8 h	[56–58]
RSPCA Approved	Yes	Yes	8 h	[59–61]
Scotch Assured	Yes	No	8 h	[62]
Bienestar Animal Welfair	No	Yes (system of integration of measures for the welfare assessment protocol for cattle [88]/sheep and goats [89] at the slaughterhouse)	Not mentioned	[63–67]
Animal Welfare Certified by GAP	Yes	Yes	All meat tiers 5+ on farm slaughter, beef cattle: tier 1: 25 h, tiers 2–4: 16 h, tier 5: 8 h, sheep: 18 h, goats: 12 h, dairy 1–4: 8 h, tier 5, 5+: 2 h	[68–71]
Animal Welfare Approved by AGW	Yes	Yes (guidelines for red meat slaughter facilities [87])	8 h	[72–77]

3.4. Environmental Enrichment and Positive Welfare Indicators

A holistic, comprehensive approach to evaluating animal welfare includes resource, management-, and animal-based indicators. Among these, animal-based indicators are the main category used in animal welfare assessment schemes and considered the most reliable, since they focus directly on the animal, evaluating its welfare through direct observation [90]. Moreover, all five aspects of animal welfare should be quantified and assessed for a holistic welfare evaluation [91]: nutrition, natural environment, health, behavior, and mental state. To ensure an acceptable welfare level, it is necessary not only to protect the animals from negative experiences by keeping these experiences below a minimum, but also to actively promote positive affective states in animals' lives [92,93].

Table 7 summarizes the positive welfare indicators identified in the selective protocols. The most common indicators are access to pasture and comfort. Access to pasture not only fulfills the strong behavioral motivational need of ruminants to graze and/or browse, but also promotes exploration, behavioral synchronization, and comfort [94,95]. Comfort is further promoted in certain standards through mandatory provision of grooming brushes and/or rubbing devices (Svensk Sigill [35,36], Anbefalet af Fyrenrs Beskyttelse [39,40], Beter Leven [42–47], Meadow Milk [48], AMA Gütesiegel [51], Animal Welfare Certified by GAP [68,69]). Bedding material requirements are often described in detail (e.g., RSPCA Approved [61]), and animals are provided with additional space through pasture or outdoor access (Table 5). Under these conditions, positive social interactions, positive affiliative behaviors, and play in young animals are also promoted [94]. In addition, structural environmental enrichment is promoted for sheep and goats, primarily by elevated platforms (Biosuisse [55], Animal Welfare Certified by GAP [71,72]). Another positive welfare indicator included in the schemes is positive human–animal interaction. While generally mentioned across standards, Welfair has gone a step further by measuring it at the farm level using the human approach test [63–67]. Finally, mother–young bonding is gaining attention, as described previously (Table 4), along with the provision of places for seclusion (KRAV [34], Animal Welfare Certified by GAP [68–71]).

Table 7. Positive welfare indicators and requirements for environmental enrichment included in the selected standards.

Labeling Scheme	Specific Environmental Enrichment	Positive Welfare Indicators Promoted	Positive Human–Animal Interaction	References
KRAV	Not specified	Access to pasture, comfort, calving in seclusion	General	[34]
Svensk Sigill	All free-range cattle in loose systems must have access to a grooming brush or similar	Access to pasture, comfort	General	[34–37]
Bedre Dyrevelfaerd	Not specified	Access to pasture, comfort	General	[38]
Anbefalet af Fyrenrs Beskyttelse	1 stationary brush per 50 cows in stables; scratching post, trees, or similar on pasture; detailed description of shelter, shade outdoors	Access to pasture, comfort, social interactions, behavioral synchronization, mother–young bonding	General	[39–41]
Beter Leven	Rubbing devices permanently in each pen	Access to pasture, comfort, mother–young bonding, social interactions	For all tiers, the farm manager must have followed a human–animal interaction workshop approved by the Society for the Protection of Animals	[42–47]

Table 7. *Cont.*

Labeling Scheme	Specific Environmental Enrichment	Positive Welfare Indicators Promoted	Positive Human–Animal Interaction	References
Meadow Milk	All cows should be able to graze simultaneously, partial herd grazing not permitted	Access to pasture, comfort, social interactions, behavioral synchronization	Not mentioned	[48]
EKO	Bedding comfort, at least 10 cm thick and brushing facilities for all cattle	Access to pasture, comfort, social interactions	Not mentioned	[49,50]
AMA Gütesiegel	Not specific, but "enhanced animal welfare" for beef requires 40% more space and scratching posts than basic requirements	Access to pasture, comfort, social interactions	Not mentioned	[51–53]
Demeter	Systems must allow animals free contact with their natural environment (sun, rain, earth underfoot, etc.).	Access to pasture, comfort, social interactions, behavioral synchronization	General	[54]
Bio Suisse	For goats, outdoor pens and pastures should be appropriately structurally enriched, e.g., with elevated areas, places of retreat	Access to pasture, comfort, play, positive social interactions, places for animals to rest and hide	General	[55]
Red Tractor	Not specified—emphasis on adequate, dry bedding	Comfort, social interactions	Analyzed in detail (appendix on good handling)	[56–58]
RSPCA Approved	Emphasis on roughage provision, explained and described in detail as bedding and enrichment	Access to pasture, comfort, positive social interactions, play, behavioral synchronization	General	[59–61]
Scotch Assured	Enough space for behavioral synchronization	Access to pasture, comfort, behavioral synchronization	General	[62]
Bienestar Animal Welfair	Not specified	Exploration, access to pasture, comfort, play, QBA, positive social interactions/affiliative behaviors, behavioral synchronization, cohesion	Analyzed and measured as indicator (avoidance distance test)	[63–67]
Animal Welfare Certified by GAP	For tier 2, objects on which beef cattle can scratch or groom, detailed environmental enrichment appendix for all species, detailed bedding description and score for calves, criterion for bullied animals' protection, one enrichment item for meat sheep/goats (1/50 animals) when moved from pasture	Access to pasture, comfort, mother–young bonding, social interactions and synchronization, play, places where animals can hide	Analyzed in detail	[68–71]
Animal Welfare Approved by AGW	Not specific, but houses conditions analyzed in detail, e.g., bedding material	Access to pasture, comfort, social interactions, mother–young bonding	Analyzed in detail, mentions that efforts must be made to develop positive relationships between the farmers–animals	[72–77]

3.5. *How Improved Is Animal Welfare in the European Labeling Schemes?*

Is the welfare of production animals actually enhanced according to the selective standards? To answer this question, we must set a comparison baseline as the welfare in conventional husbandry systems, as set by EU Directives 98/58/EC, 91/629/EEC and Regulations (EC) 1/2005 and (EC) 1099/2009. In this context, the answer is yes. All labeling schemes set higher welfare standards compared to conventional farming. Still, some of them just repeat the conventional regulations with slight modifications. These are mostly schemes developed for intensive and semi-intensive systems. Nevertheless, it is important that they explain in detail and precisely the management and handling practices, reducing ambiguities and minimizing the risk of misinterpretation. For example, Red Tractor has a separate appendix specifying the accepted practices for handling animals during movement and provides detailed requirements regarding bedding material and its condition [56–58]. Thus, although developed for intensive systems, it sets high standards for management and health-related animal-based indicators. The Welfair standards are also developed for intensive systems but include many animal-based indicators, including positive ones, that enhance animal welfare level [63–67]. These schemes even assess the positive emotional states of animals by QBA [63–67].

When comparing the standards of the selected schemes to those in organic husbandry systems, as defined by Regulation (EU) 2018/848, again, in this case, some schemes do not significantly differ from the Regulation. However, they set detailed criteria for various practices, such as pasture and stocking density on pasture (Meadow Milk [48], EKO [49,50], AMA Gütesiegel [51–53]). Additionally, considering that most of the labeling schemes require environmental enrichment and promote comfort, the overall animal welfare level is improved. It is the standards developed by NGOs that advocate for animal welfare, such as Anbefalet af Fyrenrs Beskyttelse [39–41], RSPCA Approved [59–61], Animal Welfare Certified by GAP [68–71], and Animal Welfare Approved by AGW [72–77], that set the highest animal welfare conditions. Likewise, standards developed by public–private initiatives, like Beter Leven [42–47] in the Netherlands and Bedre Dyrevelfaerd [38] in Denmark, establish a high animal welfare level, exceeding conventional EU regulations in the low tiers and even EU organic standards in the higher tiers. Notably, these standards focus on promoting animal welfare and do not issue other sustainability aspects. Table 8 summarizes the key points regarding how stringent each labeling scheme is compared to conventional and organic systems and whether they address certain animal welfare husbandry issues.

Table 8. Summary of the main points that indicate whether each standard is stricter than Regulation (EU) 2018/848 on organic production and sets a relatively high animal welfare standard, compared to both organic and conventional EU regulations.

Labeling Scheme	Is It Stricter Than EU Organic as a Standard and in General Does It Set a High Animal Welfare Level?	References
KRAV	Promotes access to pasture but also allows tethering on farms with less than 50 cows, good welfare standards, strict regarding feedstuff.	[34]
Svensk Sigill	Svensk Sigill "naturbeteskött" is strict regarding pasture criteria. All three standards have a good welfare level for animals but focus a lot on environment, biodiversity, and use of energy.	[35–37]
Bedre Dyrevelfaerd	If the animals are already organic, no conversion period is needed. Tiers 2 and 3 set a good welfare level. Max mortality level is established. Dairy calves stay with the mother 24 h after birth. Calves are in individuals pens up to 1 week in all tiers.	[38]
Anbefalet af Fyrenrs Beskyttelse	Requires an enhanced animal welfare level, with environmental complexity and positive indicators, castration of male rams prohibited, and dairy calves staying with mothers 24 h after birth.	[39–41]
Beter Leven	All tiers analyze the criteria in detail, with a higher animal welfare level for dairy cattle in all tiers (separation after mother licks the calf, as of 2030 groups kept together for 5 days, cubicle systems prohibited starting in 2040), and disbudding paste not permitted.	[42–47]
Meadow Milk	Yes, regarding pasture.	[48]
EKO	Yes, regarding pasture, but requires also a few other provisions for animals, like brushes.	[49,50]
AMA Gütesiegel	The basic AMA Gütesiegel certificates are for more intensive systems. For dairy plus and plus outdoors and for beef "enhanced animal welfare," the animal welfare is higher.	[51–53]

Table 8. *Cont.*

Labeling Scheme	Is It Stricter Than EU Organic as a Standard and in General Does It Set a High Animal Welfare Level?	References
Demeter	Although giving emphasis to the environment and feedstuffs, it is strict regarding pasture and promotes the expression of natural behaviors and keeping animals loose and horned.	[54]
Bio Suisse	It is more complex that EU organic standards, also emphasizing the environment, but sets a good animal welfare standard with environmental enrichment.	[55]
Red Tractor	Mainly for intensive systems, including mainly management-based and animal-based measures that have to do mostly with health. Still, there are very detailed criteria for management and high standards for conditions for this type of system.	[56–58]
RSPCA Approved	Sets high animal welfare standards, with complex, enriched environments, particularly for dairy calves.	[59–61]
Scotch Assured	Developed mainly to promote meat products from Scotland, not animal welfare, but has a good animal welfare level.	[62]
Bienestar Animal Welfair	Developed for intensive systems, has a holistic and high animal welfare level, and many animal-based indicators, including positive ones.	[63–67]
Animal Welfare Certified by GAP	Very high animal welfare standards starting from tier 2 and especially in tier 5,5+. Tier 5+ requires on-farm slaughter, access to pasture for the duration of the animals' lives, natural weaning for all animals, and breastfeeding by their mothers—even dairy cattle, and castration is not permitted. Very detailed way to measure animal-based indicators. The environmental enrichment analyzed is detailed in the appendix.	[68–71]
Animal Welfare Approved by AGW	High animal welfare standards, various animal-based indicators, explained in detail. Separate detailed section for transport and standards for slaughter. "Grassfed" standards require natural grassfed feeding for a minimum of one year.	[72–77]

4. Discussion

Considering the results of our review, it can be argued that the retrieved European animal welfare labeling schemes differ in various aspects, but all require an improved animal welfare level, not only compared to conventional but in many cases also compared to organic husbandry. A crucial factor contributing to this welfare amelioration is the detailed analysis of the criteria included in each scheme, which leaves no space for misinterpretation. The more specific and detailed the standard, the easier it is for the farmers to understand and implement, reducing the likelihood of practices that could negatively impact welfare. For example, it is not enough to mention that analgesia is needed after a treatment; the length of time must be precisely described [39–47,68–77]. Similarly, it is not enough to merely mention the requirement for bedding material; the standard should provide clear guidelines regarding its thickness and cleanliness [69]. Moreover, it is recommended that the standards be accompanied by detailed manuals for farmers and consumers, as well as self-evaluation checklists or a farm-specific information scheme in the form of notes. These resources should explain not only the application of each criterion but also how and why it enhances the well-being of the animals [96]. Additionally, mandatory training for the personnel, potentially performed by the initiator of the standard, can educate the farmers further, ensuring that they fully understand the welfare objectives [51–53].

When it comes to husbandry practices that enhance animal welfare, almost all labeling schemes prohibit tail docking and tethering, and promote access to pasture and day–night rhythm and daylight provision. However, castration of males is either allowed or not explicitly addressed in most standards [48–50]. The welfare of male animals is often compromised in husbandry systems. Reproduction is primarily carried out through artificial insemination, and only a few males are retained in the herd, as nearly all male calves, lambs, and kids are slaughtered for meat after weaning. To improve welfare, it is important to use castration as an animal welfare indicator in all standards and develop and incorporate more indicators that evaluate the welfare of male ruminants. Another important indicator for animal welfare assessment, though only incorporated in a few standards, is mortality [34,38,63–71]. In the future, animal welfare schemes should establish accepted mortality levels for both young and adult animals. Additionally, to further enhance the welfare of animals, it would be beneficial for labeling schemes to incorporate maximum acceptable levels for assisted calvings and caesarean sections [39–47,56–58,68–77].

Husbandry practices concerning the welfare of dairy newborn ruminants should be given greater emphasis in animal welfare schemes. In a farming system, it is often the animals in the productive stage that experience the highest welfare level. Attention must also be focused on unweaned animals, especially dairy ruminants, which are often separated from their mothers immediately after birth and, in intensive systems, kept in individual pens. Animal welfare labeling schemes should mandate a minimum mother–young contact period after calving, even if it is just for one day [34,38]. Further to this, the duration for keeping calves in separate pens should be minimized, and the provision of clean, dry bedding material of adequate height should be made a strict requirement. Based on the authors' personal observations, inadequate height of bedding material and insufficient cleanliness, which compromise comfort and thermoregulation, are common issues in dairy calf management.

Also, we propose that standards developed for labeling and communicating animal welfare in farming systems should define not only the timeline but also the specific time intervals for interventive handling practices that pose significant physical and mental stress to the animals. For example, RSPCA Approved requires that weaning from colostrum not coincide with castration and/or disbudding since both practices are significantly stressful for the animals [61]. Additionally, animal welfare standards should set a maximum number of individual animals that can be kept in a group [38,42–47,68–71] and impose stricter criteria on the regrouping of animals, as both factors negatively impact welfare.

Another aspect of animal welfare that should be thoroughly evaluated in the labeling schemes is thermal comfort. Animal Welfare Certified by GAP has incorporated a heat stress [69] and Welfair a panting score [66,67]. While the standards provide extensive analysis of shelter and shed requirements, especially for pasture-based systems [34–47,63–71], thermal comfort is often not assessed in dept within the facilities. This is particularly crucial for Mediterranean countries when the temperature is high. It also underscores the importance of performing audits during different parts of the year to cover all seasonal phases of a farming system [39–41,62,68–71] and always during a period when welfare is at risk.

According to our analysis, all animal welfare assessment schemes include positive welfare indicators, primarily access to pasture and comfort, and demand environmental enrichment, which is described in detail. However, a positive human–animal relationship is usually only generally mentioned in the standards, with Welfair being the exception by measuring it with a human approach test [63–67]. A positive human–animal relationship is valuable both for animal welfare and for productivity [95,97–100], and should be promoted in the labeling schemes more systematically. Incorporating a human-approach test as a tool for measuring animals' fearfulness towards humans would be both feasible and effective [97]. Another positive welfare indicator that should be emphasized more in the standards is behavioral synchronization, as it not only indicates positive affective states but also the adequate availability of space and resources. A practical way to measure it during an audit would be to assess how many animals in a herd can lie down in a natural body posture, eat, or drink simultaneously [39–41,63–67].

We also recommend the use of scoring tools during audits to measure indicators. Scoring scales are practical and feasible methods for assessing welfare, as they consist of a limited number of levels, accompanied by photos and explanations. Body condition and lameness scoring scales are commonly used in most standards. Notably, Animal Welfare Approved by GAP sets a nesting scoring for calves and a simple evaluation of air quality for all animals [69]. Scoring tools enhance repeatability and inter- and intra-observer reliability, making welfare assessment more effective. Moreover, scoring sheets with pictures explaining each score can serve as a feasible way to train auditors.

It is also important to develop standards for assessing animal welfare during transport and slaughter, as these production stages are not covered by most labeling schemes. Additionally, there is lack of standards for dairy sheep in intensive systems. Red Tractor [70] evaluates welfare only for sheep raised for meat. The AWIN Welfare assessment protocol for sheep is the only scheme developed for intensive dairy farming [66]. Furthermore, there is a need for animal welfare labeling schemes for buffalos and bison. No dedicated animal welfare labeling schemes were retrieved for these species, except for one standard for bison [101] developed by GAP and one for water buffaloes [102] developed by Certified Humane, an NGO based in the USA but with an international presence. According to the FAO, 15% of the world's milk production comes from buffaloes [16], so the population of buffaloes in Europe is growing and milk production is becoming more intensive.

Nowadays, consumers are willing to pay a premium price for products originating from improved animal welfare farming systems. Although this willingness to pay is complicated and the exact premium consumers are willing to pay even more so [10–15], various studies indicate that consumers do indeed pay more [13–15,21–29]. Keeping the labeling standards open to the public strengthens consumer's trust and transparency, which can influence both the willingness to pay and the economical sustainability of enhanced animal welfare farming systems. Credibility can also be reinforced by incorporating a minimum required conversion period for the farms before they are allowed to label their products under a given standard [34,35,52,54,55]. Further to this, auditing all farms rather than just a sample in a company [68–71] would strengthen this credibility, since group certification in most standards requires only sample auditing.

Moreover, we should emphasize that it is essential to communicate the farming criteria set by each labeling system to consumers through strategic promotion by the system initiators and even certification bodies. Achieving this requires collaboration between private and public entities and can be achieved through consumer and farmer demand, stricter animal welfare legislation from the state, and cooperation between private (both for-profit and non-profit) and public organizations [103]. It is also crucial to link animal welfare with the quality of animal-derived products. A notable example is the Scotch Assured labeling scheme, developed through collaboration between commercial and animal welfare NGOs, which promotes locally produced red meat of high quality [83]. Animal welfare credibility is strengthened by the fact that the Prevention of Cruelty to Animals can be present during the audits and has been involved in the development of the standard. Additionally, economic studies can further support this type of initiative, studying the economical sustainability and profitability of this type of enhanced animal welfare products.

Further to this, many labeling schemes, such as the Dutch Beter Leven [42–47] and the Danish Bedre Dyrevelfærd [38], focus solely on improving animal welfare conditions rather than addressing other aspects of sustainability. As a result, their requirements exceed the

EU legislation for conventional farming but remain lower than those of organic production. However, at the third and highest level in both systems, the welfare requirements for animals reach or even surpass those of organic farming [38,43,45]. Since these systems prioritize animal welfare, it is generally easier for a producer to adapt its husbandry practices to meet certification standards compared to the more extensive changes required for organic certification. This means that a farm can enhance animal welfare with a lower financial investment than would be needed for a full transition to organic farming.

However, certification requirements vary significantly, making it challenging to compare different protocols both from a scientific perspective and from a consumer standpoint [12–15,20]. To prevent consumer confusion, the development of a pan-European labeling system has been proposed to ensure a unified policy across countries [20,104]. This approach is supported by non-profit organizations [20,105]. Nonetheless, in practice, it could pose challenges, as not all countries would be able to meet the common requirements while remaining economically viable and competitive in the European market [20]. Additionally, the existence of various protocols—and, consequently, different farming management practices—results in animal-based products of a wide cost range, and thus being affordable for a wide group of consumers.

Finally, elaboration on how these schemes compare in terms of real-world effectiveness or consumer impact is beyond the scope of this review. Still, future studies focusing on these aspects could contribute significantly to the development of animal welfare labeling schemes not only by improving welfare assessment but also by enhancing their credibility among consumers.

Despite the variation in the requirements of animal welfare standards, the trend toward raising animals under better welfare conditions—and the corresponding consumer demand in Europe—is clear and is even supported by state-led labeling initiatives. Targeted consumer and farmer education, strategic promotional efforts, and research supporting the quality of higher-welfare products of animal origin can contribute to advancing this movement.

5. Conclusions

Animal welfare labeling serves as an important tool for informing consumers and enhancing the credibility of the production chain. Globally, consumers are increasingly concerned about the rearing conditions of production animals, driving the development of more and more labeling systems that certify improved welfare conditions. These systems vary in their structure, content, and the level of welfare they ensure. However, they all require conditions superior to those of conventional and, in many cases, even to organic farming, especially those developed by NGOs. The criteria included in these impose improved husbandry practices, particularly regarding tail docking for sheep and management of dairy calves. Dairy calves are often allowed to suckle for a period, and the maximum permitted duration for keeping calves in individual pens is shorter compared to conventional systems. Positive welfare indicators are also incorporated into the protocols, with the most common being access to pasture and comfort. The improvement in animal welfare in these schemes is largely due to the detailed and specific analysis of required practices and provisions, which minimizes the risk of misinterpretation. Nonetheless, certain limitations do remain. Specific parameters like thermal comfort, mortality, and positive human–animal relationships are not incorporated into all standards. Furthermore, there is a need for more comprehensive animal welfare evaluation during transport and slaughter. Effectively communicating the standards to consumers and strengthening their credibility can enhance consumer willingness to pay, thereby improving the economic viability of systems with improved animal welfare standards.

Author Contributions: Conceptualization, M.P. and P.S.; methodology, M.P.; writing—original draft preparation, M.P.; writing—review and editing, A.K., O.T. and P.S.; supervision, P.S. All authors have read and agreed to the published version of the manuscript.

Funding: This research received no external funding.

Conflicts of Interest: The authors declare no conflicts of interest.

References

1. Buller, H.; Blokhuis, H.; Jensen, P.; Keeling, L. Towards farm animal welfare and sustainability. *Animals* **2018**, *8*, 81. [CrossRef] [PubMed]
2. Broom, D.M. Farm animal welfare: A key component of the sustainability of farming systems. *Vet. Glas.* **2021**, *75*, 145–151. [CrossRef]
3. United Nations Committee. *Proposed Draft Recommendations on Sustainable Agricultural Development for Food Security and Nutrition Including the Role of Livestock*; Animal Health and Welfare: Easton, PA, USA, 2016; Article VIII.
4. Dwyer, C.M. Can improving animal welfare contribute to sustainability and productivity? *BSJ Agric.* **2020**, *3*, 61–65.
5. Marchant-Forde, J.N. The science of animal behavior and welfare: Challenges, opportunities, and global perspective. *Front. Vet. Sci.* **2015**, *2*, 16. [CrossRef]
6. Nurse, A. Beyond the property debate: Animal welfare as a public good. Contemp. *Justice Rev.* **2016**, *19*, 174–187. [CrossRef]
7. Alonso, M.E.; González-Montaña, J.R.; Lomillos, J.M. Consumers' Concerns and Perceptions of Farm Animal Welfare. *Animals* **2020**, *10*, 385. [CrossRef]
8. Lagerkvist, C.J.; Hess, S. A meta-analysis of consumer willingness to pay for farm animal welfare. *Eur. Rev. Agric. Econ.* **2011**, *38*, 55–78. [CrossRef]
9. Eurobarometer. European-Commission. *Attitudes of EU Citizens towards Animal Welfare, Report; Special Eurobarometer 442*; European Commission: Brussels, Belgium, 2016; p. 84.
10. Montossi, F.; Font-i-Furnols, M.; del Campo, M.; Julián, R.S.; Brito, G.; Sañudo, C. Sustainable sheep production and consumer preference trends: Compatibilities, contradictions, and unresolved dilemmas. *Meat Sci.* **2013**, *95*, 772–789. [CrossRef]
11. Clark, B.; Stewart, G.B.; Panzone, L.A.; Kyriazakis, I.; Frewer, L.J. Citizens, consumers and farm animal welfare: A meta-analysis of willingness-to-pay studies. *Food Policy* **2017**, *68*, 112–127. [CrossRef]
12. Napolitano, F.; Girolami, A.; Braghieri, A. Consumer liking and willingness to pay for high welfare animal-based product. *Trends Food Sci. Technol.* **2010**, *21*, 537–543. [CrossRef]
13. Mandolesi, S.; Latvala, T.; Naspetti, S.; Arsenos, G.; Martin-Collado, D.; Orsini, S.; Caramelle-Holtz, E.; Ozturk, E.; Zanoli, R. Consumer attitudes, motivations and barriers towards sheep and goat dairy products. *Int. J. Gastron. Food Sci.* **2024**, *36*, 100917. [CrossRef]
14. Waldrop, E.M.; Roosen, J. Consumer acceptance and willingness to pay for cow housing systems in eight European countries. *Q Open* **2021**, *1*, qoab001. [CrossRef]
15. Risius, A.; Hamm, U. The effect of information on beef husbandry systems on consumers' preferences and willingness to pay. *Meat Sci.* **2017**, *124*, 9–14. [CrossRef] [PubMed]
16. FAOSTAT. Dairy Production and Products: Products (fao.org). Available online: https://openknowledge.fao.org/server/api/core/bitstreams/fa4ada95-20fa-42c3-9c34-783be759fee9/content#:~:text=This%20chapter%20describes%20market%20developments%20and%20medium-term%20projections,cheese,%20skim%20milk%20powder%20and%20whole%20milk%20powder (accessed on 31 January 2025).
17. FAO. Dairy Market Review. Overview of Global Market Developments in 2024. Available online: https://openknowledge.fao.org/server/api/core/bitstreams/44217723-f8e7-431a-b3b2-47f10f8fbc71/content (accessed on 25 April 2025).
18. FAOSTAT. Meat (fao.org). Available online: https://openknowledge.fao.org/server/api/core/bitstreams/09e88a46-b005-4d65-8753-9714506afc38/content (accessed on 31 January 2025).
19. European Commission. Farm to Fork Strategy for a Fair, Healthy and Environmentally-Friendly Food System. 2020. Available online: https://food.ec.europa.eu/document/download/472acca8-7f7b-4171-98b0-ed76720d68d3_en?filename=f2f_action-plan_2020_strategy-info_en.pdf (accessed on 10 February 2025).

20. European Parliament Ex-Post Evaluation Unit. Animal Welfare on the Farm—Ex-Post Evaluation of the EU Legislation: PROSPECTS for Animal Welfare Labelling at EU Level. Europeam Implementation Assessment. 2021. Available online: https://www.europarl.europa.eu/thinktank/en/document/EPRS_STU(2021)662643 (accessed on 10 February 2025).
21. Vargas-Bello-Pérez, E.; Tajonar, K.; Foggi, G.; Mele, M.; Simitzis, P.; Mavrommatis, A.; Tsiplakou, E.; Habib, M.R.; Gonzalez-Ronquillo, M.; Toro-Mujica, P. Consumer attitudes toward dairy products from sheep and goats: A cross-continental perspective. *J. Dairy Sci.* **2022**, *105*, 8718–8733. [CrossRef]
22. Nalon, E.; Contiero, B.; Gottardo, F.; Cozzi, G. The Welfare of Beef Cattle in the Scientific Literature from 1990 to 2019: A Text Mining Approach. Front. *Vet. Sci.* **2021**, *7*, 588749. [CrossRef]
23. Sonoda, Y.; Oishi, K.; Chomei, Y.; Hirooka, H. How do human values influence the beef preferences of consumer segments regarding animal welfare and environmentally friendly production? *Meat Sci.* **2018**, *146*, 75–86. [CrossRef]
24. Pérez-Barea, J.J.; Espantaleón-Pérez, R.; Šedík, P. Evaluating the Perception of Socially Responsible Consumers: The Case of Products Derived from Organic Beef. *Sustainability* **2020**, *12*, 10166. [CrossRef]
25. Jiang, R.; Sharma, C.; Bryant, R.; Mohan, M.S.; Al-Marashdeh, O.; Harrison, R.; Torrico, D.D. Animal welfare information affects consumers' hedonic and emotional responses towards milk. *Food Res. Int.* **2021**, *141*, 110006. [CrossRef]
26. Marchini, A.; Riganelli, C.; Diotallevi, F.; Polenzani, B. Label information and consumer behaviour: Evidence on drinking milk sector. *Agric. Food Econ.* **2021**, *9*, 8. [CrossRef]
27. Napolitano, F.; Pacelli, C.; Girolami, A.; Braghieri, A. Effect of Information About Animal Welfare on Consumer Willingness to Pay for Yogurt. *J. Dairy Sci.* **2008**, *91*, 910–917. [CrossRef]
28. Ellis, K.; Billington, K.; McNeil, B.; McKeegan, D. Public opinion on UK milk marketing and dairy cow welfare. *Anim. Welf.* **2009**, *18*, 267–282. [CrossRef]
29. Placzek, M.; Christoph-Schulz, I.; Barth, K. Public attitude towards cow-calf separation and other common practices of calf rearing in dairy farming—A review. *Org. Agric.* **2021**, *11*, 41–50. [CrossRef]
30. Zwolińska, J.; Żakowska-Biemans, S. Is an Animal Welfare Label Enough? Role of Farm Animal Welfare Voluntary Labelling Schemes in the Development of Sustainable Livestock Production. *Proceedings* **2021**, *73*, 8.
31. Miele, M.; Blokhuis, H. Editorial: Animal welfare labelling. *Front. Anim. Sci.* **2023**, *3*, 1108111. [CrossRef]
32. Ingenbleek, P.T.M.; Krampe, C. The End of Animal Welfare Labelling as We Know It? Persisting Problems at the Consumer Level and PLF-Based Solutions. *Front. Anim. Sci.* **2022**, *3*, 819893. [CrossRef]
33. Stygar, A.H.; Krampe, C.; Llonch, P.; Niemi, J.K. How Far Are We From Data-Driven and Animal-Based Welfare Assessment? A Critical Analysis of European Quality Schemes. *Front. Anim. Sci.* **2022**, *3*, 874260. [CrossRef]
34. KRAV. Standards for KRAV-Certified Production—2024/25 Edition. Available online: https://wwwkravse.cdn.triggerfish.cloud/uploads/sites/2/2023/12/krav-standards2024-25-1702048556.pdf (accessed on 2 January 2025).
35. Sigill Kvalitetessystem AB. IP Nöt & Mjölk Grundcertifiering. Version 2020:1. 2020. Available online: https://cdn.sanity.io/files/skgc670h/production/1b7743657d6496e0ca809c47134922ad2ffbf93a.pdf (accessed on 2 January 2025).
36. Sigill Kvalitetessystem AB. IP Sigill Nöt & Mjölk Standard för Kvalitetssäkrad nöt-och Mjölkproduktion med Tillval för Naturbeteskött och Klimatcertifering. Version 2025:1. 2025. Available online: https://www.sigill.se/producent-och-foretag/not-och-mjolk (accessed on 2 January 2025).
37. Sigill Kvalitetessystem AB. IP Sigill Lamm. Version 2022:1. 2022. Available online: https://cdn.sanity.io/files/skgc670h/production/1b7743657d6496e0ca809c47134922ad2ffbf93a.pdf (accessed on 2 January 2025).
38. European Commission. Order on the Voluntary Animal Welfare Labelling Scheme. Available online: https://technical-regulation-information-system.ec.europa.eu/en/notification/25529 (accessed on 2 January 2025).
39. Anbefalet af Dyrenes Beskyttelse. Specifikation. Mærkekrav Kødkvæ. Version 17. 2025. Available online: https://www.dyrenesbeskyttelse.dk/sites/dyrenesbeskyttelse.dk/files/AADB/Maerkekrav/2025/maerkekrav-koedkvaeg-1-januar-2025.pdf (accessed on 2 January 2025).
40. Anbefalet af Dyrenes Beskyttelse. Specifikation Mærkekrav Malkekvæg. Version 10. 2024. Available online: https://www.dyrenesbeskyttelse.dk/sites/dyrenesbeskyttelse.dk/files/M%C3%A6rkekrav%20malkekv%C3%A6g%202018.pdf (accessed on 2 January 2025).
41. Anbefalet of Dyrenes Beskyttelse. Specifikation. Mærkekrav får och Lam. Version 2020. Available online: https://www.dyrenesbeskyttelse.dk/sites/dyrenesbeskyttelse.dk/files/M%C3%A6rkekrav/M%C3%A6rkekrav%20f%C3%A5r%20og%20lam.pdf (accessed on 2 January 2025).
42. Beter Leven. Beef Cattle 1 Star. Version 2.2. 2024. Available online: https://beterleven.dierenbescherming.nl/zakelijk/wp-content/uploads/sites/2/2021/07/Beef-cattle-1-star-version-2.1-d.d.-01-01-2019.pdf (accessed on 3 February 2025).
43. Beter Leven. Beef Cattle 3 Stars. Version 1.2. 2024. Available online: https://beterleven.dierenbescherming.nl/zakelijk/wp-content/uploads/sites/2/2023/09/Beef-cattle-3-stars-version-1.2-d.d.-01-09-2024.pdf (accessed on 3 February 2025).

44. Beter Leven. Dairy Cattle 1 Star. Version 2.1. 2025. Available online: https://beterleven.dierenbescherming.nl/zakelijk/wp-content/uploads/sites/2/2025/02/Dairy-Cattle-1-star-version-2.1-dd.-03-02-2025.pdf (accessed on 3 February 2025).
45. Beter Leven. Dairy Cattle 3 Stars. Version 1.0. 2021. Available online: https://beterleven.dierenbescherming.nl/zakelijk/wp-content/uploads/sites/2/2021/08/EN-Melkrund-3-ster-versie-1.0-ZW-d.d.-24-05-2021.pdf (accessed on 3 February 2025).
46. Beter Leven. Calves 2 stars. Version 2.1. 2019. Available online: https://beterleven.dierenbescherming.nl/zakelijk/wp-content/uploads/sites/2/2022/06/Calves-2-stars-version-2.0-d.d.-01-12-2022.pdf (accessed on 3 February 2025).
47. Beter Leven. Calves 3 stars. Version 1.0. 2022. Available online: https://beterleven.dierenbescherming.nl/zakelijk/wp-content/uploads/sites/2/2022/06/Calves-3-stars-version-1.0-d.d.-01-12-2022.pdf (accessed on 3 February 2025).
48. Weidemelk. Manulas. Available online: https://www.weidemelk.nl/en/conditions.html (accessed on 10 January 2025).
49. EKO NL. Normen Geiten Zuivel. 2023. Available online: https://eko-keurmerk.nl/assets/downloads/EKO_Ambities_GeitenZuivel.pdf (accessed on 10 January 2025).
50. EKO NL. Aanvullende Normen Biologische Zuivel. 2023. Available online: https://eko-keurmerk.nl/assets/downloads/EKO_Normen_Rundveezuivel.pdf (accessed on 10 January 2025).
51. AMA Gütesiegel. Richtlinie. (AMA-Produktionsbestimmungen) Rindehaltung Kälberhaltung Rinder-und Kälbermast Mutterkuhhaltung. Version 2021/1. 2021. Available online: https://media.hendriks.amainfo.at/662a439a30545b7f0ef90dc2/AMA-Guetesiegel_Richtlinie_Rinderhaltung.pdf (accessed on 20 January 2025).
52. AMA Gütesiegel. Richtlinie (AMA-Produktionsbestimmungen) Haltung von Kühen. 2024. Available online: https://media.hendriks.amainfo.at/668536ef729da3c1c053f1e8/AMA_Guetesiegel_Richtlinie_Haltung_von_Kuehen_Version-2024.pdf (accessed on 20 January 2025).
53. AMA Gütesiegel. Keeping of sheep and Goats in Austria. Available online: https://amainfo.at/tiere/nutztierhaltung-haltungsformen/schafe-ziegen (accessed on 20 January 2025).
54. Biodynamic Federation Demeter. Production, Processing and Labelling. Available online: https://demeter.net/wp-content/uploads/2024/01/2024_Int_Dem_Bio_Standard_eng.pdf (accessed on 20 January 2025).
55. Biosuisse. Standards for the Production, Processing and Trade of "bud" Products Effective as of 1 January 2023. Available online: https://international.bio-suisse.ch/dam/jcr:47dc9f0b-3dae-4936-aed8-2f844ab88497/Bio%20Suisse%20Standards%20 2023%20-%20EN.pdf (accessed on 20 January 2025).
56. Red Tractor. Beef & Lamb. Version 5. 2021. Available online: https://redtractorassurance.org.uk/wp-content/uploads/2021/11/RTStandardsV5_BeefLamb_V2.pdf (accessed on 12 January 2025).
57. Red Tractor. Dairy standards. Version 5. 2022. Available online: https://redtractorassurance.org.uk/wp-content/uploads/2023/06/RTStandards_Dairy_Streamlined.pdf (accessed on 12 January 2025).
58. Red Tractor. Dairy Goats Standards. Version 5. 2025. Available online: https://redtractorassurance.org.uk/wp-content/uploads/2025/01/Dairy-Goats-Standards-Feb-2025.pdf (accessed on 12 January 2025).
59. RSPCA. Welfare Standards for Dairy Cattle. 2021. Available online: https://science.rspca.org.uk/documents/d/science/rspca-welfare-standards-for-dairy-cattle-reformatted-final-version-december-2023 (accessed on 12 January 2025).
60. RSPCA. Welfare Standards for Beef Cattle. 2023. Available online: https://business.rspcaassured.org.uk/media/asnjl40q/beef_welfare_standards_2023.pdf (accessed on 12 January 2025).
61. RSPCA. RSPCA Approved Farming Scheme Standard Dairy Calves. Version 1.1. 2021. Available online: https://rspcaapproved.org.au/wp-content/uploads/2022/03/2021-01_DAIRYCALVES_Standard.pdf (accessed on 12 January 2025).
62. Quality Meat Scotland. Cattle & Sheep Standards. 2022. Available online: https://s3.eu-west-2.amazonaws.com/quality-meat-scotland/documents/Standards/qms_cattle_and_sheep_standards_2022_spreads_set-up.pdf (accessed on 1 February 2025).
63. Welfare Quality. Welfare Quality Assessment Protocol for Dairy Cows. Version 3.0. 2023. Available online: https://www.welfarequalitynetwork.net/media/1319/dairy-cattle-protocol.pdf (accessed on 17 December 2024).
64. Welfare Quality. Welfare Quality Assessment Protocol for Fattening Cattle. Version 3.0. 2023. Available online: https://www.welfarequalitynetwork.net/media/1320/fattening-cattle-protocol.pdf (accessed on 12 December 2024).
65. Welfare Quality. Welfare Quality Assessment Protocol for Veal Calves. Version 3.0. 2023. Available online: https://www.welfarequalitynetwork.net/media/1321/veal-calves-protocol.pdf (accessed on 12 December 2024).
66. Ruiz, R.; Dwyer, C.M. AWIN Welfare Assessment Protocol for Sheep. Version 1.1. 2015. Available online: https://www.researchgate.net/publication/275887069_AWIN_Welfare_Assessment_Protocol_for_Sheep (accessed on 12 December 2024).
67. Battini, M.; Mattiello, S.; Stilwell, G.; Vieira, A. AWIN Welfare Assessment Protocol for Goats. Version 1.1. 2015. Available online: https://air.unimi.it/retrieve/dfa8b992-3d0f-748b-e053-3a05fe0a3a96/AWINProtocolGoats.pdf (accessed on 12 December 2024).
68. Global Animal Partnership. 5-Step® Animal Welfare Standards for Beef Cattle. Version 1.1. 2020. Available online: https://globalanimalpartnership.org/wp-content/uploads/2020/05/G.A.P.s-Animal-Welfare-Pilot-Standards-for-Beef-Cattle-v1.1.pdf (accessed on 17 January 2025).

69. Global Animal Partnership. 5-Step® Animal Welfare Standards for Dairy Cattle. Version 1.0. 2021. Available online: https://www.globalanimalpartnership.org/wp-content/uploads/2021/07/G.A.P.-5-Step-Standards-for-Dairy-Cattle-v1.0-20210707.pdf (accessed on 17 January 2025).
70. Global Animal Partnership. 5-Step® Animal Welfare Standards for Meat Goats. Version 1.2. 2020. Available online: https://globalanimalpartnership.org/wp-content/uploads/2020/05/G.A.P.s-Animal-Welfare-Pilot-Standards-for-Meat-Goats-v1.2.pdf (accessed on 17 January 2025).
71. Global Animal Partnership. 5-Step® Animal Welfare Standards for Meat Sheep. Version 1.4. 2020. Available online: https://www.globalanimalpartnership.org/wp-content/uploads/2020/05/G.A.P.s-Animal-Welfare-Pilot-Standards-for-Meat-Sheep-v1.4.pdf (accessed on 17 January 2025).
72. Animal Welfare Approved by AGW. Certified Animal Welfare Approved by AGW Standards for Beef Cattle. Version 1. 2024. Available online: https://agreenerworld.org/wp-content/uploads/2024/09/AWA-Beef-Cattle-Standards-2024-v1.pdf (accessed on 19 January 2025).
73. Animal Welfare Approved by AGW. Certified Animal Welfare Approved by AGW Standards for Dairy Cattle. Version 1. 2024. Available online: https://agreenerworld.org/wp-content/uploads/2024/09/AWA-Dairy-Cattle-Standards-2024-v1.pdf (accessed on 19 January 2025).
74. Animal Welfare Approved by AGW. Certified Animal Welfare Approved by AGW Standards for Dairy Goats. Version 1. 2024. Available online: https://agreenerworld.org/wp-content/uploads/2024/09/AWA-Dairy-Goat-Standards-2024-v1.pdf (accessed on 19 January 2025).
75. Animal Welfare Approved by AGW. Certified Animal Welfare Approved by AGW Standards for Goats. Version 1. 2024. Available online: https://agreenerworld.org/wp-content/uploads/2024/09/AWA-Goat-Standards-2024-v1.pdf (accessed on 19 January 2025).
76. Animal Welfare Approved by AGW. Certified Animal Welfare Approved by AGW Standards for Dairy Sheep. Version 1. 2024. Available online: https://agreenerworld.org.uk/certifications/animal-welfare-approved/standards/dairy-sheep-standards/ (accessed on 19 January 2025).
77. Animal Welfare Approved by AGW. Certified Animal Welfare Approved by AGW Standards for Sheep. Version 1. 2024. Available online: https://agreenerworld.org/wp-content/uploads/2024/09/AWA-Sheep-Standards-2024-v1.pdf (accessed on 19 January 2025).
78. Animal Welfare. Welfair Certificate. 2024. Available online: https://www.animalwelfair.com/en/bienestar-animal/protocolos-de-evaluacion/ (accessed on 15 March 2025).
79. RSPCA Approved Farming. Available online: https://rspcaapproved.org.au (accessed on 15 March 2025).
80. Anbefalet af Fyrenrs Beskyttelse. Available online: https://www.dyrenesbeskyttelse.dk/anbefalet-af-dyrenes-beskyttelse (accessed on 15 March 2025).
81. Global Animal Partneship. Available online: https://globalanimalpartnership.org/standards (accessed on 15 March 2025).
82. A Greener World. Available online: https://agreenerworld.org (accessed on 15 March 2025).
83. Quality Meat Scotland. Available online: https://qmscotland.co.uk/integrity-assurance/quality-assurance/standards-schemes/cattle-sheep-standards (accessed on 15 March 2025).
84. Biodynamic Federation Demeter. Animal Welfare. 2021. Available online: https://demeter.net/animal-welfare/#:~:text=Only%20Demeter%20standards%20consistently%20exclude%20dehorning%20and%20genetic,to%20the%20respiratory%20system%20of%20the%20ruminant.%20[1] (accessed on 25 March 2025).
85. A Greener World. Certified Grassfed by AGW. Available online: https://agreenerworld.org/certifications/certified-grass-fed/ (accessed on 25 March 2025).
86. Red Tractor Certified Standards. Meat & Poultry Processing Scheme Standards. Version 4.0. 2023. Available online: https://redtractorassurance.org.uk/wp-content/uploads/2023/05/RTStandards_ON-FARM_May23_HR.pdf (accessed on 1 March 2025).
87. Animal Welfare Approved by AGW. Animal Welfare Approved Guidelines for Red Meat Slaughter Facilities. Version 1. 2021. Available online: https://agreenerworld.org/wp-content/uploads/2024/02/AWA-Slaughter-Guidelines-for-Red-Meat-2021.v1.pdf (accessed on 25 April 2025).
88. IRTA. System of Integration of Measures for the Animal Welfare Assessment Protocol for Cattle in Slaughterhouses. 2023. Available online: https://www.animalwelfair.com/docs/sacrificio-vacuno/protocolo/Protocol_WQ_Cattle-EN-final.pdf (accessed on 1 March 2025).
89. IRTA. System of Integration of Measures for the Animal Welfare Assessment Protocol for Sheep and Goats in Slaughterhouses. 2023. Available online: https://animalwelfair.com/docs/sacrificio-ovino-caprino/protocolo/Protocolo_ovino_caprino_matadero_en.pdf (accessed on 1 March 2025).

90. Fraser, D. Understanding Animal Welfare. *Acta Vet. Scand.* **2008**, *50*, S1. [CrossRef]
91. Mellor, D.J.; Reid, C.S. Concepts of animal well-being and predicting the impact of procedures on experimental animals. Improving the well-being of animals in the research environment. *WellBeing Int.* **1994**, 3–18.
92. Lawrence, A.; Sandoe, P.; Vigors, B. What is so positive about positive animal welfare?- A critical review of the literature. *Animals* **2019**, *9*, 783. [CrossRef]
93. Mellor, D.J. Updating animal welfare thinking: Moving beyond the five freedoms towards a life worth living. *Animals* **2016**, *6*, 21. [CrossRef]
94. Mattiello, S.; Battini, M.; De Rosa, G.; Napolitano, F.; Dwyer, C. How can we assess positive welfare in ruminants? *Animals* **2019**, *9*, 758. [CrossRef] [PubMed]
95. Papageorgiou, M.; Simitzis, P.E. Positive Welfare Indicators in Dairy Animals. *Dairy* **2022**, *3*, 814–841. [CrossRef]
96. RSPCA. RSPCA Approved Farming Schemes Standard Dairy Calves Information Notes. 1.1. 2021. Available online: https://rspcaapproved.org.au/wp-content/uploads/2022/03/2021-01_DAIRYCALVES_InformationNotes.pdf (accessed on 12 January 2021).
97. Rault, J.-L.; Waiblinger, S.; Boivin, X.; Hemsworth, P. The Power of a Positive Human—Animal Relationship for Animal Welfare. *Front. Vet. Sci.* **2020**, *7*, 590867. [CrossRef]
98. Lange, A.; Franzmayr, S.; Wisenöcker, V.; Futschik, A.; Waiblinger, S.; Lürzel, S. Effect of different stroking styles on bahaviour and cardiac parameters in heifers. *Animals* **2020**, *10*, 426. [CrossRef]
99. Tamioso, P.R.; Rucinque, D.S.; Taconeli, C.A.; Da Silva, G.P.; Maiolino Molento, C.F. Behavior and body surface temperature as welfare indicators in selected sheep regularly brushed by a familiar observer. *J. Vet. Behav.* **2017**, *19*, 27–34. [CrossRef]
100. Briefer, F.E.; Tettamanti, F.; McElliigott, A.G. Emotions in goats: Mapping physiological, behavioral and vocal profiles. *Anim. Behav.* **2015**, *99*, 131–143. [CrossRef]
101. Global Animal Partneship. 5-Step Animal Welfare Standards for Bison. Version 1.1. 2020. Available online: https://globalanimalpartnership.org/wp-content/uploads/2020/05/GAP-Bison-Standards-v2.0.pdf (accessed on 3 March 2025).
102. Humane Farm Animal Care. Dairy Water Buffaloes. Version 2. 2020. Available online: https://certifiedhumane.org/wp-content/uploads/Std21_-Dairy-Water-Buffaloes-21v2_updated.pdf (accessed on 28 March 2025).
103. Saatkamp, H.W.; Luuk, W.; Vissers, S.M.; van Horne, P.L.M.; de Jong, E.C. Transition from Conventional Broiler Meat to Meat from Production Concepts with Higher Animal Welfare: Experiences from The Netherlands. *Animals* **2019**, *9*, 483. [CrossRef]
104. Heinola, K.; Latvala, T.; Niemi, J.K. Consumer trust and willingness to pay for establishing a market-based animal welfare assurance scheme for broiler chickens. *Poult. Sci.* **2023**, *102*, 102765. [CrossRef]
105. Eurogroup for Animals. Animal Welfare and Food Labeling: Initiating the Transition Through High Quality Consumer Information. 2020. Available online: https://www.eurogroupforanimals.org/files/eurogroupforanimals/2021-05/E4A-AW-Food_Labeling-2020-web-version.pdf (accessed on 3 March 2025).

ruminants

MDPI

Systematic Review

Slow-Release Urea Technologies: Effects on Nitrogen Utilization and Small Ruminant Performance

Leilson R. Bezerra [1,2,*], Pedro H. S. Mazza [3], Kevily H. de O. S. de Lucena [1], Maria Andrea F. Amorim [1], José Fabio P. de Moura [1], Juliana Paula F. de Oliveira [4], Ronaldo L. Oliveira [5], Analívia M. Barbosa [5], Mozart A. Fonseca [2] and José Morais Pereira Filho [1]

1 Department of Animal Science, Federal University of Campina Grande, Patos 58708110, Paraíba, Brazil; hkevily@gmail.com (K.H.d.O.S.d.L.); andreaamorimjp@gmail.com (M.A.F.A.); jose.fabio@ufcg.edu.br (J.F.P.d.M.); jmorais@cstr.ufcg.edu.br (J.M.P.F.)
2 Department of Animal and Range Sciences, Clayton Livestock Research Center, New Mexico State University, Clayton, NM 88415, USA; mozart@nmsu.edu
3 Animal Science Department, Federal University of Northern Tocantins, Araguaína 77804970, Tocantins, Brazil; pedro.mazza@ufnt.edu.br
4 Animal Science Department, Federal University of Sergipe, Nossa Senhora da Gloria 49680000, Sergipe, Brazil; jupaula.oliv@academico.ufs.br
5 Animal Science Department, Federal University of Bahia, Salvador 40170110, Bahia, Brazil; ronaldooloiveira@ufba.br (R.L.O.); analiviabarbosa@gmail.com (A.M.B.)
* Correspondence: leilson@ufpi.edu.br

Simple Summary: Urea is a common and inexpensive supplement in the diet of ruminants because it provides nitrogen that rumen microbes use to grow. However, when urea dissolves too quickly in the stomach, it can release ammonia at toxic levels and reduce the efficiency of nutrient use. To address this problem, scientists have developed "slow-release urea," a technology that controls the speed at which urea is broken down. By covering the urea with protective materials such as waxes, oils, or plant-based substances, it is possible to better match nitrogen release with the energy that animals obtain from their feed, improving animal performance and providing a promising way to make livestock production more efficient, safe, and environmentally friendly.

Abstract: Urea is one of the most widely used sources of non-protein nitrogen (NPN) in ruminant diets due to its low cost and high availability. However, its rapid solubilization in the rumen can result in abrupt ammonia release, leading to toxicity risks and low nitrogen utilization efficiency. In this context, slow-release technologies, particularly microencapsulation, have been developed to synchronize NPN release with fermentable carbohydrate availability, thereby enhancing microbial protein synthesis, improving animal performance, and reducing environmental impacts. This review compiles recent advances in urea microencapsulation, emphasizing different wall materials such as waxes, lipids, polysaccharides, and fatty acids, as well as drying techniques and formulation strategies. Slow-release urea (SRU) addition in small ruminants' diet may increase nutrient intake and digestibility, improve N balance, and reduce urinary excretion losses. Regarding performance, positive responses are observed when nitrogen release is properly synchronized with energy availability, although the results may vary depending on the encapsulant type, forage-to-concentrate ratio, and ruminal passage rate. Additionally, effects on meat quality and environmental parameters indicate that this technology holds not only zootechnical but also socio-environmental potential. It is concluded that urea microencapsulation can represent a promising alternative to optimize NPN use efficiency in ruminant production systems, though greater methodological standardization, long-term evaluations, and comparative economic analyses are required to encourage its broader adoption.

Keywords: ruminant nutrition; non-protein nitrogen; microencapsulation; ruminal efficiency

1. Introduction

Urea is a widely used source of non-protein nitrogen (NPN) in ruminant nutrition, primarily due to its low cost per unit of nutrient. It represents a viable alternative to partially or completely replace true protein from commercial meals, which are expensive, and from oilseed cakes, which often contain antinutritional factors that may compromise nutrient digestibility [1,2].

However, the high solubility of urea makes it highly susceptible to rapid hydrolysis in the rumen, resulting in accelerated ammonia release [3]. When supplied in excess, this ammonia may exceed the liver's capacity for conversion back to urea, impairing recycling through saliva or urinary excretion [4]. This overload increases blood ammonia concentration, potentially leading to ruminal alkalosis and acute toxicity. Therefore, the safe use of urea requires continuous adaptation of the ruminal microbiota to optimize its conversion into microbial protein and to reduce the risk of toxicity [2].

Microbial protein synthesis in the rumen strongly depends on ammonia (NH_3-N) availability. Classical studies demonstrate that microbial synthesis efficiency is maximized when ruminal NH_3-N concentrations are approximately 5–11 mmol/L ($\approx$7–15 mg·dL^{-1}) [4,5]; substantially lower concentrations can support only limited microbial growth, while rapid and excessive releases of N into the rumen increase the risk of ammonia release into portal and systemic blood [5,6]. In episodes of overload (e.g., acute urea ingestion), ruminal ammonia spikes from $\approx$9.7 to $\approx$32.0 mg/100 mL within minutes have been observed, with corresponding elevations in portal blood (up to $\approx$8.0 mg/100 mL) and subsequent arterial rise, illustrating that ruminal spikes can rapidly exceed hepatic detoxification capacity. Clinical signs of hyperammonemia typically appear around $\approx$1.0 mg/dL plasma ammonia, and associated mortality is generally reported when blood ammonia exceeds $\approx$2.0 mg/dL, emphasizing the narrow threshold between high N availability and systemic toxicity [6]. Although the ruminant liver has the adaptive capacity to increase ureagenesis, and a significant portion of hepatic urea is recycled to the gastrointestinal tract (recent modeling indicates that a substantial fraction of urea-N may return to the lumen, model estimates even suggest that a large portion of hepatic N is recycled to the digestive tract), this recycling has limits. It does not eliminate the risk of acute ruminal peaks [7]. Therefore, nutritional strategies that reduce rapid ruminal amino acid degradation (e.g., microencapsulated lysine) can improve N use efficiency and reduce nitrogen losses and hyperammonemia risk [8].

In this context, strategies have been developed to control the rate of urea release in the rumen environment. By promoting a more gradual release of NPN, it becomes possible to synchronize nitrogen supply with the availability of fermentable energy, thereby enhancing microbial growth and microbial protein synthesis [1,2]. Technologies such as microencapsulation have been investigated for this purpose, using different materials and methods. Alternatives tested over the past few decades include starch-based systems [1], formaldehyde treatments [2], lipid protection [3,9], liquid urea with calcium chloride [10], and the use of polymers [11].

This type of approach is particularly relevant in arid and semi-arid regions worldwide, where forage scarcity or the predominance of low-quality feedstuffs, often with high fiber content and slow ruminal degradation, limits animal performance. In such environments, the lack of synchronization between nitrogen release and the availability of fermentable carbohydrates further restricts the utilization of free urea, compromising ruminal fermentation efficiency and microbial protein production [12,13].

Given this scenario, research efforts have intensified to develop and improve slow-release urea technologies aimed at enhancing the compatibility between nutrient degradation rates and maximizing the efficiency of non-protein nitrogen utilization [14]. Therefore, this review summarizes advances in encapsulation techniques and evaluates their effects on nitrogen utilization and small ruminant performance.

2. Protein Metabolism in Ruminants

Protein is an essential macromolecule for the functioning and maintenance of animal organisms, playing roles in cellular structure, hormonal regulation, enzymatic activity, stimulus reception, and the storage of genetic information [2,14]. Protein metabolism in ruminants plays a central role in sustaining growth, reproduction, milk, and meat production. It is a complex process influenced by the interaction between dietary proteins and the rumen microbial ecosystem. Microorganisms in the rumen degrade dietary proteins into peptides, amino acids, and ammonia, which are later used for microbial protein synthesis [11,13]. This microbial protein is a major source of amino acids for the host animal once digested in the small intestine. The balance between rumen degradable and undegradable proteins strongly affects nutrient utilization efficiency. Thus, efficient protein metabolism reduces nitrogen losses, contributing to animal performance and environmental sustainability [10].

To facilitate the understanding of these processes, an infographic (Figure 1) was prepared to schematically represent protein metabolism in sheep, highlighting the interaction between dietary proteins, ruminal microbial activity, nitrogen recycling through the urea cycle, and the contribution of slow-release urea (SRU) to synchronizing nitrogen availability with energy supply.

Understanding these mechanisms is essential for developing feeding strategies to optimize productivity in ruminant systems. In ruminants, dietary proteins are initially degraded in the rumen by microorganisms into peptides and amino acids, which are subsequently used for microbial protein synthesis [9]. The extent and rate of this degradation vary widely depending on the type of feed: plant proteins, particularly those from legumes, tend to be more degradable than those of animal origin and usually have higher protein content compared to grasses [13].

Given this dynamic, modern ruminant feeding systems adopt the concept of metabolizable protein (MP), defined as the total amount of amino acids truly absorbable in the small intestine, originating from both microbial protein and rumen undegraded dietary protein [14–17]. However, not all amino acids reaching the intestine are efficiently absorbed and utilized. The fraction of MP effectively incorporated into body tissues, milk synthesis, or metabolic maintenance is what determines the efficiency of metabolizable protein utilization [18].

In this context, Galvani et al. [19] suggested an average efficiency of 62% for MP utilization in sheep, a value considered safe to ensure adequate protein supply while preventing negative impacts on rumen fermentation. According to Van Soest [20], diets with protein levels below 8% of dry matter (DM) may impair microbial growth and, consequently, reduce diet digestibility. Under conditions of low protein intake, physiological adaptations occur, such as increased renal urea reabsorption [21] and enhanced ruminal urea transport [22], allowing this recycled urea to be used as a nitrogen source by ruminal microorganisms.

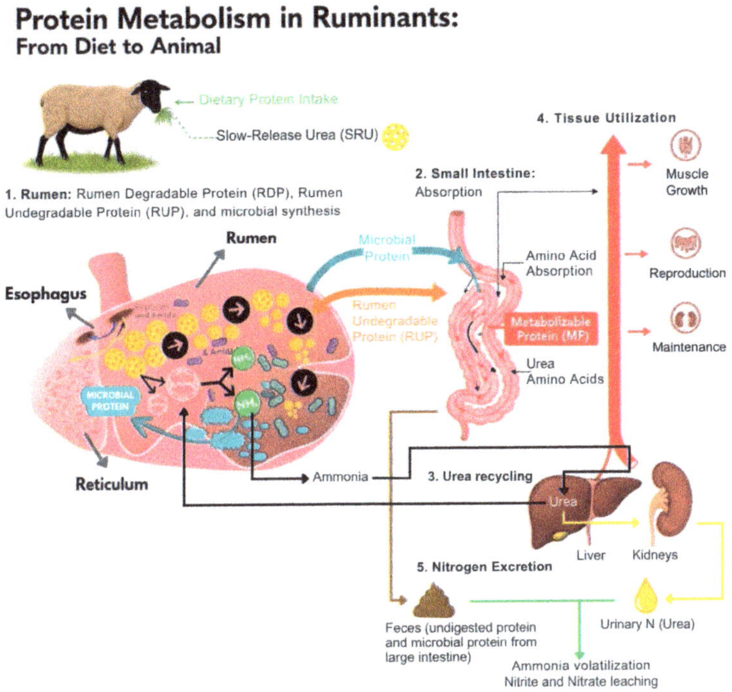

Figure 1. Schematic representation of protein metabolism in sheep. Dietary proteins are degraded in the rumen into rumen degradable protein (RDP), releasing peptides, amino acids, and ammonia (NH_3), which support microbial protein synthesis. Rumen undegradable protein (RUP) escapes ruminal degradation and, together with microbial protein, contributes to metabolizable protein (MP), effectively absorbed in the small intestine. Slow-release urea (SRU) provides controlled nitrogen availability synchronized with carbohydrate fermentation. The diagram also illustrates nitrogen recycling through the urea cycle and nitrogen excretion (urinary and fecal).

On the other hand, excessive crude protein (CP) supplementation does not lead to greater productive gains and brings both economic and environmental drawbacks. Nitrogen excess requires additional energy for excretion and reduces CP utilization efficiency [23]. Moreover, it increases N excretion, mainly as urinary urea, contributing to ammonia volatilization and nitrate leaching into groundwater [24], thus raising the environmental impact of livestock systems.

Although ruminants can utilize NPN sources such as urea, the efficiency of this process remains limited. Nitrogen balance studies in dairy and beef systems have shown that only 27% and 14% of ingested N, respectively, were converted into milk or live weight gain [25]. The remaining N is excreted via feces or urine, contributing to environmental pollution through soil and water contamination and atmospheric emissions of nitrogenous compounds [26]. Importantly, urinary N is more susceptible to rapid losses via volatilization and leaching than fecal N, representing a mitigation opportunity through nutritional management [27].

Fecal N excretion, largely derived from undigested microbial protein and proteins fermented in the large intestine, is directly influenced by diet quality. Diets with higher digestibility, in addition to supplying more fermentable energy to rumen microorganisms, tend to enhance microbial protein production. However, part of this protein may escape digestion in the small intestine and be excreted in feces [28], particularly in diets with high CP content.

The efficiency of nitrogen metabolism in ruminants can also be assessed through nitrogen retention, defined as the difference between ingested N and total excretion. High retention rates are associated with adequate fermentable energy supply in the rumen, which favors microbial synthesis, reduces urinary urea excretion, and increases the proportion of N excreted in feces [29]. Pfeffer et al. [30] estimated that well-formulated diets should provide approximately 16 g of N/kg of dry matter, considering inevitable losses: 11 g in feces, 4 g as non-urea N in urine, and 1 g as urinary urea. This requirement can be met by either true protein sources or NPN, provided there is proper synchronization with the energy available to microorganisms.

3. Non-Protein Nitrogen (NPN)

The use of urea [$(NH_2)_2CO$] as a source of NPN has been widely adopted in ruminant nutrition, primarily because it represents an effective strategy to reduce production costs. Urea is rapidly hydrolyzed by ureolytic bacteria present in the rumen, releasing ammoniacal nitrogen (NH_3-N), which can be incorporated by ruminal microorganisms into microbial protein, essential for ruminant growth and maintenance [31].

Among ruminal microorganisms, ureolytic bacteria play a key role in dietary urea hydrolysis since they produce the enzyme urease, which catalyzes the conversion of urea into ammonia (NH_3) and carbon dioxide [32,33]. However, the rate of urea hydrolysis often exceeds the capacity of ruminal bacteria to assimilate NH_3 into microbial protein, thereby reducing the efficiency of urea-N utilization by ruminants [34]. Furthermore, the isolation and cultivation of these bacteria are complex, resulting in limited knowledge of the ureolytic ruminal microbiota [35]. This knowledge gap hampers the development of targeted strategies to regulate the rate of urea hydrolysis [36].

The NH_3-N peak in the rumen following urea supplementation usually occurs between 1 and 2 h after feeding. In contrast, when true protein sources are supplied, this peak is observed between 3 and 5 h after intake [37,38]. This difference reflects the rapid metabolism of urea in the ruminal environment. However, such rapid hydrolysis also raises concerns, such as the low incorporation rate of NH_3-N by the microbiota and the risk of ammonia intoxication when excessive amounts are provided [20].

Urea remains an attractive alternative because of its low cost per nutrient unit and high protein equivalent, reaching up to 281% [39]. This allows for its partial replacement of plant protein sources from meals and cakes. Nevertheless, its rapid degradation rate can lead to the sudden release of large amounts of NH_3-N, accumulating in the rumen. When absorption exceeds the liver's capacity to convert ammonia into urea, it accumulates in the bloodstream, potentially leading to ruminal alkalosis, intoxication, and, in severe cases, animal death.

Ammonia production in ruminants is multifactorial. It can originate from dietary amino acids, glutamine catabolism in enterocytes, and peripheral tissues such as skeletal muscle. In mammals, at least 20 metabolic reactions produce ammonia, with glutaminase, glutamate dehydrogenase, and the purine nucleotide cycle being the most relevant [40]. In the gastrointestinal tract (GIT), ammonia arises mainly through two mechanisms: microbial degradation of nitrogenous compounds in the lumen and microbial hydrolysis of urea diffusing across the intestinal wall from the bloodstream [41].

In diets with very low protein levels, urea becomes the main NH_3 source in the rumen. Rumen undegradable protein and indigestible protein pass into the duodenum without influencing ruminal NH_3 production. Conversely, degradable NPN, such as urea, is rapidly converted into NH_3. Dietary nucleic acids are also degraded by bacterial peptidases and deaminases, generating ammonia [42]. In animals with high protein intake, 10% to 15% of dietary protein may be deaminated to generate energy, leading to additional NH_3 production [43].

Ammonia absorbed through the ruminal epithelium, small intestinal mucosa, and large intestine is transported via the portal vein to the liver, where it is converted into urea. This urea may then be recycled back to the rumen, intestine, or saliva; excreted by the kidneys; or secreted into milk or sweat [44]. Hepatic metabolism, therefore, plays a central role in maintaining nitrogen homeostasis.

Urea intoxication in ruminants is an acute and often fatal syndrome that results from the excessive conversion of NPN into ammonia within the ruminal environment. The pathophysiological mechanism begins with the hydrolysis of urea by the enzyme urease, produced by ruminal microorganisms, leading to the release of ammonia and carbon dioxide. Under normal conditions, the ammonia generated is either incorporated by ruminal microbes for microbial protein synthesis or absorbed through the ruminal wall and converted to urea in the liver. However, when excessive amounts of urea are ingested, or when ammonia is released too rapidly, hepatic detoxification mechanisms become overwhelmed, resulting in the accumulation of ammonia in the bloodstream [45].

Ammonia toxicity exerts its effects through multiple deleterious cellular mechanisms. Ammonia inhibits the citric acid cycle, induces lactic acidosis, decreases ATP production, and disrupts the Na^+/K^+-ATPase pump, severely impairing cellular energy metabolism. These effects are particularly pronounced in the central nervous system, where ammonia readily crosses the blood–brain barrier and directly interferes with neurotransmission. Critical blood ammonia thresholds for toxicity are well established in the literature. In cattle, clinical signs such as ataxia occur when blood ammonia-nitrogen exceeds 20 mg/L, with death usually occurring above 49 mg/L. In sheep, fatalities have been reported when blood ammonia-nitrogen surpasses 32 mg/L, indicating a slightly different sensitivity between species [46]. Other studies report that death commonly occurs when blood ammonia reaches approximately 5 mg per 100 mL (equivalent to 50 mg/L) [45].

The interval between urea ingestion and the onset of clinical signs varies markedly across species, generally ranging from 20 to 60 min in cattle and 30 to 90 min in sheep, and can be even longer in horses [21]. This temporal variation reflects differences in urea hydrolysis rate, ammonia absorption, and detoxification capacity among species. In sheep, the progression of intoxication may be extremely rapid, with some animals found dead without overt clinical signs.

Early clinical signs of urea intoxication include muscle tremors, especially of the face and ears, abdominal pain, frothy salivation, polyuria, and bruxism. As intoxication progresses, animals may display restlessness, labored breathing, motor incoordination, bloat, and tetany. In severe cases, mydriasis, tetanic contractions, lateral recumbency, and convulsions develop, rapidly progressing to coma and death. In sheep specifically, experimental studies have documented distinctive clinical manifestations, including pronounced muscle fasciculations, tremors, teeth grinding, ataxia, lateral recumbency, bloat, regurgitation, hyperesthesia, mydriasis, and seizures. Anuria and absence of salivation have also been described as late signs [47].

The practical aspects of urea intoxication in sheep involve both predisposing factors and preventive measures. Major risk factors include abrupt introduction of urea into the diet without proper adaptation, poor mixing of feed leading to localized areas of high urea concentration, administration of urea on an empty stomach when ruminal pH is elevated and microbial populations are reduced, and provision of urea without readily fermentable carbohydrates to balance ammonia release. Toxicity is further exacerbated under stress, dehydration, or conditions of low energy intake, which compromise the microbial capacity to assimilate ammonia.

From a diagnostic perspective, urea intoxication can be confirmed through clinical findings in association with specific laboratory alterations. Ruminal fluid pH becomes alkaline, typically ranging from 7.1 to 7.9 (normal 6.0–7.0), with elevated ruminal ammonia concentrations often exceeding 500–800 mg/L compared with normal values of 60–680 mg/L. The presence of white froth in the airways at necropsy and the characteristic odor of ammonia are supportive post-mortem findings. Blood sampling for ammonia determination must be performed immediately after collection and processed without delay, since values may artificially increase if samples are not properly preserved.

Treatment of urea intoxication is largely symptomatic and supportive, as no specific antidote exists. Therapeutic measures include the oral administration of weak acids such as vinegar (acetic acid) or diluted hydrochloric acid to neutralize ruminal alkalinity and reduce ammonia absorption, provision of easily fermentable carbohydrates to stimulate microbial growth and ammonia utilization, fluid therapy to correct dehydration and enhance renal excretion of nitrogenous compounds, and administration of sedatives or anticonvulsants when seizures occur. Prognosis depends critically on the rapidity of intervention, and animals in advanced stages of intoxication usually have a poor to grave prognosis.

Prevention of urea intoxication in sheep requires the implementation of strict nutritional management protocols. Urea introduction into the diet should be gradual, beginning with low doses and increasing progressively over 7 to 14 days to allow for ruminal microbiota adaptation. Uniform mixing of urea into feed is essential, and direct administration of powdered or granulated urea to animals must be avoided. The simultaneous provision of readily fermentable carbohydrates, such as molasses or grains, is crucial to synchronize ammonia release with energy availability for ruminal microbes. Additionally, urea should not exceed 1% of total dietary dry matter or 3% of concentrate, and animals must have continuous access to high-quality water.

To mitigate the risks associated with the rapid release of ammonia in the rumen, technological strategies such as urea microencapsulation have been explored. This technique aims to control the release of the compound, ensuring gradual and safer availability of ammonia, optimizing synchronization with fermentable carbohydrate supply [20]. Controlled release improves nitrogen utilization efficiency and reduces toxicity risks. Freire et al. [48] emphasized that the rate of ammonia release in the rumen is crucial both for microbial protein synthesis and for the adaptation of urea to the ruminal environment. Additionally, urea can improve the digestibility and utilization of low-quality roughages, such as hays and straws [49], supporting animal performance even under nutritional restrictions.

A comprehensive understanding of the pathophysiological mechanisms of urea intoxication, coupled with knowledge of critical blood ammonia thresholds and practical aspects of prevention and management, is fundamental for the safe and effective use of urea as a source of non-protein nitrogen in sheep production systems. Proper implementation of these practices allows producers to benefit from the economic advantages of urea supplementation while minimizing the risks of intoxication, thereby contributing to the sustainability and efficiency of ruminant production.

Therefore, the use of urea as an NPN source remains a viable and economically attractive strategy, provided that practices are adopted to ensure metabolic safety and maximize microbial efficiency, particularly through technologies such as microencapsulation.

4. Microencapsulation Technique

Microencapsulation is a technology designed to package solids, liquids, or gases into small capsules, whose main function is to enable controlled release of their core content at defined rates, over time, and under specific conditions [50]. In animal nutrition, this approach has gained increasing interest due to its potential to improve nutrient utilization efficiency, such as urea, while also favoring the modulation of ruminal microbiota and enhancing ruminant productive performance [51,52].

Historically, the microencapsulation technique was first described in the 1950s by Green and Schleicher, who patented capsules containing dyes initially developed for copying purposes [51]. Since then, this technology has evolved, being applied to the encapsulation of liquids, solids, or gases within wall-forming materials, resulting in microcapsules with the capacity for controlled release of the core [50]. This feature is particularly relevant in ruminant nutrition, as certain compounds, such as urea and methionine, undergo rapid transformations in the rumen, namely hydrolysis and deamination, respectively, thereby reducing their utilization efficiency [53].

Although relatively recent in zootechnical applications, microencapsulation was already described in the 1970s as a process involving polymeric coatings applied to solid particles, liquid droplets, or gases, generating microcapsules capable of releasing their core under specific conditions of time and environment [54]. This conceptual evolution reinforces the potential of the technique as a strategy to overcome ruminal constraints.

In ruminant nutrition, two main objectives justify the use of microencapsulation: (i) the controlled release of nitrogenous compounds, such as urea, in the rumen, and (ii) the protection of amino acids, fatty acids, and probiotic microorganisms against ruminal degradation [55–57]. As highlighted by Comunian et al. [58], several techniques have been investigated and employed for ingredient encapsulation, which can be grouped into three broad categories: physical (lyophilization, extrusion, spray-chilling, fluidized bed, co-crystallization, among others), chemical (polymerization, interfacial polymerization), and physicochemical (coacervation, solvent evaporation, liposomes, fusion-emulsion).

The choice of an appropriate technique depends on multiple factors, including the site and mode of application of the final product, the desired particle size, release mechanisms, and the physicochemical properties of both the core and the encapsulating material [51]. Moreover, cost and safety of materials, particularly in the context of feed ingredients, are fundamental considerations [58].

When lipid-based wall materials are used, the most common method is fusion-emulsification. This method consists of melting the lipids that compose the encapsulant, followed by core incorporation through dispersion or dissolution. To stabilize the emulsion, an emulsifier such as soy lecithin is added to the molten lipid [59]. The emulsion is then dispersed into an aqueous phase containing an oil-in-water surfactant. Subsequently, the system is left at room temperature, allowing for lipid solidification and aqueous fraction evaporation [60].

Another widely used method, especially for preserving heat-sensitive compounds, is lyophilization. According to Azeredo [61], this technique involves freezing the material and subsequently removing water by sublimation under vacuum, resulting in a high-quality product with minimal structural changes and good stability. In microencapsulation, lyophilization can be applied to emulsions previously formed between the core and encapsulant, favoring content preservation and encapsulation efficiency.

Following encapsulation, one of the most relevant aspects for success in ruminant nutrition is understanding the mechanisms that regulate the release of the core in the ruminal environment. In this context, both the rate and mode of release are strongly influenced by the properties of the wall material, the nature of the core, and the ruminal physicochemical conditions.

In the rumen, the main identified release mechanisms are diffusion and biodegradation (or enzymatic erosion), both mediated by microorganisms and their digestive enzymes [62,63]. Diffusion occurs when the coating remains intact, allowing ruminal fluid to penetrate, dissolve the core, and subsequently release it through pores or channels within the encapsulant matrix [64]. This process depends on factors such as solubility of the core, wall permeability, and microcapsule morphology [65,66]. Biodegradation, on the other hand, results from the action of ruminal enzymes, such as proteases and lipases, that hydrolyze wall constituents, particularly when composed of proteins or lipids, thereby exposing the core to the external environment and facilitating its release [67].

Although some encapsulating materials, such as waxes or saturated fats, are considered relatively inert in the ruminal environment, evidence from electron micrographs suggests that the structure of these walls can be partially removed or disrupted by microbial action. Scanning electron microscopy (SEM) allows for the evaluation of surface morphology in both isolated phases and microencapsulated systems. For the preparation of micrographs, samples were fixed on aluminum stubs using double-sided adhesive tape and subsequently coated with gold. According to Medeiros et al. [63], carnauba wax exhibits an irregular yet intact surface, sealed, non-porous, and marked by distinctive features. In contrast, urea displays a rough, porous surface with visible fissures [52,63]. When urea was encapsulated in carnauba wax, the microstructure retained the characteristic features of the wax, confirming its protective role. The observation that carnauba wax formed the only visible external phase supports the success of the microencapsulation process [63]. The effectiveness of carnauba wax for material incorporation has also been reported in the literature, where it was employed to hydrophobize bentonite, vermiculite, and diatomite. Similarly, Carvalho Neto et al. [53] demonstrated its potential as a microencapsulating agent for valine.

According to Carvalho et al. [52], bee wax micrographs show a slightly irregular, intact, sealed, non-porous, and smooth surface, which contrasts with the more irregular morphology of carnauba wax [63]. In the case of SRU encapsulated with beeswax, the surface appeared very similar to that of isolated bee wax, indicating that urea was located inside the microparticle, although minor surface exposure was detected. However, Carvalho et al. [52] further observed that when sulfur was incorporated into the bee wax encapsulant, the microstructure became clearly rough and aggregated, suggesting a significant presence of exposed urea. This effect was likely associated with the sulfur source, which reduced the protective efficiency of the matrix and led to nitrogen exposure.

Medeiros et al. [63] demonstrated from the scanning electron micrographs obtained before and after the in vitro ruminal digestion process of microcapsules formulated with carnauba wax as wall material, the capacity of ruminal microorganisms to promote physical and/or enzymatic modifications of microcapsules, potentially as an adaptive strategy to access metabolically valuable internal contents. Thus, interactions between capsules and ruminal microbiota appear more dynamic than previously assumed, with implications for release kinetics and compound bioavailability.

In practice, multiple mechanisms often act simultaneously or sequentially, and release may also be conditioned by specific triggers, such as pH, temperature, or the presence of particular enzymes, offering greater precision in nutrient delivery [67,68].

It is important to emphasize that the core purpose of microencapsulation is not only to protect the core but also to ensure its controlled release at the appropriate time and site. Without such control, the technique loses its functionality [67]. When correctly applied, these systems provide a viable solution to challenges faced in ruminant nutrition, particularly in replacing soybean meal with alternative sources such as non-protein nitrogen compounds, protected amino acids, and functional lipids. Therefore, microencapsulation emerges as a promising technological tool to enhance feed efficiency and sustainability in animal production.

5. Methodology

This research followed the PRISMA (Preferred Reporting Items for Systematic Reviews and Meta-Analyses) framework, structured as a systematic and descriptive review with a quantitative analytical approach. The review aimed to compile and critically assess the scientific evidence on slow- and controlled-release urea (SRU/CRU) technologies applied to sheep feeding systems, emphasizing their impact on nitrogen metabolism, feed efficiency, and animal performance.

5.1. Literature Search Strategy

A comprehensive and systematic search was conducted from September 2016 to April 2025 across three major academic databases: SciELO, ScienceDirect (Elsevier), and the National Library of Medicine (PubMed). The search strategy was developed using the Boolean method, combining controlled descriptors and free-text terms to maximize retrieval sensitivity and specificity.

The search string was formulated as follows: ("slow-release urea" OR "controlled-release urea") AND ("sheep" OR "lamb") AND ("feedlot" OR "intensive feeding") AND ("nitrogen metabolism" OR "protein efficiency").

After a pilot search, descriptions were selected to ensure adequate indexing under the Agricultural and Biological Sciences category. Only peer-reviewed articles published in English were considered. The reference management software Mendeley 1.19.3 (Elsevier®) was used to organize records, identify duplicates, and facilitate screening.

5.2. Eligibility Criteria

Inclusion and exclusion criteria were defined a priori to ensure methodological transparency and reproducibility. As inclusion criteria, studies were included if they: were original research articles published between 2016 and 2025; reported experimental data on the development, evaluation, or application of SRU or CRU in sheep; were conducted under feedlot or controlled feeding conditions using complete diets; and provided at least one quantitative response variable, such as nutrient intake, digestibility, ruminal or nitrogen metabolism, growth, or productive performance.

Priority was given to studies presenting novel encapsulation materials (e.g., lipid, wax, polymeric, or polysaccharide matrices), detailed characterizations of protection technologies, and trials assessing synchrony between nitrogen release and carbohydrate degradation.

In contrast, as an exclusion criterion, publications were excluded if they: were review papers, theses, dissertations, or book chapters; presented duplicated data across databases; lacked precise or reproducible methodological descriptions; reported results unrelated to SRU/CRU or nitrogen metabolism; or did not address the guiding research question of this review.

5.3. Study Selection Process

The selection process was conducted in three consecutive phases: Identification: Initial screening based on titles and keywords retrieved from the databases. Screening: Abstracts were evaluated to determine alignment with the review objectives. Eligibility and Inclusion:

Full texts were read and assessed against all inclusion and exclusion criteria. From an initial pool of 246 studies (76 from SciELO, six from PubMed, and 164 from ScienceDirect), 23 manuscripts fulfilled all eligibility criteria and were retained for in-depth analysis.

A PRISMA [69] flow diagram (Figure 2) summarizes the article selection process, including the number of records identified, screened, excluded, and finally included in the synthesis.

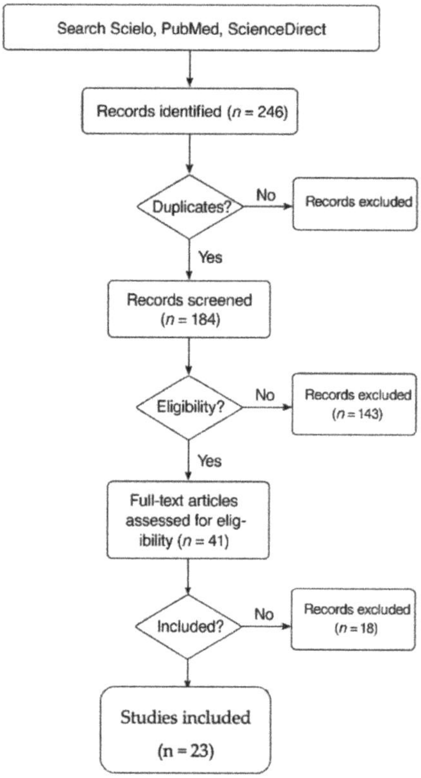

Figure 2. PRISMA [69] flow diagram summarizes the article selection process.

5.4. Data Extraction and Synthesis

Two reviewers independently extracted and verified data from the selected studies to minimize bias. Extraction fields included: Author(s) and year of publication; Experimental design and animal category; Type of SRU/CRU and encapsulation material; Analytical and evaluation methods (chemical, metabolic, or performance-based); Main quantitative outcomes (intake, digestibility, ruminal parameters, nitrogen balance, growth rate, or carcass traits).

Information was organized into comparative tables and summary figures, focusing on the technological characteristics and biological responses of SRU systems. To enhance visual interpretation, results were synthesized into heatmaps and trend graphs highlighting relationships between the degradation rate of carbohydrate sources, nitrogen release kinetics, and ruminal synchronization efficiency.

5.5. Guiding Research Question

This review was structured around the question: "Which slow- or controlled-release urea technologies demonstrate effective applicability and productive advantages in sheep nutrition under feedlot or intensive feeding conditions?".

5.6. Quality Assessment and Bias Control

Each included study was critically evaluated for methodological consistency, clarity of experimental design, replication, and statistical robustness. The risk of bias was assessed based on sample size, replication, data transparency, and the adequacy of control treatments. Studies with unclear reporting or methodological limitations were included only when their findings were essential to technological comparisons, with their limitations explicitly discussed in the results and discussion below.

6. New Systems of Slow-Release Urea in the Rumen

The use of protected urea is already a well-established practice in ruminant nutrition. However, recent advances in encapsulation processes have expanded the possibilities of this technology, enabling the development of more efficient and specific systems for controlled release. This has encouraged its adoption on a larger scale and stimulated the registration of new patents in the sector.

One of the main challenges in urea microencapsulation is the choice of the encapsulating material. Various industrial techniques have been employed to reduce the degradation rate of urea in the rumen. Among the pioneering strategies are "amireia" [1], formaldehyde-treated urea [2], lipid protection [3], liquid urea associated with calcium chloride [5], biuret [9], and polymer encapsulation [11]. To be effective, the encapsulating material must provide high protection and retention capacity, commercial availability, and safety for animals.

The wall material, or encapsulant, coats the core to form microparticles (Figure 3). Compounds such as carbohydrates, mainly polymer [37,70,71], lipids [52,53,63,72,73], and cellulose [49,74] can be used, provided they are compatible with the core, stable at different pH values, capable of film formation, possess good emulsifying properties, and adequate porosity for the desired release. These characteristics must be evaluated in conjunction with the properties of the core and the environmental conditions at the release site.

To promote the slow release of urea in the rumen, the ideal encapsulant should be chemically inert, water-insoluble, and highly hydrophobic. Waxes such as carnauba and beeswax meet these criteria, offering good moisture resistance and a melting point close to 65 °C [63,74]. Carnauba wax, abundant in the Caatinga biome, characteristic of Brazil's semiarid region, is a viable alternative because it is biodegradable, inert to ruminal microbiota, and hydrophobic [63]. Composed of long-chain saturated lipids extracted from the palm *Copernicia cerifera*, it is considered safe for animals [75]. In addition, carnauba wax has environmental and economic importance for local communities in the Caatinga, being an important source of extractive income [76].

Medeiros et al. [63] evaluated carnauba wax as an encapsulating matrix in different core-to-encapsulant ratios using the lyophilization technique. All tested systems demonstrated satisfactory performance, reinforcing the potential of this wax in urea encapsulation. Beeswax also stands out as a lipidic encapsulant, with desirable characteristics such as hydrophobicity, thermal stability, and moisture resistance [52,74]. It also contains between 12 and 15% of free fatty acids, predominantly tetracosanoic acid, being inert to ruminal microorganisms [77].

Figure 3. Different urea encapsulation systems using distinct wall materials. (**a**) [53,63] Carnauba wax: (**a.1**) pure wax flakes; (**a.2**) emulsion obtained by melting carnauba wax with urea and lecithin; (**a.3**) dried and ground material, showing a flour-like structure. (**b**) [52] Beeswax: (**b.1**) pure wax blocks; (**b.2**) emulsion obtained by melting beeswax with urea and lecithin; (**b.3**) dried and ground material, showing a flour-like structure. (**c**) [71,72] Vegetable fat: (**c.1**) emulsion prepared with vegetable fat, aqueous urea solution, and lecithin; (**c.2**) material after drying; (**c.3**) homogenized material with a pasty appearance. (**d**) [70] Sodium alginate: (**d.1**) beads formed by mixing urea with sodium alginate and immersing in calcium solution; (**d.2**) capsules after the gelification process; (**d.3**) dried material, showing a circular and solid structure. (Source: Photos by the author).

Different wall materials used for urea encapsulation exhibit distinct properties that directly influence microencapsulation efficiency, the protection of urea, and the release profile of nitrogen in the rumen (Table 1).

Carnauba wax is initially available in flakes and is melted together with urea and lecithin to form an emulsion [53,63]. After drying and grinding, it results in a flour-like material. This matrix provides a continuous hydrophobic barrier without apparent porosity, offering efficient protection against the rapid solubilization of urea [78,79]. Its irregular but sealed surface suggests strong potential for encapsulation and controlled release.

Beeswax, typically sold in blocks, follows a process similar to that of carnauba wax, also yielding an emulsion and, after drying, a flour-like material. Its particles, however, display a smoother and less irregular surface compared to carnauba. While beeswax is capable of encapsulating urea, studies indicate a higher likelihood of superficial crystal exposure, particularly when combined with sulfur, which reduces its protective efficiency [52,80].

Vegetable fat is used to prepare emulsions with aqueous urea solutions and lecithin. After drying, it produces a homogeneous material with a pasty texture. This matrix is more flexible and less structured than waxes, which may compromise the physical resistance of the encapsulated product. On the other hand, its malleability allows for adjustments in the release rate through formulation strategies [71].

Sodium alginate, in contrast to lipid matrices, forms solid beads through ionic gelation in calcium solutions. After drying, it produces circular and rigid capsules with high structural integrity. Being a hydrophilic system, it favors the production of uniform and resistant particles, but it may provide less protection against rapid diffusion of urea in aqueous environments, depending on the crosslinking density of the gel [70].

In comparative terms, waxes (carnauba and beeswax) are hydrophobic and create effective physical barriers against the immediate solubilization of urea, but they differ in texture: carnauba produces a more sealed and irregular matrix, whereas beeswax forms a smoother surface with a greater chance of urea exposure [81]. Vegetable fat offers flexibility and ease of processing but yields less robust structures, which may limit long-term protection [71,72]. Sodium alginate, due to its polysaccharide nature, forms well-defined solid capsules and represents a non-lipid alternative that combines simplicity of production with the potential for controlled release, although it is more susceptible to rapid hydration [70].

However, not all encapsulating matrices are efficient in protecting urea. For example, Nayohan et al. [82] attempted to encapsulate urea using chitosan, but this material proved ineffective. In vitro ruminal fermentation assays confirmed this inefficiency, as the degradation of urea in the rumen was not altered compared to the unprotected form, demonstrating that the choice of encapsulating material is crucial to ensuring the functionality of slow-release urea systems.

Table 1. Comparison of different materials used as urea encapsulants.

Encapsulant	Form/Structure Obtained	Surface Characteristics	Advantages	Limitations
Carnauba wax [53,63,78,80,83]	Flakes → emulsion with urea and lecithin → dried and ground, flour-like material	Irregular but sealed surface, intact and non-porous	Hydrophobic barrier, good protection against rapid solubilization, controlled release, high structural integrity	More rigid processing; requires higher thermal energy during preparation
Beeswax [52,81,83]	Blocks → emulsion with urea and lecithin → dried and ground, flour-like material	Smooth, slightly irregular surface, sealed but may expose urea crystals	Hydrophobic, effective encapsulation, less irregular morphology	Higher risk of exposed urea; reduced efficiency when combined with sulfur
Vegetable fat [71,72]	Emulsion with aqueous urea solution + lecithin → dried homogeneous material, pasty texture	Less rigid, homogeneous structure, flexible consistency	Easy processing, flexible matrix, adjustable release rate	Lower physical resistance; less durable protection
Sodium alginate [70]	Beads formed by gelification in Ca^{2+} → circular and solid capsules after drying	Rigid, circular, uniform, hydrophilic structure	Simple production, resistant capsules, particle uniformity	Lower hydrophobic barrier, higher risk of rapid diffusion in aqueous medium

Overall, each encapsulant presents specific advantages and limitations: waxes stand out for their strong physical barrier, vegetable fat for its simplicity and adaptability, and sodium alginate for its structural robustness, albeit with a greater tendency toward quick hydration.

Several experimental studies have been conducted to evaluate the performance of different encapsulating matrices and methodologies applied to the production of SRU, focusing on process yield, encapsulation efficiency, product stability, controlled nitrogen release, and the impact on animal metabolism.

In Figure 4, the microencapsulation yield (MY) is presented, calculated as MY = (M_{final}/ $M_{inicial}$) × 100, which compares the total mass of materials used in the encapsulation process with the mass obtained after processing. In Figure 5, the microencapsulation efficiency (EM) is shown, determined as EM = ($U_{retained}$/$U_{inserted}$) × 100, which relates the amount of urea effectively retained in the final product to the amount initially used in the process. These parameters provide important insights for selecting the most effective combinations of core and wall materials, as well as for identifying operational and technological limitations, considering different formulations, urea proportions, drying methods, and the effects of sulfur addition.

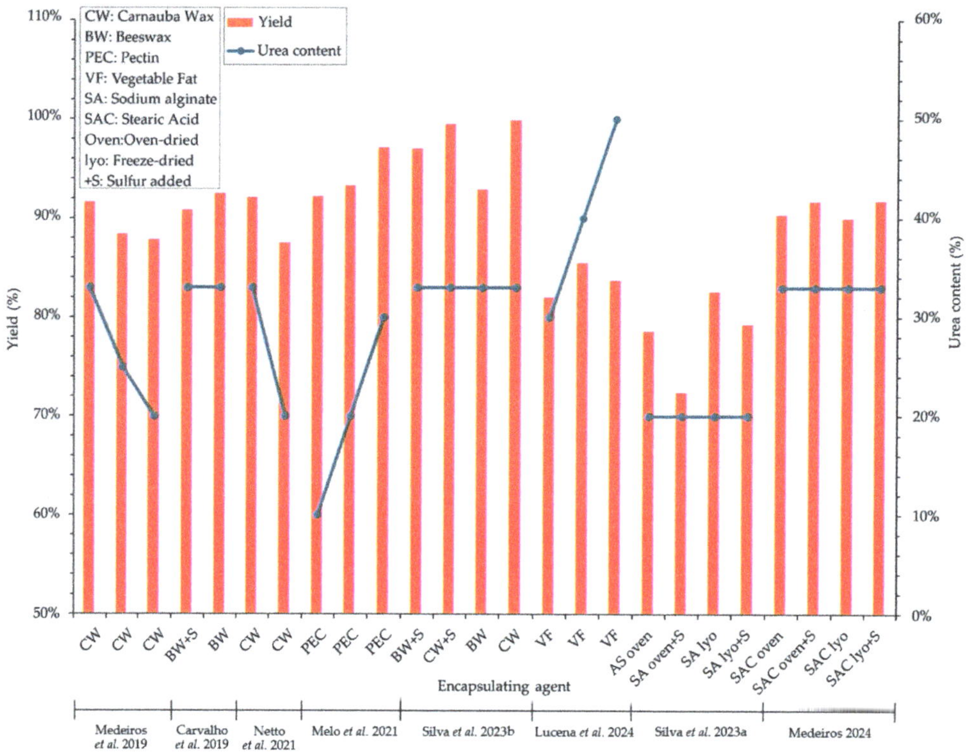

Figure 4. Microencapsulation yield (MY) of slow-release urea systems, calculated as MY = (M_{final}/ $M_{inicial}$) × 100. The figure compiles data from several studies using different wall materials: CW = Carnauba Wax; BW = Beeswax; PEC = Pectin; VF = Vegetable Fat; SA = Sodium Alginate; SAC = Stearic Acid. Some formulations were produced with different drying processes (Oven = Oven-dried; Lyo = Freeze-dried) and others with or without sulfur addition (+S = Sulfur added). Blue dots indicate the level of urea incorporated into each encapsulation system, and connected dots belong to the same study [52,63,69–71,80,83,84].

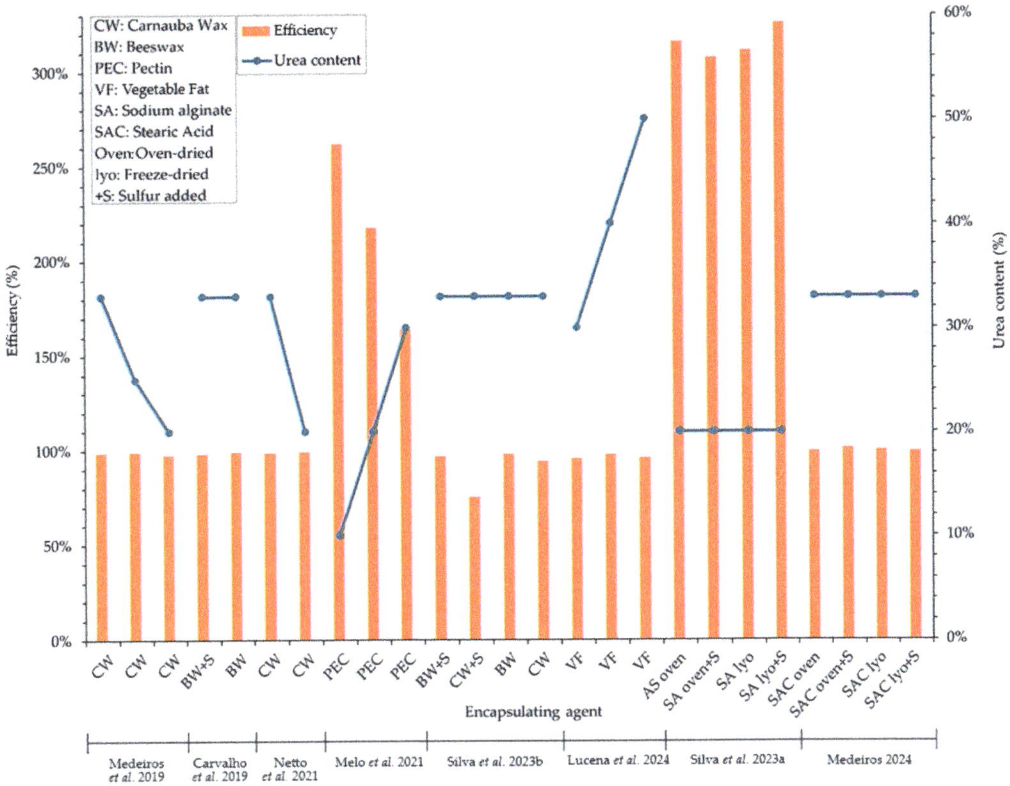

Figure 5. Microencapsulation efficiency (ME) of slow-release urea systems, calculated as ME = $(U_{retained}/U_{inserted}) \times 100$. The figure compiles data from several studies using different wall materials: CW = Carnauba Wax; BW = Beeswax; PEC = Pectin; VF = Vegetable Fat; SA = Sodium Alginate; SAC = Stearic Acid. Some formulations were produced with different drying processes (Oven = Oven-dried; Lyo = Freeze-dried) and others with or without sulfur addition (+S = Sulfur added). Blue dots indicate the level of urea incorporated into each encapsulation system, and connected dots belong to the same study [52,63,69–71,80,83,84].

Carvalho et al. [52] used beeswax as an encapsulant, observing its effectiveness in reducing urea hygroscopicity, which favored homogenization with diet ingredients and prevented solidification during storage. The 33% urea proportion showed the best yield and urea content. However, sulfur addition impaired the quality of the formulation.

Another matrix evaluated was citrus pectin [69], a natural polysaccharide successfully used with the external ionic extrusion/gelation technique. Pectin concentrations between 3 and 5% (*w*/*v*) and urea concentrations between 10 and 30% (*w*/*w*) were tested. Results indicated that pectin favored the slow release of non-protein nitrogen, optimizing its utilization and reducing nitrogen excretion.

When diets with high non-protein nitrogen levels, such as urea, are provided to ruminants, adequate sulfur supplementation is necessary to meet ruminal microbial requirements for sulfur amino acid synthesis. In this context, incorporating sulfur directly into encapsulated urea systems would logically optimize the simultaneous supply of these two essential nutrients for microbial metabolism. However, several studies have shown that sulfur addition during the encapsulation process can negatively affect the efficiency and quality of the formed microparticles.

Because it is a polar molecule, sulfur, especially in the form of magnesium sulfate, can alter the interaction between the core and encapsulating materials, affecting emulsion stability, microcapsule shape, and final encapsulation yield. Silva et al. [70], for example, evaluated systems encapsulated with 3% (*w*/*v*) sodium alginate containing 20% (*w*/*w*) urea, using the external ionic extrusion/gelation technique and testing two drying methods (oven and freeze-drying). Results indicated that oven drying performed better; however, magnesium sulfate addition significantly compromised urea encapsulation efficiency. Similarly, Silva et al. [83] compared carnauba and beeswax as encapsulating matrices using emulsification followed by lyophilization at a 2:1 (encapsulant:core) ratio. Again, sulfur addition reduced system efficiency, highlighting the need for caution in formulating encapsulated products with multiple components.

Medeiros [84], evaluating stearic acid at a 2:1 ratio as wall material, also investigated the effect of sulfur addition to the urea core and compared two dehydration methods (oven drying and lyophilization). Results showed the same negative effect of sulfur on encapsulation. Still, under all tested conditions, stearic acid demonstrated high urea encapsulation efficiency, emphasizing the relevance of its physicochemical properties, such as high melting point (65 °C) and saturated nature, as determining factors for lipid application in protecting molecules in the ruminal environment. It is worth noting that stearic acid, in addition to being metabolically inert in the rumen, can be absorbed in the small intestine and used as an energy source by the animal.

Most SRU systems use about 33% urea in the formulation, considered the upper limit to maintain microparticle integrity and quality. Encapsulation yields (Figure 4) are generally above 90%, with Silva et al. [83] reporting nearly 100% when using carnauba wax. Conversely, systems with only 20% urea showed average yields of 87% [63,80], although SRU in pectin reached 97% [69].

Values above 100%, such as those reported by Silva et al. [83] with carnauba wax with and without sulfur, may be attributed to moisture retention during drying, due to wax hardness and the hygroscopicity of magnesium sulfate heptahydrate. Silva et al. [70] also obtained yields exceeding 300% with sodium alginate, as urea concentration was calculated based on the hydrated solution, which concentrated after drying.

Particle size also influences urea degradation rate: larger particles degraded less and released more slowly [63]. Microencapsulation efficiency was close to 100% for waxes and vegetable fat and exceeded 150% for pectin [69] and sodium alginate [70] systems.

The analyzed systems showed low water activity (~0.5 at 25 °C), an important parameter for the stability of powdered feeds, whose residual moisture must be below 4% [85]. All formulations met this criterion. Results obtained with vegetable fat as encapsulant [63] differed from those reported by Medeiros et al. [63], Carvalho et al. [52], and Netto et al. [80], as increasing fat proportion did not improve yield, unlike other encapsulants. Lucena et al. [71] highlighted a relevant finding regarding the ratio between wall material and core: higher wall content is not always associated with greater yields and efficiencies. This occurs because the interaction between urea, soy lecithin, and vegetable fat during emulsion formation can significantly influence this relationship, especially during dehydration. At this stage, vegetable fat may hinder water evaporation, compromising emulsion stability and, consequently, final yield and material stability.

7. Slow-Release Urea in Small Ruminant Feeding

The use of urea in ruminant nutrition is limited by the risk of ammonia intoxication, especially when there is a rapid release of NPN in the rumen, surpassing the microbial assimilation capacity. This abrupt release can cause an excessive increase in ruminal ammonia concentration, leading to systemic absorption and hepatic overload in the conversion to urea, with a consequent rise in serum urea levels and higher urinary excretion. As an alternative, slow-release urea has been proposed with the aim of synchronizing nitrogen release with carbohydrate fermentation in the rumen, favoring microbial growth and improving nitrogen use efficiency.

Based on bench-scale laboratory tests, it is possible to verify that the protection of urea through encapsulation with waxes and carbohydrates is effective, ensuring gradual release of NPN. However, it is crucial to understand how these systems behave in the actual ruminal environment, where variables such as passage rate, pH, and diet composition may directly influence the efficiency of encapsulated nitrogen utilization. Diets with high energy concentration and different forage-to-concentrate ratios, for instance, can affect the release pattern and utilization of protected urea.

Beyond the technical efficiency of encapsulation, it is essential to assess how these systems affect nutrient intake and digestibility, parameters directly related to diet acceptance and ingredient utilization. Some wall materials, such as carnauba wax, may present odors and specific sensory characteristics that influence palatability, potentially impacting voluntary intake. In this context, evaluating DM and CP intake allows for verification not only of diet acceptance but also of potential adjustments in the synchronization between NPN release and the availability of fermentable carbohydrates. Such synchronization is decisive to maximize NPN use efficiency by fibrolytic bacteria, whose activity is fundamental for fiber degradation and, consequently, energy utilization of the diet, reflecting in changes in the apparent digestibility of DM, CP, and NDF.

Table 2 presents the effects of SRU inclusion on DM, CP, and NDF intake and digestibility. Table 3 compiles the data related to nitrogen metabolism, while Figure 6 shows the animals' average daily gain (ADG). A joint analysis of these results makes it possible to identify whether SRU contributes to a better balance between nutrient supply and utilization, optimizing ruminal fermentation and enhancing productive performance.

The inclusion of microencapsulated urea in sheep diets has promoted greater DM intake, especially with increasing levels of SRU [52,63,80,83], in addition to reducing urinary urea excretion, indicating better ruminal nitrogen utilization. However, Geron et al. [38], using an established commercial SRU product, did not observe changes in intake.

The encapsulation process of urea can reduce or even eliminate the negative effect that free urea exerts on voluntary intake, even though wall materials may contain high ether extract (EE) levels. Lucena et al. [71], by including up to 3% encapsulated urea using vegetable fat as wall material, raised the total dietary EE content to 6.25%, without reducing intake or altering ingestive behavior. Despite the high EE content, there was also no negative effect on fiber digestibility. Similar results were reported by Medeiros [84] with the inclusion of 4.5% encapsulated urea with stearic acid.

Waxes also demonstrated positive effects on microbial activity on fiber [52], possibly due to greater ruminal pH stability [69,71,72]. This stability favors protozoal growth, whose contribution to fermentation and nutrition remains debated [86]. Gradual ammonia release from SRU favors microbial nitrogen utilization and improves DM digestibility [63,69].

Regarding animal performance, responses to encapsulated urea vary according to wall material, dose, encapsulation technique, and animal category (Figure 6). The best weight gain results were obtained with SRU encapsulated in carnauba wax [78], reinforcing the superiority of this system over others. Systems with vegetable fat [71] or combined waxes [69] also proved viable, though with a more modest impact on performance.

SRU also contributes to the reduction in ruminal ammonia and serum urea peaks, promoting better nitrogen balance without compromising ruminal conditions [70,71]. The inclusion of SRU (2% of the diet) encapsulated with oven-dried sodium alginate proved efficient in nitrogen retention and reduction in urinary excretion [70].

Table 2. Dry matter intake (DMI) and crude protein (CP) intake, and apparent digestibility of dry matter (DM), crude protein (CP), and neutral detergent fiber (NDF) in sheep fed diets containing different nitrogen sources (soybean meal, free urea, and slow-release urea—SRU), using distinct encapsulating agents and inclusion levels in the total diets.

References	Encapsulating Agent	Levels of Inclusion	Intake (g/d)		Apparent Digestibility Coefficient (g/100 g Ingested)		
			DM	CP	DM	CP	NDF
Geron et al. [31,38]	Soybean Meal	-	946	129	66.8 b	52.1 b	40.7
	Commercial Protected urea	0.4	956	132	66.8 b	59.2 ab	45.1
		0.8	972	134	71.8 a	65.4 a	52.2
		1.2	836	116	63.2 c	54.7 b	41.3
Carvalho et al. [52]	Soybean Meal	-	785 c	107	58.6	64.9	49.5 c
	Beeswax	1.5	788 c	108	60.2	59.8	57.0 b
		3.0	905 a	126	66.1	64.2	59.2 b
		4.5	813 b	109	64.3	65.1	62.2 a
Netto et al. [80]	Soybean Meal	-	1000 b	132 c	67.3	74.1 b	48.4
	Carnauba Wax	1.5	1004 b	135 c	67.6	74.5 b	50
		3.0	1071 ab	141 b	68.3	78.0 a	49.3
		4.5	1167 a	151 a	67.5	79.3 a	49.5
Lucena et al. [71]	Free Urea	0.5	843	120	73.3	63.2	53.9
	Vegetal Fat	1.25	813	119	73.2	66.9	50.4
		2.0	856	125	74.8	65.5	48.3
		3.0	871	129	73.4	67.5	45.4
Silva et al. [70]	Free Urea	0.5	972 b	137 b	64.5 ab	72.5	51.9 b
	Sodium alginate	1.0	1183 a	166 a	66.9 a	71.9	51.8 b
		1.5	1011 ab	152 ab	61.2 ab	67.9	55.8 ab
		2.0	1100 ab	153 ab	58.9 b	68.9	62.1 a
Batista et al. [78]	Free Urea	0.4	1686	250	-	-	-
	Carnauba Wax	1.0	1684	239	-	-	-
		2.0	1641	236	-	-	-
		3.0	1653	237	-	-	-
		4.0	1573	223	-	-	-
Medeiros [84]	Free Urea	1.5	1583	226	74.81	69.69	66.98
	Stearic Acid	4.5	1430	199	74.22	71.24	66.92
	Commercial Protected urea	1.5	1527	215	72.36	66.7	64.09

abc Means followed by different letters in the same column are statistically different by Tukey's test at 5%.

Netto et al. [80] observed linear increases in DM, CP, and total digestible nutrient intake with SRU encapsulated in carnauba wax. Similarly, Silva et al. [83] reported improved DM and NDF digestibility with sodium alginate-encapsulated SRU. Lucena et al. [71] observed increased TDN intake and nitrogen retention, although Netto et al. [80] did not identify changes in excretion despite higher nitrogen intake.

Urea encapsulation also showed positive effects on ruminal environment. SRU inclusion promotes greater ruminal pH stability, improving conditions for fiber fermentation and favoring cellulolytic microbial growth. These effects were more pronounced with sodium alginate and pectin encapsulation [69,70], which maintained pH within the optimal range for microbial enzymatic activity.

Ruminal ammonia nitrogen concentration was also positively influenced by SRU. Gradual release provides continuous ammonia supply, optimizing microbial assimilation and reducing losses through excessive absorption or urinary excretion [52,71]. These dynamic favors microbial protein synthesis, leading to greater dietary nitrogen use efficiency [31,63].

Table 3. Nitrogen metabolism in sheep fed diets containing soybean meal, free urea, and slow-release urea (SRU), using different encapsulating agents and inclusion levels. This table compiles data from several studies evaluating parameters such as nitrogen balance, urinary and fecal nitrogen excretion, nitrogen use efficiency, and body nitrogen retention, comparing different nitrogen sources.

References	Encapsulating Agent	Levels of Inclusion	Nitrogen Metabolism (g/d)			
			Intake	Urinary	Fecal	Retained
Geron et al. [31,38]	Soybean Meal	-	20.7 ab	1.77	9.91 a	9.25 c
	Commercial protected urea	0.4	21.2 a	1.08	8.24 ab	11.23 b
		0.8	21.5 a	0.86	7.64 b	13.75 a
		1.2	18.6 b	1.11	8.11 ab	9.22 c
Carvalho et al. [52]	Soybean Meal	-	18.0	6.8	4.79	6.44
	Beeswax	1.5	18.4	7.27	5.45	5.66
		3.0	21.3	6.53	4.69	10.1
		4.5	19.6	6.83	5.67	7.11
Netto et al. [80]	Soybean Meal	-	21.1 b	5.18	5.34	10.6 b
	Carnauba Wax	1.5	21.7 b	5.52	5.58	10.6 b
		3.0	22.6 ab	5.93	5.86	10.8 b
		4.5	24.2 a	5.81	5.88	12.5 a
Lucena et al. [71]	Free Urea	0.5	19.3	5.52 a	7.11 a	6.67 d
	Vegetal Fat	1.25	19.0	4.85 b	6.42 b	7.75 c
		2.0	20.0	4.63 b	6.96 ab	8.41 b
		3.0	20.7	3.76 c	6.77 ab	10.2 a
Silva et al. [70]	Free Urea	0.5	21.8 b	4.75 b	6.88	10.2 c
	Sodium alginate	1.0	26.6 a	5.22 ab	6.85	14.5 a
		1.5	24.3 ab	5.42 ab	6.85	12.0 b
		2.0	24.8 ab	5.92 a	6.89	12.0 b
Medeiros [84]	Free Urea	1.5	34.1	11.1	3.16	19.9 a
	Stearic Acid	4.5	33.6	9.38	2.81	21.4 a
	Commercial protected urea	1.5	31.0	11.4	3.45	16.1 b

abcd Means followed by different letters in the same column are statistically different by Tukey's test at 5%.

Lucena et al. [71], although not observing significant effects of vegetable fat encapsulation on the ADG of feedlot lambs compared to free urea, reported that all treatments achieved expressive gains, with values above 200 g/day. A relevant aspect of this study is that SRU inclusion with vegetable fat allowed soybean meal to be reduced from 14% to 8% of the diet DM, without compromising animal performance. This represents a substantial reduction in the use of a costly and seasonally available ingredient, while demonstrating the efficiency of encapsulated NPN in partially replacing true protein sources without impairing ruminal fermentation or animal growth.

The effectiveness of SRU also depends on its interaction with forage sources and effective fiber levels. Its efficiency may be compromised by factors such as particle size, forage-to-concentrate ratio, and ruminal passage rate, which directly affect the retention time of encapsulated urea in the rumen and synchronization with fermentable energy availability.

In this context, carnauba wax-encapsulated SRU has shown contrasting results, possibly due to diet composition and physical structure. For example, Netto et al. [80] reported a quadratic increase in ADG of feedlot lambs fed *Cynodon* spp. hay, ranging from 90 to 148 g/day. Conversely, Batista et al. [78], evaluating carnauba wax SRU in high-concentrate diets (60:40) with low physically effective forage, found no improvement in weight gain or carcass traits compared to free urea. Moreover, inclusions above 27.5 g/kg DM impaired performance. The authors attributed this to the small particle size, which accelerated passage rate, reducing ruminal exposure time of encapsulated urea and limiting nitrogen assimilation. These findings reinforce the importance of integrating diet formulation, ingredient structure, and synchronization of NPN release with carbohydrate availability.

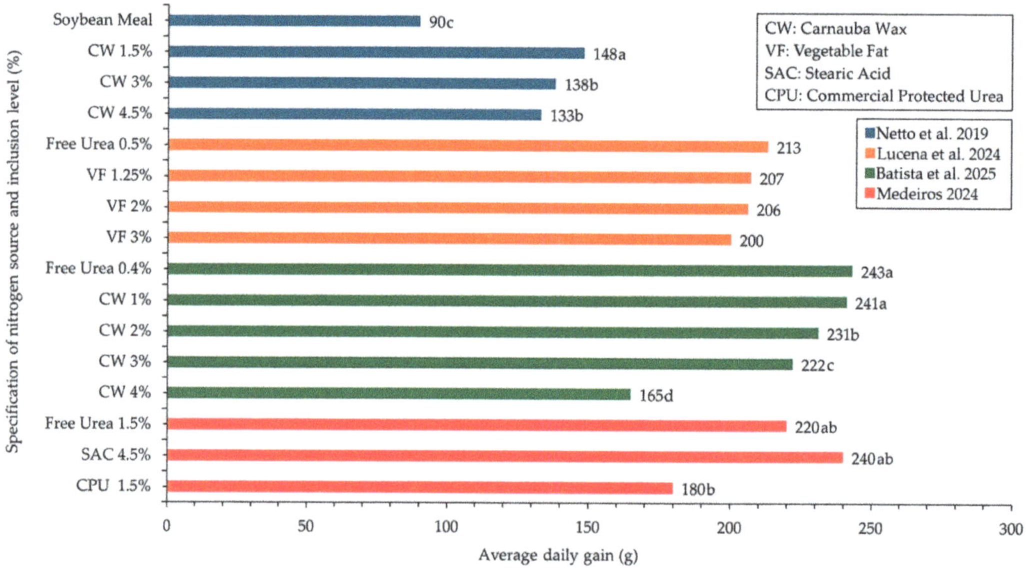

Figure 6. Average daily gain (ADG) of sheep fed diets containing different nitrogen sources: soybean meal, free urea, and slow-release urea (SRU) encapsulated with distinct wall materials (CW: carnauba wax; VF: vegetable fat; SAC: stearic acid; CPU: commercial protected urea), included at different levels in the total diets. (abcd Means followed by different letters in the same column are statistically different by Tukey's test at 5%.) [63,70,72,75].

Complementary in vitro evidence has been provided by Inácio et al. [87], who evaluated SRU protected with carnauba wax associated with cassava starch and sugarcane molasses powder. The authors observed improved metabolic activity of ruminal microorganisms and linked these effects to changes in gas production dynamics, which were associated with the utilization of simple carbohydrates such as fructose (from molasses) and starch (from cassava), compared with fibrous carbohydrates (cellulose and hemicellulose) derived from forages. These results highlight the potential of combining SRU with readily fermentable carbohydrate sources to enhance microbial fermentation efficiency and modulate ruminal fermentation patterns.

Positive results were also obtained by Medeiros [84], who, using corn silage, observed that stearic acid-encapsulated urea yielded ADG of 240 g/day in sheep, a 25% increase compared to commercial SRU. This superior performance may be linked to the physicochemical properties of stearic acid, such as its high melting point and saturated nature, which confer greater resistance to ruminal degradation and more controlled nitrogen release.

Besides ruminal metabolism and performance, it is also relevant to evaluate the impact of encapsulation systems on meat quality. Diet composition strongly influences lipid profile and nutritional value of meat, potentially yielding benefits for human health. Thus, it is crucial to assess whether encapsulating materials affect fatty acid deposition in body tissues, impacting not only productivity but also product quality and value.

Lucena et al. [71] found that SRU protected with vegetable fat improved ruminal microbial efficiency. Although this fat, due to its high saturation degree, is considered inert in the rumen, it can be digested in the small intestine as a source of energy and fatty acids, potentially influencing meat lipid composition. According to Mazza et al. [72], low-trans-fat SRU increased fatty acid and energy intake in lambs without altering carcass traits or physicochemical meat composition. However, including this encapsulate at 1.25% of diet DM increased trans-MUFA, n-6 PUFA, and CLA concentrations in meat compared to free urea. These findings suggest that vegetable fat-protected SRU may serve not only to optimize nitrogen utilization but also to improve meat lipid profile, potentially enhancing consumer health benefits.

Environmental parameters also benefit from SRU use. Reduced urinary urea excretion and greater nitrogen retention lower the risk of nitrous oxide emissions and soil/water contamination. Moreover, reduced ruminal ammonia peaks indirectly mitigate methane production by improving fiber fermentation and microbial efficiency [80].

8. Conclusions

From a technological standpoint, advances in urea encapsulation have been driven by improved production techniques (extrusion, emulsification, freeze-drying), as well as careful selection of wall materials. Promising systems, such as those based on carnauba wax, sodium alginate, and pectin, offer excellent stability, core protection, and gradual NPN release, with potential for large-scale application.

It should be emphasized, however, that system efficiency also depends on diet composition, adaptation time, and feeding management. Diets with adequate fermentable carbohydrates and effective fiber are essential to ensure ammonia utilization and avoid ruminal disorders.

Based on available studies, it is concluded that slow-release encapsulated urea may represent a viable alternative to improve NPN use efficiency in small ruminants, and it has potential to improve nitrogen utilization and contribute to more efficient feeding strategies, though the results are variable. Developing systems with regional raw materials, such as carnauba wax, further adds social and environmental value to the technology, promoting sustainable resource use and strengthening local production chains.

Future research should focus on standardizing encapsulation efficiency evaluation methods, optimizing release according to ruminal passage rate, and assessing long-term effects on health, metabolism, and productivity. In addition, comparative economic studies between different slow-release technologies and traditional NPN sources may encourage broader adoption of this strategy in livestock production systems.

9. Limitations and Future Perspectives

Despite the promising advances in slow-release urea technologies for small ruminants, several limitations currently hinder their widespread adoption and optimal implementation. Importantly, the findings presented here should not be extrapolated to other species, such as goats or cattle, since the inclusion criteria restricted the review to sheep meat production.

One of the main technical challenges is the variability in coating materials. Waxes, lipids, and polysaccharides often show inconsistent release profiles under different ruminal conditions, making standardization across production systems difficult. Manufacturing complexity adds to this issue, as microencapsulation requires specialized equipment and expertise that are not always available to smaller feed producers, limiting scalability. In addition, the long-term stability of encapsulated products under varying temperature and humidity conditions during storage and transport remains poorly characterized, raising concerns about product efficacy over time.

Methodological gaps represent another critical limitation in the current state of slow-release urea research. The absence of universally accepted protocols for evaluating release kinetics and efficacy across different research groups has led to difficulties in comparing results and establishing clear performance benchmarks. Most existing research focuses on short-term trials, with insufficient data on long-term effects on animal health, reproduction, and overall performance, leaving important questions about chronic exposure and sustained benefits unanswered. The variable response factors observed in different studies, including inconsistent results due to differences in forage-to-concentrate ratios, ruminal passage rates, and individual animal variations, further complicate the development of universal application guidelines.

Economic and practical barriers also pose significant challenges to the broader implementation of slow-release urea technologies. Limited comprehensive economic evaluations comparing the costs of slow-release urea production and application versus the benefits obtained from conventional urea supplementation make it difficult for producers to make informed investment decisions. The scale-up of laboratory techniques to commercial production remains challenging, and producers are often reluctant to adopt new feed technologies without extensive field validation, creating market resistance.

Knowledge limitations persist in several key areas, including an incomplete understanding of the precise mechanisms governing nitrogen-energy synchronization in different ruminant species, which hampers the development of optimized formulations. Limited research on species-specific responses means that optimization for specific small ruminant breeds and production systems remains underdeveloped, while insufficient long-term studies on environmental benefits and potential ecological consequences leave important sustainability questions unanswered.

Looking toward the future, several promising research perspectives emerge that could address these limitations and unlock the full potential of slow-release urea technologies. Technological innovations hold particular promise, with the development of smart release systems that incorporate pH-responsive and time-controlled mechanisms capable of adapting to individual animal ruminal conditions representing a significant advancement opportunity. Nanotechnology applications, including the exploration of nanoencapsulation techniques, could provide improved precision in nitrogen release control, while research into biodegradable coating materials derived from agricultural by-products could offer environmentally sustainable and cost-effective alternatives to current synthetic options.

Methodological advancements will be crucial for the field's progression, particularly through standardization initiatives that establish international protocols for testing and validating slow-release urea products. The development of in vivo monitoring technologies capable of real-time ruminal monitoring could enable precise optimization of nitrogen release patterns, while the creation of mathematical models to predict optimal formulations based on diet composition and animal characteristics could revolutionize formulation development processes.

Applied research priorities should focus on conducting extended trials lasting more than six months to evaluate chronic effects on animal health, reproduction, and productivity, providing the long-term data currently lacking in the literature. Comprehensive cost–benefit evaluations across different production systems and geographic regions are essential for demonstrating economic viability, while life-cycle analyses of slow-release urea technologies, including carbon footprint assessments and waste reduction potential evaluations, will be critical for establishing environmental credentials.

Integration and implementation strategies will require the development of precision nutrition approaches that combine slow-release urea with precision feeding systems for individualized supplementation strategies. Extension and education programs targeting veterinarians, nutritionists, and producers will be essential for ensuring optimal implementation, while collaboration with regulatory agencies to establish safety and efficacy standards for commercial products will facilitate market acceptance.

Emerging applications present exciting opportunities for the technology's expansion beyond traditional supplementation roles. Investigation of slow-release urea's potential in reducing greenhouse gas emissions from ruminant production could position it as a climate change mitigation tool, while exploration of incorporating urea derived from organic waste streams into slow-release formulations could contribute to circular economy initiatives. Research into simultaneous delivery of other nutrients, including vitamins, minerals, and probiotics through co-encapsulation technologies, could create value-added products that provide multiple benefits simultaneously.

To realize these future perspectives, research efforts should prioritize immediate goals including the establishment of standardized protocols for release kinetics evaluation, comprehensive economic feasibility studies, and the development of quality control standards for commercial products. Medium-term objectives should focus on completing long-term safety and efficacy trials in diverse production systems, optimizing formulations for specific small ruminant species and production stages, and validating environmental benefits through comprehensive life-cycle assessments. The long-term vision encompasses achieving commercial-scale production with cost-competitive pricing, integrating slow-release urea into precision livestock farming systems, and establishing global adoption frameworks with regulatory approval across major livestock-producing regions. Success in addressing these limitations and pursuing these future perspectives will ultimately determine whether slow release urea technologies can fulfill their promise as a transformative approach to optimizing nitrogen utilization in small ruminant production systems.

Author Contributions: Conceptualization, L.R.B., K.H.d.O.S.d.L., J.F.P.d.M., J.P.F.d.O., R.L.O., A.M.B., M.A.F. and J.M.P.F.; Data curation, K.H.d.O.S.d.L.; Formal analysis, P.H.S.M., J.P.F.d.O. and M.A.F.; Investigation, L.R.B., K.H.d.O.S.d.L., J.F.P.d.M., R.L.O. and A.M.B.; Methodology, P.H.S.M., M.A.F.A. and J.M.P.F.; Project administration, L.R.B.; Software, P.H.S.M.; Supervision, L.R.B. and J.F.P.d.M.; Validation, L.R.B., J.P.F.d.O. and J.M.P.F.; Visualization, L.R.B.; Writing—original draft, K.H.d.O.S.d.L. and M.A.F.A.; Writing—review and editing, L.R.B., R.L.O., A.M.B., M.A.F. and J.M.P.F. All authors have read and agreed to the published version of the manuscript.

Funding: This study was partially supported by the National Council for Scientific and Technological Development (Brazil) agency from CNPq (Grant 306920/2023-9) and MCTIC/CNPq (Grant #406734/2022-4).

Institutional Review Board Statement: Not applicable.

Informed Consent Statement: Not applicable.

Acknowledgments: The authors acknowledge the contribution of all researchers whose work was included in this systematic review. The authors thank Research Support Foundation of the State of Paraiba (FAPESq), CAPES-Brazil, CNPq-Brazil, "INCT-Cadeia produtiva da Carne" and "INCT-Polissacarideos", and UFCG/CSTR.

Conflicts of Interest: The authors declare no conflicts of interest.

References

1. Fonnesbeck, P.V.; Kearl, L.C.; Harris, L.E. Feed Grade Biuret as a Protein Replacement for Ruminants. A Review. *J. Anim. Sci.* **1975**, *40*, 1150–1184. [CrossRef]
2. Niazifar, M.; Besharati, M.; Jabbar, M.; Ghazanfar, S.; Asad, M.; Palangi, V.; Eseceli, H.; Lackner, M. Slow-release non-protein nitrogen sources in animal nutrition: A review. *Heliyon* **2024**, *10*, e33752. [CrossRef] [PubMed]
3. Forero, O.; Owens, F.N.; Lusby, K.S. Evaluation of slow-release urea for winter supplementation of lactating range cows. *J. Anim. Sci.* **1980**, *50*, 532–538. [CrossRef]
4. Satter, L.D.; Slyter, L.L. Effect of ammonia concentration on rumen microbial protein production in vitro. *Br. J. Nutr.* **1974**, *32*, 199–208. [CrossRef] [PubMed]
5. Bartley, E.E.; Avery, T.B.; Nagaraja, T.G.; Watt, B.R.; Davidovich, A.; Galitzer, S.; Lassman, B. Ammonia toxicity in cattle. V. Ammonia concentration of lymph and portal, carotid and jugular blood after the ingestion of urea. *J. Anim. Sci.* **1981**, *53*, 494–498. [CrossRef]
6. Haliburton, J.C.; Morgan, S.E. Nonprotein nitrogen-induced ammonia toxicosis and ammoniated feed toxicity syndrome. *Vet. Clin. N. Am. Food Anim. Pract.* **1989**, *5*, 237–249. [CrossRef]
7. Li, M.M.; Titgemeyer, E.C.; Hanigan, M.D. A revised representation of urea and ammonia nitrogen recycling and use in the Molly cow model. *J. Dairy Sci.* **2019**, *102*, 5109–5129. [CrossRef]
8. Hristov, A.N.; Bannink, A.; Crompton, L.A.; Huhtanen, P.; Kreuzer, M.; McGee, M.; Nozière, P.; Reynolds, C.K.; Bayat, A.R.; Yáñez-Ruiz, D.R.; et al. Invited review: Nitrogen in ruminant nutrition: A review of measurement techniques. *J. Dairy Sci.* **2019**, *102*, 5811–5852. [CrossRef]
9. Löest, C.A.; Lambert, B.D.; Trater, A.M.; Titgemeyer, E.C. Branched-chain amino acids for growing cattle limit-fed diets based on soybean hulls. *J. Anim. Sci.* **2001**, *79*, 2747–2753. [CrossRef]
10. Cass, J.L.; Richardson, C.R. *In Vitro Ammonia Release from Urea/Calcium Compounds as Compared to Urea and Cottonseed Meal*; Technical Report No. T-5-342; Texas Tech University: Lubbock, TX, USA, 1994.
11. Galo, E.; Emanuele, S.M.; Sniffen, C.J.; White, J.H.; Knapp, J.R. Effects of a polymer-coated urea product on nitrogen metabolism in lactating holstein dairy cattle. *J. Dairy Sci.* **2003**, *86*, 2154–2162. [CrossRef]
12. Santos, F.A.P. Metabolismo de proteínas. In *Nutrição de Ruminantes*; Berchielli, T.T., Pires, A.V., Oliveira, S.G.d., Eds.; Funep: Jaboticabal, Brazil, 2006.
13. Mcdonald, P.; Edwards, R.A.; Greenhalgh, J.F.D.; Morgan, C.A.; Sinclair, L.A.; Wilkinson, R.G. *Animal Nutrition*, 5th ed., Longman Scientific and Technical: New York, NY, USA, 1995; 607p.
14. AFRC (Agricultural and Food Research Council). *Energy and Protein Requirements of Ruminants*; CAB International: Wallingford, UK, 1993. [CrossRef]
15. CSIRO. *Nutrient Requirements of Domesticated Ruminants*; CSIRO Publishing: Collingwood, Australia, 2007.
16. NRC (National Research Council). *Nutrient Requirements of Small Ruminants: Sheep, Goats, Cervids, and New World Camelids*; The National Academies Press: Washington, DC, USA, 2007. [CrossRef]
17. Tedeschi, L.O.; Cannas, A.; Fox, D.G. A nutrition mathematical model to account for dietary supply and requirements of energy and other nutrients for domesticated small ruminants: The development and evaluation of the Small Ruminant Nutrition System. *Small Rum. Res.* **2010**, *89*, 174–184. [CrossRef]
18. Oldham, J.D. Efficiencies of amino acid utilization. In *Feed Evaluation and Protein Requirement Systems for Ruminants*; Jarrige, R., Alderman, G., Eds.; CEC: Luxembourg, 1987.
19. Galvani, D.B.; Pires, A.V.; Susin, I.; Gouvêa, V.N.; Berndt, A.; Abdalla, A.L.; Tedeschi, L.O. Net protein requirements and metabolizable protein use for growing ram lambs fed diets differing in concentrate level and roughage source. *Small Rum. Res.* **2018**, *165*, 79–86. [CrossRef]
20. Van Soest, P.J. *Nutritional Ecology of the Ruminant*, 2nd ed.; Cornell University Press: Ithaca, NY, USA, 1994. [CrossRef]

21. Starke, S.; Muscher, A.S.; Hirschhausen, N.; Pfeffer, E.; Breves, G.; Huber, K. Expression of urea transporters is affected by dietary nitrogen restriction in goat kidney. *J. Anim. Sci.* **2012**, *90*, 3889–3897. [CrossRef] [PubMed]
22. Muscher, A.S.; Schröder, B.; Breves, G.; Huber, K. Dietary nitrogen reduction enhances urea transport across goat rumen epithelium. *J. Anim. Sci.* **2010**, *88*, 3390–3398. [CrossRef] [PubMed]
23. *NRC-Nutrient Requirements of Dairy Cattle*, 6th ed.; National Academy Press: Washington, DC, USA, 1989.
24. Tamminga, S. Nutrition management of dairy cows as a contribution of pollution control. *J. Dairy Sci.* **1992**, *75*, 345–357. [CrossRef]
25. Hristov, A.N.; Hanigan, M.; Cole, A.; Todd, R.; Mcallister, T.A.; Ndegwa, P.M.; Rotz, A. Ammonia emissions from dairy farms and beef feedlots: A review. *Can. J. Anim. Sci.* **2011**, *91*, 1–35. [CrossRef]
26. Cameron, K.C.; Di, H.J.; Moir, J.L. Nitrogen losses from the soil/plant system: A review. *Ann. Appl. Biol.* **2013**, *162*, 145–173. [CrossRef]
27. Dijkstra, J.; Bannink, A.; Bosma, P.M.; Lantinga, E.A.; Reijs, J.W. Modelling the effect of nutritional strategies for dairy cows on the composition of excreta nitrogen. *Front. Sustain. Food Syst.* **2018**, *2*, 63. [CrossRef]
28. Lukas, M.; Südekum, K.-H.; Rave, G.; Friedel, K.; Susenbeth, A. Relationship between fecal crude protein concentration and diet organic matter digestibility in cattle. *J. Anim. Sci.* **2005**, *83*, 1332–1344. [CrossRef]
29. Schuba, J.; Südekum, K.-H.; Pfeffer, E.; Jayanegara, A. Excretion of faecal, urinary urea and urinary non-urea nitrogen by four ruminant species as influenced by dietary nitrogen intake: A meta-analysis. *Liv. Sci.* **2017**, *198*, 82–88. [CrossRef]
30. Pfeffer, E.; Schuba, J.; Südekum, K.-H. Nitrogen supply in cattle coupled with appropriate supply of utilisable crude protein at the duodenum, a precursor to metabolisable protein. *Arch. Anim. Nut.* **2016**, *70*, 293–306. [CrossRef]
31. Geron, L.J.V.; Garcia, J.; De Aguiar, S.C.; Da Costa, F.G.; Da Silva, A.P.; Sousa Neto, E.L.; De Carvalho, J.T.H.; Roberto, L.S.; Coelho, K.S.M.; Santos, I.S. Effect of slow release urea in sheep feed on nitrogen balance. *Semin. Ciências Agrárias* **2018**, *39*, 683–696. [CrossRef]
32. Wanapat, M. *Feeding of Ruminants in the Tropics Based on Local Feed Resources*; Khon Kaen Publishing Company Ltd.: KhonKaen, Thailand, 1999; p. 236.
33. Owens, F.N.; Lusby, K.S.; Mizwicki, K.F.; Orero, O. Slow ammonia release from urea: Rumen and metabolism studies. *J. Anim. Sci.* **1980**, *50*, 527–531. [CrossRef] [PubMed]
34. Patra, A.K. Urea/Ammonia Metabolism in the Rumen and Toxicity in Ruminants. In *Rumen Microbiology: From Evolution to Revolution*; Puniya, A., Singh, R., Kamra, D., Eds.; Springer: New Delhi, India, 2015; pp. 329–341. [CrossRef]
35. Kim, M.; Morrison, M.; Yu, Z. Status of the phylogenetic diversity census of ruminal microbiomes. *FEMS Microbiol. Ecol.* **2011**, *76*, 49–63. [CrossRef]
36. Firkins, J.; Yu, Z. Characterisation and quantification of the microbial populations in the rumen. In *Ruminant Physiology, Digestion, Metabolism and Impact of Nutrition on Gene Expression, Immunology and Stress*; Sejrsen, K., Hvelplund, T., Nielsen, M.O., Eds.; Wageningen Academic Publishers: Wageningen, The Netherlands, 2006; pp. 19–54.
37. Calomeni, G.D.; Gardinal, R.; Venturelli, B.C.; De Freitas Júnior, J.E.; Vendramini, T.H.A.; Takiya, C.S.; De Souza, H.N.; Rennó, F.P. Effects of polymer-coated slow-release urea on performance, ruminal fermentation, and blood metabolites in dairy cows. *Rev. Bras. Zootec.* **2015**, *44*, 327–334. [CrossRef]
38. Geron, L.J.V.; Aguiar, S.C.; Carvalho, J.T.H.; Juffo, G.D.; Silva, A.P.; Neto, E.L.S.; Coelho, K.S.M.; Garcia, J.; Diniz, L.C.; Paula, E.J.H. Effect of slow release urea in sheep feed on intake, nutriente digestibility, and ruminal parameters. *Semin. Ciências Agrárias* **2016**, *37*, 2793–2806. [CrossRef]
39. Castañeda-Serrano, R.D.; Ferriani-Branco, A.; Teixeira, S.; Garcia-Diaz, T.; Diego-SofiatI, A. Slow release urea in beef cattle diets: Digestibility microbial synthesis and rumen kinetic. *Agrociencia* **2013**, *47*, 13–24.
40. Cooper, A.; Plum, F. Biochemistry and phiysiology of the brain ammonia. *Physiol. Rev.* **1987**, *67*, 446–519. [CrossRef]
41. Pisulewski, P.M.; Okorie, A.U.; Buttery, P.J.; Haresign, W.; Lewis, D. Ammonia concentration and protein synthesis in the rumen. *J. Sci. Food Agric.* **1981**, *32*, 759–766. [CrossRef]
42. Tan, Z.; Murphy, M.R. Ammonia production, ammonia absorption, and urea recycling in ruminants. *J. Anim. Feed Sci.* **2004**, *13*, 389–904. [CrossRef]
43. Meijer, A.J.; Lamers, W.H.; Chamuleau, R.A. Nitrogen metabolism and ornithine cycle function. *Physiol. Rev.* **1990**, *70*, 701–748. [CrossRef] [PubMed]
44. Alio, A.; Theurer, C.B.; Lozano, O.; Huber, J.T.; Swingle, R.S.; Delgado-Elorduy, A.; Cuneo, P.; DeYoung, D.; Webb, K.E. Splanchnic nitrogen metabolism by growing beef steers fed diets containing sorghum grain flaked at different densities. *J. Anim. Sci.* **2000**, *78*, 1355–1363. [CrossRef] [PubMed]
45. Lieske, C.L.; Volmer, P.A. Nonprotein nitrogen toxicosis. In *Clinical Veterinary Toxicology*; Plumlee, K.H., Ed.; Mosby: St. Louis, MO, USA, 2003; pp. 231–236.

46. Rogers, A. Urea toxicity in ruminants: Ammonia blood levels and clinical implications. *Vet. Clin. N. Am. Food Anim. Pract.* **2023**, *39*, 287–301. [CrossRef]
47. Gimelli, A.; Pupin, R.C.; Guizelini, C.C.; Gomes, D.C.; Franco, G.L.; Vedovatto, M.; Gaspar, A.O.; Lemos, R.A.A. Urea poisoning in cattle: A brief review and diagnostic approach. *Pesq. Vet. Bras.* **2023**, *43*, e07228. [CrossRef]
48. Freire, R.D.L.; Moreira, S.B.; Pereira, M.M.; Ferreira, N.D.; Santos, J.R.; Pedreira, S.M.; Silva, O.G.H.; Peixoto, M.A.C. Ureia de liberação lenta em dietas de ovinos confinados: Consumo e digestibilidade dos nutrientes. *Rev. Cient. Prod. Anim.* **2012**, *14*, 77–80. [CrossRef]
49. Cardoso, G.S.; Pereira, L.B.; Martini, A.P.M.; De Moura, A.F.; Da Silva, M.A.; Cattelam, P.M.M.; Klein, J.L.; Druzian, A.S.; Brondani, I.L.; Alves Filho, D.C. Non-carcass components of finished feedlot steers fed with slow release or agricultural urea in substitution of soybean meal. *Semin. Ciências Agrárias* **2018**, *39*, 2761–2770. [CrossRef]
50. Nedovic, V.; Kalusevica, A.; Manojlovicb, V.; Levica, S.; Bugarskib, B. An overview of encapsulation technologies for food applications. *Procedia Food Sci.* **2011**, *1*, 1806–1815. [CrossRef]
51. Ghosh, S.K. Functional coatings and microencapsulation: A general perspective. In *Functional Coatings*; Wiley: Hoboken, NJ, USA, 2006; pp. 1–28. [CrossRef]
52. Carvalho, A.B.; Silva, A.L.; Silva, A.M.A.; Netto, A.J.; Medeiros, T.T.B.; Araujo Filho, J.M.; Agostini, D.L.S.; Oliveira, D.L.V.; Mazzetto, S.E.; Kotzebue, L.R.V.; et al. Effect of slow-release urea microencapsulated in beeswax and its inclusion in ruminant diets. *Small Rum. Res.* **2019**, *179*, 56–63. [CrossRef]
53. Carvalho Neto, J.P.; Bezerra, L.R.; Da Silva, A.L.; Moura, J.F.P.; Pereira Filho, J.M.; Da Silva Filho, E.C.; Guedes, A.F.; Araújo, M.J.; Edvan, R.L.; Oliveira, R.L. Methionine microencapsulated with a carnauba (*Copernicia prunifera*) wax matrix for protection from degradation in the rumen. *Liv. Sci.* **2019**, *228*, 53–60. [CrossRef]
54. Todd, R.D. Microencapsulation and flavour industry. *Flavour Ind.* **1970**, *1*, 768–771.
55. Risch, S.J. Encapsulation: Overview of uses and techniques. In *Encapsulation and Controlled Release of Food Ingredients*; American Chemical Society: Washington, DC, USA, 1995. [CrossRef]
56. Sobel, R.; Versic, R.; Gaonkar, A.G. Introduction to microencapsulation and controlled delivery in foods. In *Microencapsulation in the Food Industry*; Academic Press: Cambridge, MA, USA, 2014; pp. 3–12. [CrossRef]
57. Jafari, S.M. *Nanoencapsulation of Food Bioactive Ingredients: Principles and Applications*; Academic Press: Cambridge, MA, USA, 2017.
58. Comunian, T.A.; Favaro-Trindade, C.S. Microencapsulation using biopolymers as an alternative to produce food enhanced with phytosterols and omega-3 fatty acids: A review. *Food Hydrocoll.* **2016**, *61*, 442–457. [CrossRef]
59. Fávaro-Trindade, C.S.; Pinho, S.C.; Rocha, G.A. Microencapsulação de ingredientes alimentares. *Braz. J. Food Technol.* **2008**, *11*, 103–112.
60. Siddiqui, A.; Patwardhan, G.A.; Liu, Y.Y.; Nazzal, S. Mixed backbone antisense glucosylceramide synthase oligonucleotide (MBO-asGCS) loaded solid lipid nanoparticles: In vitro characterization and reversal of multidrug resistance in NCI/ADR-RES cells. *Int. J. Pharm.* **2010**, *400*, 251–259. [CrossRef] [PubMed]
61. Azeredo, H.M.C. *Encapsulação: Aplicação à Tecnologia de Alimentos*; Alimentos e Nutrição: Araraquara, Brasil, 2005; pp. 89–97.
62. Lopes, C.M.; Lobo, J.M.S.; Costa, P. Formas farmacêuticas de liberação modificada: Polímeros hidrifílicos. *Braz. J. Pharm. Sci.* **2005**, *41*, 143–154. [CrossRef]
63. Medeiros, T.T.B.; Silva, A.M.A.; Silva, A.L.; Bezerra, L.R.; Agostini, D.L.S.; Oliveira, D.L.V.; Mazzetto, S.E.; Kotzebue, L.R.V.; Oliveira, J.R.; Souto, G.S.B.; et al. Carnauba wax as a wall material for urea microencapsulation. *J. Sci. Food Agric.* **2019**, *99*, 1078–1087. [CrossRef]
64. Gunder, W.; Lippold, B.H.; Lippold, B.C. Release of drugs from ethyl cellulose microcapsules (diffusion pellets) with pore formers and pore fusion. *Eur. J. Pharm. Sci.* **1995**, *3*, 203–214. [CrossRef]
65. Higuchi, T. Mechanism of sustained-action medication. Theoretical analysis of rate of release of solid drugs dispersed in solid matrices. *J. Pharm. Sci.* **1963**, *52*, 1145–1149. [CrossRef]
66. Siepmann, J.; Siepmann, F. Modeling of diffusion-controlled drug delivery. *J. Control. Release* **2012**, *161*, 351–362. [CrossRef]
67. Dhakal, S.P.; He, J. Microencapsulation of vitamins in food applications to prevent losses in processing and storage: A review. *Food Res. Int.* **2020**, *137*, 109326. [CrossRef]
68. Singh, M.N.; Hemant, K.S.Y.; Ram, M.; Shivakumar, H.G. Microencapsulation: A promising technique for controlled drug delivery. *Res. Pharm. Sci.* **2010**, *5*, 65–75.
69. Page, M.J.; McKenzie, J.E.; Bossuyt, P.M.; Boutron, I.; Hoffmann, T.C.; Mulrow, C.D.; Shamseer, L.; Tetzlaff, J.M.; Akl, E.A.; Brennan, S.E.; et al. The PRISMA 2020 statement: An updated guideline for reporting systematic reviews. *BMJ* **2021**, *71*, 372. [CrossRef]

70. Melo, M.; Da Silva, A.; Silva Filho, E.; Oliveira, R.; Silva Junior, J.; Oliveira, J.P.; Vaz, A.; Moura, J.; Pereira Filho, J.; Bezerra, L. Polymeric microparticles of calcium pectinate containing urea for slow release in ruminant diet. *Polymers* **2021**, *13*, 3776. [CrossRef] [PubMed]
71. Silva, A.; Pereira Filho, J.M.; Oliveira, J.; Lucena, K.; Mazza, P.; Silva Filho, E.; Nascimento, A.; Pereira, E.; Vaz, A.; Barbosa, A.; et al. Effect of slow-release urea on intake, ingestive behavior, digestibility, nitrogen metabolism, microbial protein production, blood and ruminal parameters of sheep. *Trop. Anim. Health Prod.* **2023**, *55*, 414. [CrossRef] [PubMed]
72. Lucena, K.H.D.O.; Mazza, P.H.; Oliveira, R.L.; Barbosa, A.M.; Perreira Filho, J.M.; Bessa, R.J.; Alves, S.P.; Edvan, R.L.; Pereira, E.S.; Fonseca, M.; et al. Slow-releasing urea coated with low-trans vegetable lipids: Effects on lamb performance, nutrient digestibility, nitrogen balance, and blood parameters. *Anim. Feed Sci. Technol.* **2024**, *310*, 115925. [CrossRef]
73. Mazza, P.H.S.; Bezerra, L.R.; de Lucena, K.H.O.S.; Pereira Filho, J.M.; Barbosa, A.M.; Bessa, R.J.B.; Alves, S.P.; Araújo, M.J.; Parente, M.O.M.; Pereira, E.S.; et al. Effect of dietary inclusion of urea encapsulated in low-trans vegetable fat microspheres on fatty acids intake, carcass traits, quality, and fatty acid composition of lamb meat. *Small Rumin. Res.* **2024**, *236*, 107289. [CrossRef]
74. Souza, C.; De Freitas, L.A.P.; Maia, C.; Gonçalves, P.M.B. Topical formulation containing beeswax-based nanoparticles improved in vivo skin barrier function. *AAPS PharmSciTech* **2017**, *18*, 2505–2516. [CrossRef]
75. Rodrigues, D.C.; Caceres, C.A.; Ribeiro, H.L.; Abreu, R.F.A.; Cunha, A.P.; Azeredo, H.M.C. Influence of cassava starch and carnauba wax on physical properties of cashew tree gum-based films. *Food Hydrocoll.* **2014**, *38*, 147–151. [CrossRef]
76. Varela, M.C.; Carvalho, R.G. Viabilidade ambiental para a criação de unidades de conservação na Ilha da Coroa, Mossoró—RN. *Soc. Nat.* **2009**, *21*, 7–21. [CrossRef]
77. Barros, A.I.R.N.; Nunes, F.H.F.M.; Da Costa, M.M.F. *Manual de Boas Práticas na Produção de Cera de Abelha: Princípios Gerais*; FNAP—Federação Nacional dos Apicultores de Portugal: Bragança, Portugal, 2009.
78. Batista, T.S.; Bezerra, L.R.; Mazza, P.H.S.; Barbosa, A.M.; de Castro, B.S.C.; Ferrer, M.D.; Ferreira, A.C.; de Morais, J.S.; da Silva Filho, E.C.; Oliveira, R.L. Effects of dietary supplementation with slow-release urea coated with carnauba wax on intake, ingestive behavior, selectivity, ruminal parameters, performance and carcass traits of lambs. *Small Rum. Res.* **2025**, *246*, 107468. [CrossRef]
79. Inô, C.F.A.; Pereira Filho, J.M.; Oliveira, R.M.T.; Oliveira, J.F.P.; Silva Filho, E.C.; Nascimento, A.M.S.S.; Oliveira, R.L.; Nascimento, R.R.; Lucena, K.H.O.S.; Bezerra, L.R. New technology of rumen-protected bypass lysine encapsulated in lipid matrix of beeswax and carnauba wax and natural tannin blended for ruminant diets. *Animals* **2024**, *14*, 2895. [CrossRef]
80. Netto, A.J.; Silva, A.M.D.A.; Bezerra, L.R.; Carvalho, A.D.B.; Agostini, D.L.D.S.; De Oliveira, D.L.V.; Mazzetto, S.E.; Kotzebue, L.R.V.; Oliveira, J.R.; Oliveira, R.L. Lipid microspheres containing urea for slow release of non-protein N in ruminant diets. *Anim. Prod. Sci.* **2021**, *62*, 191–200. [CrossRef]
81. Feitosa, T.; Bezerra, L.; Silva, A.C.; Oliveira, D.; Mazzetto, S.; Cavalcanti, M.; Pereira Filho, J.M.; Oliveira, R.; Oliveira, J.; Fonseca, M.; et al. New manufacturing technology for rumen-protected fat using Apis mellifera beeswax: Yield, characterization, and ruminant feeding. *Trop. Anim. Health Prod.* **2025**, *57*, 6. [CrossRef]
82. Nayohan, S.; Wiryawan, K.G.; Jayanegara, A. Effect of urea coating by chitosan on the dynamics of ammonia concentration and rumen fermentation in vitro. *IOP Conf. Ser. Earth Environ. Sci.* **2022**, *951*, 012004. [CrossRef]
83. Silva, A.C.C.; Bezerra, L.R.; de Lima, J.A.M.; Oliveira, J.R.; de Oliveira, D.L.; Mazzetto, S.E.; Cavalcanti, M.T.; Filho, J.M.P.; Oliveira, J.P.F.; da Silva, A.L. Lipid microspheres of different waxes containing urea enriched with a sulfur source for slow release in the rumen. *Small Rum. Res.* **2023**, *228*, 107104. [CrossRef]
84. Medeiros, E.C. Stearic Acid Microspheres Containing Urea: Production, Characterization and In Vivo Evaluation. Master's Thesis, Federal University of Campina Grande, Patos, Brazil, 2024.
85. Zayed, G.; Roos, Y.H. Influence of trehalose and moisture content on survival of Lactobacillus salivarius subjected to freeze-drying and storage. *Process Biochem.* **2004**, *39*, 1081–1086. [CrossRef]
86. Williams, A.G.; Coleman, G.S. *The Rumen Protozoa*; Springer: New York, NY, USA, 1992. [CrossRef]
87. Inácio, A.G.; Ítavo, C.C.B.F.; Dias, A.M.; Difante, G.S.; Queiroz, J.F.; Oliveira, L.C.S.; Santos, G.T.; Ítavo, L.C.V. A new feed additive composed of urea and soluble carbohydrate coated with wax for controlled release in ruminal fluid. *Sci. Rep.* **2022**, *12*, 4487. [CrossRef] [PubMed]

MDPI AG
Grosspeteranlage 5
4052 Basel
Switzerland
Tel.: +41 61 683 77 34

Ruminants Editorial Office
E-mail: ruminants@mdpi.com
www.mdpi.com/journal/ruminants

www.ingramcontent.com/pod-product-compliance
Lightning Source LLC
Chambersburg PA
CBHW061511210926
51850CB00008B/1885
9783725869909